Additive Manufacturing of Fibre Reinforced Polymer Composites

Additive Manufacturing of Fibre Reinforced Polymer Composites

Guest Editors

Chengxing Yang
Kui Wang
Jianxun Zhang
Andrea Codolini

MDPI

Basel • Beijing • Wuhan • Barcelona • Belgrade • Novi Sad • Cluj • Manchester

Guest Editors
Chengxing Yang
School of Traffic &
Transportation Engineering
Central South University
Changsha
China

Kui Wang
School of Traffic &
Transportation Engineering
Central South University
Changsha
China

Jianxun Zhang
School of Aerospace
Engineering
Xi'an Jiaotong University
Xi'an
China

Andrea Codolini
CIMComp
University of Cambridge
Cambridge
UK

Editorial Office
MDPI AG
Grosspeteranlage 5
4052 Basel, Switzerland

This is a reprint of the Special Issue, published open access by the journal *Polymers* (ISSN 2073-4360), freely accessible at: https://www.mdpi.com/journal/polymers/special_issues/X3M6665SCL.

For citation purposes, cite each article independently as indicated on the article page online and as indicated below:

Lastname, A.A.; Lastname, B.B. Article Title. *Journal Name* **Year**, *Volume Number*, Page Range.

ISBN 978-3-7258-5731-9 (Hbk)
ISBN 978-3-7258-5732-6 (PDF)
https://doi.org/10.3390/books978-3-7258-5732-6

Contents

polymers

MDPI

Editorial

Additive Manufacturing of Fibre-Reinforced Polymer Composites

Chengxing Yang [1,2,3,*], Kui Wang [1], Jianxun Zhang [4] and Andrea Codolini [5]

1 Key Laboratory of Traffic Safety on Track, Ministry of Education, School of Traffic & Transportation Engineering, Central South University, Changsha 410075, China; kui.wang@csu.edu.cn
2 State Key Laboratory of Safety and Resilience of Civil Engineering in Mountain Area, East China Jiaotong University, Nanchang 330013, China
3 National Key Laboratory of Equipment State Sensing and Smart Support, National University of Defense Technology, Changsha 410073, China
4 State Key Laboratory for Strength and Vibration of Mechanical Structures, School of Aerospace Engineering, Xi'an Jiaotong University, Xi'an 710049, China; jianxunzhang@mail.xjtu.edu.cn
5 Engineering and Physical Science Research Council, Centre for Innovative Manufacturing in Composites, University of Cambridge, Cambridge CB2 1TN, UK; ac2386@cam.ac.uk
* Correspondence: chengxing_yang_hn@163.com

1. Introduction

Additive manufacturing (AM) has emerged as a transformative approach to fabricating complex geometries with tailored architectures, offering significant advantages in terms of design freedom, material efficiency, and on-demand production. Among the diverse materials applicable to AM, fibre-reinforced polymer composites (FRPCs) have garnered increasing attention due to their exceptional mechanical properties, high specific strength, and lightweight nature [1–3]. The synergistic combination of reinforcing fibres and polymer matrices significantly enhances structural integrity, stiffness, and impact resistance, rendering FRPCs suitable for a wide array of high-performance applications, including aerospace, automotive, marine, biomedical, and consumer products [4,5]. Furthermore, the incorporation of bio-based and biodegradable matrices has elevated FRPCs as promising candidates for sustainable material systems, aligning with global efforts toward carbon neutrality and circular economy strategies [6].

Recent developments in AM technologies have further expanded the potential of FRPCs by enabling the incorporation of both short and continuous fibres within thermoplastic and thermosetting matrices [7–9]. Advanced printing techniques such as fused deposition modelling (FDM), stereolithography (SLA), and selective laser sintering (SLS) have been tailored to accommodate fibre reinforcements, offering unprecedented opportunities to control fibre orientation, layer thickness, and part topology [10,11]. Moreover, novel strategies, such as curved layer deposition, voxel-level printing, spatially controlled fibre alignment, and hybrid AM processes, have been proposed to overcome traditional challenges in manufacturing fibre-reinforced components, resulting in enhanced mechanical performance and structural optimization [12,13]. High-performance engineering polymers, including polyether ether ketone (PEEK), polyamide (PA), and polylactic acid (PLA), have been successfully integrated into AM platforms, thereby extending the application scope of FRPCs to demanding environments, such as aerospace propulsion systems and medical implants [14].

Despite the substantial progress made, several challenges remain in fully realizing the potential of AM-FRPCs. Key issues include controlling the interfacial adhesion between the fibre and matrix, mitigating voids and porosity during printing, managing thermal residual stresses, and addressing the anisotropic behaviour caused by layer-by-layer deposition [15,16]. To address these limitations, recent studies have explored computational simulation, machine learning-assisted process optimization, and real-time monitoring systems to guide printing accuracy and material behaviour [17]. Meanwhile, the use of natural fibres, nanocellulose, and lignocellulosic reinforcements offers eco-friendly alternatives that retain competitive mechanical and thermal performance [18,19]. This Special Issue brings together recent innovations in materials design, processing technologies, and characterization techniques to advance the fundamental understanding and practical implementation of additive manufacturing for fibre-reinforced polymer composites.

2. Overview of the Published Articles

The Special Issue "Additive Manufacturing of Fibre-Reinforced Polymer Composites" brings together 13 impactful studies that delve into various facets of additive manufacturing (AM) applied to fibre-reinforced polymer composites (FRPCs). These contributions span innovations in mechanical property enhancement, process optimization, material sustainability, and smart sensing functionalities. Each article provides valuable insights that collectively represent the forefront of this rapidly evolving field.

The study by Wang et al. (Contribution 1) focuses on improving the interlaminar fracture toughness of aramid fibre/epoxy resin composites by incorporating laser-induced graphene (LIG) and short Kevlar fibres. A major bottleneck for aramid composites in structural applications is their poor Mode II interfacial toughness. Through Mode II and tensile tests, this research shows that LIG, combined with Kevlar fibres, significantly enhances toughness by 381.60%. More impressively, LIG offers a functional role in damage monitoring due to its piezoresistive properties, enabling the real-time resistance-based detection of delamination and crack propagation. Although LIG slightly reduces the tensile strength, this drawback is mitigated by the Kevlar additions. SEM analysis confirms enhanced surface roughness and fibre bridging, indicating improved crack deflection pathways. This multifunctional strategy represents a notable step toward intelligent FRPCs capable of both load bearing and structural health monitoring.

Zhang et al.'s (Contribution 2) work provides an in-depth evaluation of glass fibre-reinforced polymer (GFRP) main spars for wind turbine blades containing artificial delamination and wrinkle defects. By combining tensile testing with acoustic emission (AE) monitoring, the study analyses how these defects evolve under load and affect mechanical integrity. Using K-means clustering of AE data, the research successfully classifies different damage modes and correlates them with microstructural observations. Notably, the AE signature and characteristic frequencies remain consistent across different defect types, suggesting a degree of universality in GFRP damage responses. The study offers practical guidance for condition-based monitoring in renewable energy structures and contributes significantly to our understanding of defect-driven failure in FRPCs.

Andreozzi et al. (Contribution 3) investigates the buckling behaviour of isogrid structures manufactured using continuous carbon fibre-reinforced polymers via an Anisoprint Composer A3. This paper demonstrates that increasing the infill density enhances the buckling resistance and transitions failure from local to global buckling, leading to more uniform stress distribution. However, SEM and optical analyses highlight voids as a persistent issue that undermines the mechanical performance. By employing coextrusion technology and geometrically optimized structures, this study bridges the gap between high-performance aerospace-grade components and cost-effective 3D printing processes. The findings are especially relevant for aerospace and automotive industries seeking weight-efficient, load-bearing printed components.

Stepanov et al. (Contribution 4) addresses the challenges associated with 3D printing high-viscosity, glass fibre-filled PEEK composites. Using a combination of Taguchi design, FEM simulation, and artificial neural networks, this study identifies the optimal printing parameters (460 °C extruder temp, 20 mm/min travel speed, and 4 rpm screw rotation) that yield homogeneous microstructures and low porosity. Computed tomography validates these parameters, showing porosity levels under 10%. Furthermore, strategies for fibre alignment and post-printing treatment are proposed to enhance mechanical performance. This comprehensive approach to parameter tuning sets a benchmark for high-temperature engineering polymers and their applications in biomedical and aerospace components.

Wang et al. (Contribution 5) contributes an analytical and numerical framework for assessing the low-velocity impact behaviour of clamped rectangular sandwich tubes composed of fibre metal laminates (FMLs) and foam cores. Using a modified rigid plastic model, the study evaluates the effects of the metal volume fraction, fibre content, and foam strength on impact resistance. Validation against finite element simulations confirms the model's predictive accuracy. The results reveal that a higher fibre content and foam strength substantially enhance energy absorption and the load-bearing capacity. This study not only informs the structural design of impact-resistant components, but also showcases the synergy between fibre layups and hybrid core structures.

Almeida (Contribution 6) explores the development of eco-friendly FFF filaments by reinforcing polylactic acid (PLA) with cocoa husk-derived cellulose fibres. Both untreated (UCFF) and chemically treated (TCFF) fibres are investigated, with TCFF yielding an 18% improvement in tensile strength. Through SEM, TGA, and FTIR analyses, the study reveals that chemical treatment enhances fibre–matrix interactions and thermal stability. ANOVA confirms the statistical significance of these improvements, and the high R2 values underscore the robustness of the experimental design. This work highlights the viability of agricultural waste as a reinforcement for AM biocomposites, offering a sustainable path forward in reducing the reliance on petroleum-based materials.

Zhang et al. (Contribution 7) explored the potential of integrating continuous carbon fibres (CCFs), short carbon fibres (SCFs), and short glass fibres (SGFs) within polyamide (PA) and polylactide (PLA) matrices to design high-performance I-beam structures using AM. Employing a multi-objective optimization approach grounded in the NSGA-II algorithm, the study developed three I-beam types: primitive (P-type), designed (D-type), and optimized (O-type). Mechanical testing demonstrated remarkable improvements in the stiffness-to-mass and load-to-mass ratios, with optimized I-beams exhibiting increases of 30.05% and 40.59%, respectively. The study underscored the effectiveness of structural optimization in enhancing AM-fabricated composite components and emphasized the critical role of fibre type and matrix selection in tailoring mechanical performance.

Wang et al. (Contribution 8) presented a systematic investigation into the flexural behaviour of PLA-based composite panels fabricated via multi-material fused filament fabrication (FFF). Three PLA derivatives, including foam agent-modified and glass fibre-reinforced eco-friendly variants, were combined in varying sequences and ratios. The Taguchi method was used to analyze the influence of four parameters: material sequence, relative volume ratio, filling pattern, and filling density. The study identified the optimal configurations for maximizing the bending strength and modulus of elasticity. The material sequence emerged as the most influential factor, with performance varying up to 60% depending on the configuration. The validated results provide important guidelines for future multi-material 3D printing applications, particularly in lightweight structural design.

Niu et al. (Contribution 9) addressed the challenge of enhancing the toughness and durability of cement-based composites by incorporating multi-walled carbon nanotubes and polypropylene fibres. Through comprehensive mechanical testing and microstructural analysis, the study revealed that MWCNTs improve compressive strength and reduce mass loss, while PP fibres increase flexural performance and resist crack propagation. The synergistic effect of MWCNTs and PP fibres led to improved internal densification and pore refinement. Notably, the optimized composites showed a 19.1% increase in flexural strength and a reduction of more than 25% in key durability indicators. This study provides a practical pathway for developing advanced, fibre-modified, cement-based composites for construction and infrastructure.

Wang et al. (Contribution 10) examined how printing path strategies affect the structural integrity of continuous fibre-reinforced polymer honeycomb structures (CFRPHSs) fabricated using FDM. By experimenting with multiple path configurations, the team found that fibre dislocations at path corners resulted in stiffness variations and localized weaknesses. Among the tested geometries, the staggered trapezoidal path yielded the best mechanical outcomes, achieving the highest specific load capacity (68.33 N/g) and flexural stiffness (627.70 N/mm). The study concluded that printing path planning plays a vital role in ensuring mechanical uniformity and enhancing the load-bearing capacity in CFRPHSs, contributing valuable insights into structural design optimization. Zhang et al. (Contribution 11) proposed a novel approach to fabricating multilayer truss structures using continuous fibre-reinforced thermoplastic composites and spatial 3D printing. The design enabled the creation of pyramid trusses with tunable density and directionally extended layers. Experimental tests showed high specific stiffness (up to 401.91 MPa) and compressive strength (30.26 MPa), with relative densities as low as 1.45%. The flexibility of layer-by-layer design allowed for targeted performance customization based on application demands. This work highlights spatial 3D printing as a promising route for creating multifunctional structures in aerospace, civil engineering, and other sectors requiring low-weight, high-efficiency materials.

Baranowski et al. (Contribution 12) introduced a simulation-driven strategy to optimize a new laser-sintering platform for fabricating continuous carbon fibre-reinforced polymer parts. Using COMSOL Multiphysics, the authors evaluated the heat distribution and fibre–matrix integration efficiency. The optimized process significantly reduced the width and depth of the heat-affected zone (by 56% and 44%, respectively) and shortened the fibre integration time by over 230%. This contribution demonstrates the value of numerical modelling in refining the AM process parameters and enhancing throughput, especially for high-strength, low-batch components in aerospace and defence. Bogusz (Contribution 13) offered a high-resolution, experimental investigation into the in-plane shear behaviour of GFRP composites using $\pm 45^{\circ}$ off-axis tension tests and digital image correlation (DIC). The non-contact DIC technique enabled the visualization of strain fields and crack initiation in real time. The study identified microcrack formation and strain localization as contributors to shear nonlinearity. The comparison with strain gauge data validated the methodology and highlighted the importance of sensor placement in mechanical analysis. The findings provide crucial insights into failure mechanisms and serve as a benchmark for validating finite element models of fibre-reinforced composites.

Collectively, these 13 articles exemplify the current trajectory of research in the additive manufacturing of fibre-reinforced polymer composites. From structural optimization and sustainable biocomposites to simulation-enhanced process control and advanced characterization, each study contributes distinctively to the scientific and practical understanding of next-generation AM materials and methods.

3. Conclusions and Outlooks

In conclusion, the research contributions featured in this Special Issue collectively underscore the dynamic progression of additive manufacturing (AM) in the realm of fibre-reinforced polymer composites (FRPCs). From advanced structural design and process parameter optimization to the development of multifunctional and sustainable composite systems, these studies demonstrate the diverse capabilities of AM technologies to meet the evolving demands of high-performance applications. Notably, the integration of continuous and short fibres into thermoplastic matrices, the use of eco-friendly reinforcements, and the adoption of simulation-guided approaches have significantly advanced both the mechanical performance and functional sophistication of AM-FRPC components. Moreover, the contributions highlighting the role of artificial intelligence in process optimization and digital image correlation in failure analysis signal a trend toward data-driven, digitally informed manufacturing ecosystems.

Looking forward, the future of AM-FRPCs will likely be shaped by several strategic research directions. Firstly, there is a pressing need to develop scalable AM processes that ensure consistent fibre alignment, void-free structures, and robust interfacial bonding. This can be achieved by integrating real-time monitoring, closed-loop control systems, and in situ process diagnostics. Secondly, the exploration of hybrid material systems, such as bio-based polymers combined with advanced synthetic fibres, holds promise for balancing sustainability and performance. Thirdly, multi-material and functionally graded composites fabricated via voxel-level or multi-axis AM platforms could unlock next-generation components with spatially tuned properties. Additionally, machine learning and physics-informed models can further support intelligent process planning and quality assurance. Finally, standardization in testing protocols, lifecycle assessments, and the long-term performance evaluation of AM-FRPCs will be critical for accelerating their adoption in safety-critical industries such as the aerospace, energy, and biomedical sectors. By addressing these opportunities, the research community can continue to push the boundaries of what is achievable through the additive manufacturing of fibre-reinforced composites.

Author Contributions: Conceptualization, C.Y., K.W., J.Z. and A.C.; methodology, C.Y.; formal analysis, C.Y. and K.W.; investigation, C.Y. and K.W.; resources, C.Y., K.W. and J.Z; writing—original draft preparation, C.Y.; writing—review and editing, C.Y., K.W., J.Z. and A.C.; funding acquisition, C.Y. All authors have read and agreed to the published version of the manuscript.

Funding: This research was funded by the National Natural Science Foundation of China (No. 52302460), the Hunan Provincial Natural Science Foundation of China (No. 2023JJ20074), the open project of State Key Laboratory of Safety and Resilience of Civil Engineering in Mountain Area, East China Jiaotong University (No. HJGZ2024113), the open project of National Key Laboratory of Equipment State Sensing and Smart Support (No. WDZC20235250304), and the Young Elite Scientists Sponsorship Program by CAST (No. 2022QNRC001).

Acknowledgments: The Guest Editors would like to acknowledge all contributing authors for their submissions to this Special Issue. Particular gratitude is expressed to the Editors-in-Chief for their invitation to organize this Special Issue of *Polymers* and to the Section Managing Editors for their efficient and pro-active management of the editorial process.

Conflicts of Interest: The authors declare no conflicts of interest.

List of Contributions (Published Papers):

1. Wang, B.; Tian, W.; Wang, C.; Wang, Q. Research on Interlayer Toughening and Damage Detection of Laser-Induced Graphene and Short Kevlar Fibers Aramid Fiber/Epoxy Resin Composites. *Polymers* **2024**, *16*, 3380.
2. Zhang, Y.; Xue, S.; Chen, C.; Ma, T.; Zhou, B. Study on Acoustic Emission Characteristics and Damage Mechanism of Wind Turbine Blade Main Spar with Different Defects. *Polymers* **2024**, *16*, 3261.
3. Andreozzi, M.; Bruni, C.; Forcellese, A.; Gentili, S.; Vita, A. Compression Behavior of 3D Printed Composite Isogrid Structures. *Polymers* **2024**, *16*, 2747.
4. Stepanov, D.Y.; Dontsov, Y.V.; Panin, S.V.; Buslovich, D.G.; Alexenko, V.O.; Bochkareva, S.A.; Batranin, A.V.; Kosmachev, P.V. Optimization of 3D Printing Parameters of High Viscosity PEEK/30GF Composites. *Polymers* **2024**, *16*, 2601.

5. Wang, Y.; Zhang, J.; Guo, H.; Yuan, H. Low-Velocity Impact of Clamped Rectangular Sandwich Tubes with Fiber Metal Laminated Tubes. *Polymers* **2024**, *16*, 1833.
6. Almeida, V.H.M.d.; Jesus, R.M.d.; Santana, G.M.; Khan, S.; Silva, E.F.M.S.; Cruz, I.S.d.; Santos, I.d.S.; dos Anjos, P.N.M. The Development of Biocomposite Filaments for 3D Printing by Utilizing a Polylactic Acid (PLA) Polymer Matrix Reinforced with Cocoa Husk Cellulose Fibers. *Polymers* **2024**, *16*, 1757.
7. Zhang, X.; Sun, P.; Zhang, Y.; Wang, F.; Tu, Y.; Ma, Y.; Zhang, C. Design and Optimization of 3D-Printed Variable Cross-Section I-Beams Reinforced with Continuous and Short Fibers. *Polymers* **2024**, *16*, 684.
8. Wang, Z.; Wang, L.; Tang, F.; Shen, C. PLA-Based Composite Panels Prepared via Multi-Material Fused Filament Fabrication and Associated Investigation of Process Parameters on Flexural Properties of the Fabricated Composite. *Polymers* **2024**, *16*, 109.
9. Niu, X.; Chen, Y.; Li, Z.; Guo, T.; Ren, M.; Chen, Y. Study on the Properties of Multi-Walled Carbon Nanotubes (MWCNTs)/Polypropylene Fiber (PP Fiber) Cement-Based Materials. *Polymers* **2024**, *16*, 41.
10. Wang, K.; Wang, D.; Liu, Y.; Gao, H.; Yang, C.; Peng, Y. Path Planning and Bending Behaviors of 3D Printed Continuous Carbon Fiber Reinforced Polymer Honeycomb Structures. *Polymers* **2023**, *15*, 4485.
11. Zhang, D.; Tian, X.; Zhou, Y.; Wang, Q.; Yan, W.; Akmal Zia, A.; Wu, L.; Li, D. Spatial 3D Printing of Continuous Fiber-Reinforced Composite Multilayer Truss Structures with Controllable Structural Performance. *Polymers* **2023**, *15*, 4333.
12. Baranowski, M.; Shao, Z.; Spintzyk, A.; Kößler, F.; Fleischer, J. Simulation-Based Identification of Operating Point Range for a Novel Laser-Sintering Machine for Additive Manufacturing of Continuous Carbon-Fibre-Reinforced Polymer Parts. *Polymers* **2023**, *15*, 3975.
13. Bogusz, P. Digital Image Correlation Analysis of Strain Fields in Fibre-Reinforced Polymer–Matrix Composite under ±45° Off-Axis Tensile Testing. *Polymers* **2023**, *15*, 2846.

References

1. Wong, J.; Altassan, A.; Rosen, D.W. Additive Manufacturing of Fiber-Reinforced Polymer Composites: A Technical Review and Status of Design Methodologies. *Compos. Part B Eng.* **2023**, *255*, 110603. [CrossRef]
2. Jamal, M.A.; Shah, O.R.; Ghafoor, U.; Qureshi, Y.; Bhutta, M.R. Additive Manufacturing of Continuous Fiber-Reinforced Polymer Composites via Fused Deposition Modelling: A Comprehensive Review. *Polymers* **2024**, *16*, 1622. [CrossRef] [PubMed]
3. Wagmare, R.; Harshe, R.; Pednekar, J.; Patro, T.U. Additive Manufacturing of Continuous Fiber-Reinforced Polymer Composites: Current Trend and Future Directions. *Prog. Addit. Manuf.* **2024**, *10*, 1–28. [CrossRef]
4. Karaş, B.; Smith, P.J.; Fairclough, J.P.; Mumtaz, K. Additive Manufacturing of High Density Carbon Fibre Reinforced Polymer Composites. *Addit. Manuf.* **2022**, *58*, 103044. [CrossRef]
5. Chen, H.; Pilla, S.; Li, G.; Ijeoma, M.W.; Carbajales-Dale, M. Electricity Use in Big Area Additive Manufacturing of Fiber-Reinforced Polymer Composites. *Sci. Data* **2024**, *11*, 1362. [CrossRef] [PubMed]
6. Devarajan, B.; LakshmiNarasimhan, R.; Venkateswaran, B.; Mavinkere Rangappa, S.; Siengchin, S. Additive Manufacturing of Jute Fiber Reinforced Polymer Composites: A Concise Review of Material Forms and Methods. *Polym. Compos.* **2022**, *43*, 6735–6748. [CrossRef]
7. Armstrong, C.D.; Yue, L.; Kuang, X.; Roach, D.J.; Dunn, M.L.; Qi, H.J. A Hybrid Additive Manufacturing Process for Production of Functional Fiber-Reinforced Polymer Composite Structures. *J. Compos. Mater.* **2023**, *57*, 841–850. [CrossRef]
8. Goh, G.D.; Yap, Y.L.; Agarwala, S.; Yeong, W.Y. Recent Progress in Additive Manufacturing of Fiber Reinforced Polymer Composite. *Adv. Mater. Technol.* **2019**, *4*, 1800271. [CrossRef]
9. Qiao, Y.; Fring, L.D.; Pallaka, M.R.; Simmons, K.L. A Review of the Fabrication Methods and Mechanical Behavior of Continuous Thermoplastic Polymer Fiber–Thermoplastic Polymer Matrix Composites. *Polym. Compos.* **2023**, *44*, 694–733. [CrossRef]
10. Fang, G.; Zhang, T.; Huang, Y.; Zhang, Z.; Masania, K.; Wang, C.C. Exceptional Mechanical Performance by Spatial Printing with Continuous Fiber: Curved Slicing, Toolpath Generation and Physical Verification. *Addit. Manuf.* **2024**, *82*, 104048. [CrossRef]
11. Kafle, A.; Luis, E.; Silwal, R.; Pan, H.M.; Shrestha, P.L.; Bastola, A.K. 3D/4D Printing of Polymers: Fused Deposition Modelling (FDM), Selective Laser Sintering (SLS), and Stereolithography (SLA). *Polymers* **2021**, *13*, 3101. [CrossRef] [PubMed]
12. Ferreira, I.; Machado, M.; Alves, F.; Torres Marques, A. A Review on Fibre Reinforced Composite Printing via FFF. *Rapid Prototyp. J.* **2019**, *25*, 972–988. [CrossRef]

13. Long, Y.; Zhang, Z.; Yan, C.; Huang, Z.; Fu, K.; Li, Y. Multi-Objective Optimization for Improving Printing Efficiency and Mechanical Properties of 3D-Printed Continuous Plant Fibre Composites. *Compos. Commun.* **2022**, *35*, 101283. [CrossRef]
14. Ashrafi, B.; Díez-Pascual, A.M.; Johnson, L.; Genest, M.; Hind, S.; Martinez-Rubi, Y.; González-Domínguez, J.M.; Martínez, M.T.; Simard, B.; Gómez-Fatou, M.A.; et al. Processing and Properties of PEEK/Glass Fiber Laminates: Effect of Addition of Single-Walled Carbon Nanotubes. *Compos. Part A Appl. Sci. Manuf.* **2012**, *43*, 1267–1279. [CrossRef]
15. Yang, D.; Zhang, H.; Wu, J.; McCarthy, E.D. Fibre Flow and Void Formation in 3D Printing of Short-Fibre Reinforced Thermoplastic Composites: An Experimental Benchmark Exercise. *Addit. Manuf.* **2021**, *37*, 101686. [CrossRef]
16. Nishikawa, R.; Tamaki, K.; Notoya, O.; Yamaguchi, M. Carbon Nanotube Localization at Interface in Cocontinuous Blends of Polyethylene and Polycarbonate. *J. Appl. Polym. Sci.* **2020**, *137*, 48676. [CrossRef]
17. Ali, M.H.; Yerbolat, G.; Abilgaziyev, A. Modeling and Simulation of Composite Materials for SLS-Based 3D Printing. *Manuf. Technol.* **2020**, *20*, 135–142. [CrossRef]
18. Luhar, S.; Suntharalingam, T.; Navaratnam, S.; Luhar, I.; Thamboo, J.; Poologanathan, K.; Gatheeshgar, P. Sustainable and Renewable Bio-Based Natural Fibres and Its Application for 3D Printed Concrete: A Review. *Sustainability* **2020**, *12*, 10485. [CrossRef]
19. Deb, D.; Jafferson, J.M. Natural Fibers Reinforced FDM 3D Printing Filaments. *Mater. Today Proc.* **2021**, *46*, 1308–1318. [CrossRef]

polymers

MDPI

Article

Digital Image Correlation Analysis of Strain Fields in Fibre-Reinforced Polymer–Matrix Composite under ±45° Off-Axis Tensile Testing

Paweł Bogusz

Faculty of Mechanical Engineering, Military University of Technology, gen. S. Kaliskiego 2, 00-908 Warsaw, Poland; pawel.bogusz@wat.edu.pl

Abstract: This study presents an experimental investigation of an in-plane shear of a glass lamina composite using a ±45° off-axis tension test. Typically, the shear stress curve, shear modulus, and in-plane shear strength for composite lamina-type materials are identified. Previous research indicated that a loading rate affects the strength of this composite. This study extends the existing literature by utilising a non-contact optical digital image correlation (DIC) method to measure strain distribution during the test. Two cross-head displacement rates were examined. The obtained strain maps reveal an uneven distribution resembling fabric texture. As the deformation progresses, the differences in the strain pattern increase. Subsequently, a quantitative analysis of the differences between regions with extreme (minimum and maximum) strain values and regions with average values was conducted. Based on these measurements, shear stress–strain curves, indicating variations in their courses, were constructed. These differences may reach several percent and may influence the analysis of numerical simulations. The DIC results were validated using strain gauge measurements, a commonly utilised method in this test. It was demonstrated that the location of the strain gauge installation impacts the results. During the tests, the occurrence of multiple microcracks in the resin was observed, which can contribute to the nonlinearity observed in the shear stress–shear strain curve.

Keywords: fibre-reinforced polymer–matrix composite; vinyl-ester resin; in-plane shear test; experimental mechanics; digital image correlation method; strain map

1. Introduction

In recent decades, the usage of fibre-reinforced polymer–matrix (FRP) composites as structural elements in a variety of applications due to their advantageous properties has increased. The most commonly used FRP is mainly made of glass fibre (GFRP), carbon fibre (CFRP), and aramid fibre (AFRP). In general, they offer high strength, low self-weight, ease of handling, low maintenance requirements, high durability, high tensile strength, high chemical resistance, and lower rates of aqueous corrosion. A great variety of tensile properties of various FRP materials causes some limitations during the design process. GFRP composites, in general, indicate lower tensile modulus and strength than CFRP composites; however, GFRP composites exhibit higher deformability, better impact resistance, and a much lower material cost. Therefore, GFRP composites currently dominate the reinforcement components in pultruded products due to their exceptional performance at a competitive cost [1–3].

Pultruded Glass Fibre-Reinforced Polymer (GFRP) refers to a composite material that is formed in the pultrusion process. In this process, continuous glass fibres are impregnated with a polymer resin, typically a thermosetting resin such as polyester or epoxy. The impregnated fibres are then pulled through a heated die, in which the resin cures and solidifies, resulting in a continuous, reinforced profile or structural shape.

GFRP is commonly used in various industries, including construction, infrastructure, aerospace, and automotive, to produce beams, columns, rods, and other structural components.

In finite element method simulations (FEM), the laminae are typically homogenised and modelled as linearly elastic-brittle orthotropic or monotropic materials. To determine the constants of elasticity, strength, and ultimate strains for FRP composites, standard tests are commonly used, or non-standard strength tests are proposed [4]. In the case of composite materials, the in-plane shear properties can be identified via a $\pm 45^{\circ}$ off-axis tension test. This test follows the specifications outlined in ASTM D3518/D3518M-94 and PN-EN ISO 14129 standards [5,6]. The test produces a well-known shear stress–shear strain curve with a strong non-linear behaviour. The standard procedure involves displacement-controlled loading at a crosshead displacement rate of 2 mm/min and determines parameters such as an initial in-plane shear modulus, ultimate in-plane shear strain (0.05), and in-plane shear strength corresponding to shear strains ≤ 0.05.

Quasi-static $\pm 45^{\circ}$ off-axis tension/compression tests were investigated in [7–15] on various FRP composite materials with thermoset or thermoplastic matrices at different crosshead displacement ratios [4]. Only short-term displacement-controlled processes were examined. The following laminae were subjected to testing: woven glass/polyester [7], unidirectional carbon/epoxy [8–10], unidirectional graphite/polyimide [11], unidirectional glass/epoxy [12,13], unidirectional carbon/polymer matrix [14], and stitched glass/vinylester [15]. The following crosshead displacement ratios were used: 0.1 mm/min [8], 0.6 mm/min [10,14], 2 mm/min [9,12,15], and 5 mm/min [7].

In [14], the tests were conducted using a split-Hopkinson bar to investigate high strain rates. Significant increases in shear strength and yielding strength were observed along with an increase in strain rate from 5×10^{-4} 1/s to 1.3×10^{3} 1/s, while the failure strain decreased along with the increasing strain rate. This phenomenon can be attributed to the viscoelastic properties of polymers and polymer–matrix composites, as higher strength and reduced ductility were observed in epoxy compression tests conducted at higher rates. The viscoelastic properties of polymers and polymer–matrix composites were extensively discussed in numerous papers. In [16,17], the constitutive viscoelastic modelling of unidirectional glass/polyester composites was developed using a fractional exponential function.

In [4], the authors conducted a study on a Glass Fibre-Reinforced Plastic (GFRP) composite. The off-axis tensile tests of the composite were performed at four different levels of displacement rates: 0.02 mm/min, 2 mm/min, 20 mm/min, and 200 mm/min. The objective was to demonstrate that the high nonlinearity observed in the shear stress–shear strain curve is caused by the viscoelastic flow of the resin at low shear stress levels, as well as the combined effects of viscoelastic flow and plastic microcracks in the resin at high shear stress levels. To confirm this, a classic short-term (1 h) in-plane shear creep test was conducted on $\pm 45^{\circ}$ off-axis samples. These samples were subjected to in-plane shear stress equal to 67% of the average in-plane shear strength obtained in the previous test.

Contemporary numerical models require precise measurements of experimental curves and strength parameters. DIC measurement is currently the most important and constantly developed non-contact strain measurement method [18–21]. Its most significant 3D version has been extended to numerous research fields, including standard materials such as metals, plastics or thermoplastics [22], advanced polymeric composites [23], and biological materials such as tissues [24] or carotid arteries [25]. Therefore, DIC has become an important method in the field of experimental mechanics [18].

The DIC method was used by other researchers to study the strain field of composite materials. The study conducted in [26] investigated the shear properties of carbon/epoxy composites with $0^{\circ}/90^{\circ}$ textile reinforcement structures and varying layer thicknesses under $\pm 45^{\circ}$ tensile loading. To analyse the deformation process on the sample surfaces during loading, an optical 2D DIC measuring method was employed. The findings indicated that shear behaviour is influenced by the layer thickness and the weight per unit area of the reinforcement. Contrary to the authors' expectations, the optical deformation analysis of the sample surfaces revealed an uneven strain distribution.

In [27], a novel oblique end tab was designed to mitigate stress concentration and in-plane bending moments resulting from off-axis tension loading. Finite element analysis and experiments were conducted on polyetheretherketone (PEEK)/AS4 unidirectional thermoplastic composites (CFRTP) to assess the effectiveness of the proposed testing method. Simulation and test results demonstrated that the utilisation of oblique end tabs successfully eliminated stress concentration and bending movements. The digital image correlation (DIC) method was employed to investigate the deformation response, specifically tension/shear coupling, across the entire field of the off-axis specimen. The test results revealed significant non-linear behaviour and non-uniform strain distribution under combined tension/shear stresses. A fractographic examination was carried out to explore the damage mechanisms under the tension/shear combined stress state. Specimens with off-axis angles of 30°, 45°, and 60° exhibited failure in a mixed tension/shear mode.

In paper [28], DIC was utilised for tensile and bond tests on composite reinforcements with varying textiles and matrices. The results obtained from two DIC software programs were validated by comparing them with displacement and strain transducers. DIC provided additional insights into the damage pattern, such as crack location and width, as well as the load transfer mechanism between the composite and the substrate. One advantage of DIC is the ability to select multiple measurement points after the test, overcoming the limitations of traditional transducers. However, since DIC only monitors the outer surface of the specimen, direct information about the embedded textile in the matrix is not available. The combination of DIC and traditional sensors in laboratory testing allows for improving the understanding of the mechanical behaviour of composite reinforcements as well as the identification of their fundamental properties.

FRP composites are vulnerable to the presence of holes and cutouts, as several experiments have shown a significant decrease in strength. This strength degradation is primarily attributed to a stress concentration around the hole. In review [29], the objective was to understand the impact of holes on the mechanical properties of FRP laminates, with a specific focus on carbon FRP laminates and experimental findings. A comprehensive analysis of the tensile, compressive, flexural, and post-buckling properties of laminates with holes was provided to facilitate the optimal component design. The strain field near the holes obtained via DIC was compared to the output of finite element analysis, and a good agreement was observed between both sets of results.

In [30], a series of dynamic longitudinal compression tests were conducted on cross-ply IM7/8552 samples cut at different off-axis angles to produce different combinations of compression and shear stresses. Together with the results from quasi-static tests [31] performed in a similar manner, quasi-static and dynamic fibre kinking failure envelopes were obtained using a classical laminate theory. The experimental results revealed that while the leading fibre kinking theories accurately predicted the effects of a strain rate on uniaxial compression strength, they failed to account for shear effects, both in quasi-static and dynamic scenarios. The digital image correlation method was employed to validate strain gauge measurements and ensure the adequacy of the testing conditions.

FRP materials commonly employed in civil engineering exhibit significant asymmetries and heterogeneity due to manual installation procedures [32]. This study aimed to explore the applicability of an optical full-field DIC technique for strain field measurement on FRP structures in the civil engineering industry. The DIC method and more traditional measurement methods were compared. The DIC results provided valuable insights into the stress distribution around the structure. Consistent measurements demonstrated the potential of DIC in capturing rupture strains of FRP materials and understanding their capacity under the influence of these strains. Furthermore, DIC offered improved capabilities for detecting debonding effects.

The authors of [33] focused on the experimental investigation of tensile testing of CFRP-UD coupon specimens using the 10° off-axis test and DIC strain mapping to measure the full-field deformation response. A sensitivity analysis of the measured stresses, strains, and moduli was performed by varying the aspect ratio of the specimens and the location of strain measurement. Test-related uncertainties, such as the loading angle and specimen preparation details, were investigated and quantified with the use of these full-field DIC measurements.

This present study extends upon the research presented in [4]. In addition to the previous study, DIC measurements were performed on the same GFRP composite. This method enabled the measurement of strain distribution on the composite in a ±45° off-axis tension test. Similar papers can be found in the literature. However, this manuscript specifically focuses on a unique G/V composite and presents specific results. The assessment of strain maps and their comparison with the strain gauge method, which is not extensively documented in the literature, was conducted.

Two strain rates, 0.02 and 20 mm/min, were selected for this study. The measured DIC maps revealed non-uniform strain distributions. Subsequently, a quantitative analysis of the differences between regions with extreme (minimum and maximum) values and regions with average values was conducted. Based on these measurements, shear stress–strain curves, presenting variations in their courses, were constructed. These differences can reach several percent and may influence strength parameters, consequently affecting the analysis of numerical simulations. During the testing process, the formation of multiple microcracks in the resin was observed, which can contribute to the nonlinearity observed in the shear stress–shear strain curve, and, thus, supports the statement presented in [4].

2. Materials and Methods

This research involved conducting ±45° off-axis tension tests to measure shear properties of flat rectangular samples of a Glass Fibre Reinforced Polymer (GFRP) composite (producer: ROMA Co., Ltd., Grabowiec, Poland) in order to analyse strain distribution. The samples were made and tested in accordance with the guidelines contained in the PN-EN ISO 14129:2000 standard [6]. The nominal dimensions of the samples were 250 × 25 mm. The overall thickness of the plate was approximately 2.6 mm. The samples had adhesive patches, with a length of 70 mm and a thickness of 3 mm, applied on both sides of the grip sections. The scheme of a sample is presented in Figure 1.

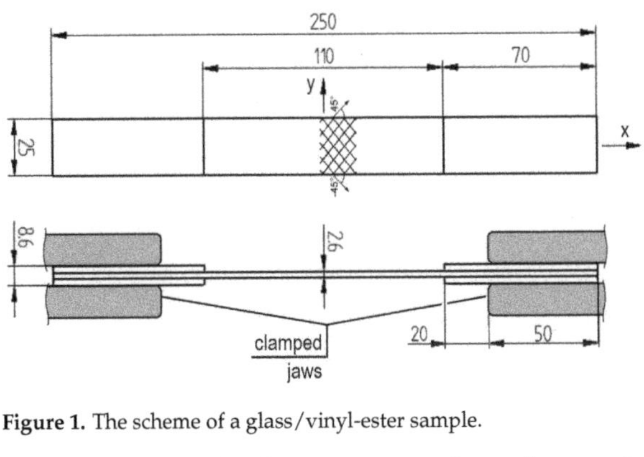

Figure 1. The scheme of a glass/vinyl-ester sample.

The investigated G/V composite is a layered composite made of Polimal VE-11M vinyl-ester resin (producer: CIECH Sarzyna S.A., Nowa Sarzyna, Poland), used as a matrix reinforced with GBX800 [±45] glass fabric (producer: DIPEX Co., Sered', Slovakia). This composite design mixture is utilised for large-scale roof coverings. Glass composite is more cost-effective than carbon-reinforced composites yet still suitable for this purpose.

The G/V composite under investigation is a layered composite consisting of Polimal VE-11M vinyl-ester resin (manufacturer: CIECH Sarzyna S.A., Nowa Sarzyna, Poland) used as a matrix and reinforced with GBX800 [±45] glass fabric (manufacturer: DIPEX Co.,

Sered', Slovakia). This composite is employed, among others, for extensive roof coverings. Glass composite is a more economical alternative to carbon-reinforced composites while remaining suitable for this specific application.

The Polimal VE-11M resin is a flame-retardant neutral vinyl-ester resin known for its high thermal and chemical resistance. The curing system used for 1 kg of this resin, at a temperature above 18 °C and low air humidity, includes Cobalt accelerator Co 1% (10 mL) and MEKP low reactive hardener (20 mL). Post-curing of the resin is required at an increased temperature of 80 °C for 4 h. Resin V is suitable for vacuum infusion technology [34]. The basic parameters of this resin before and after curing/post-curing are summarised in Table 1. Tensile strength is equal to 80 MPa, the tensile modulus is 3500 MPa, and relative elongation at break is 3.5%.

Table 1. Mean values and standard deviations for in-plane shear modulus G_{12} and in-plane shear strength R_{12} of the G/V composite [4,34].

Parameter, Standard	Unit	Value
Viscosity at 25 °C, DIN 53015	[mPa · s]	300 ÷ 500
Gelation time at 25 °C, ISO 2535	[min]	10 ÷ 30
Tensile strength, ISO 527-2	[MPa]	80
Relative elongation at break, ISO 527-2	[%]	3.5
Tensile modulus, ISO 527-2	[MPa]	3500
Heat distortion temperature under load (HDT), ISO 75-2	[°C]	85

The GBX800 fabric (referred to as G) is a stitched E-glass fabric with a nominal surface density of 800 g/m^2. It is quasi-balanced, orthogonal, and bi-directional fabric with a ±45 warp/weft orientation in relation to the longitudinal axis of a fabric strip and test samples.

Composite plates of G/V were manufactured by ROMA Co., Ltd., Grabowiec, Poland, using vacuum infusion technology and had a $[\pm 45]_{2S}$ fabric sequence. The plates were post-cured following the guidelines in [34]. The key parameters of one lamina were determined to be thickness of 0.663 mm, fibre volume fraction of 48%, and mass density of 1.70 g/cm^3.

For the measurement of shear strains, two Vishay strain gauges (producer: Vishay Precision Group, Inc., Wendell, NC, USA) with resistance of 120 Ω were installed in the central measurement zone of the samples following the T-rosette scheme, as shown in Figure 2. The strain gauges were attached along and perpendicular to the axis of the sample in accordance with standard requirements. This allowed for the evaluation of shear strains occurring in the composite fabrics oriented at a 45° angle, while the sample, itself, was subjected to axial tension.

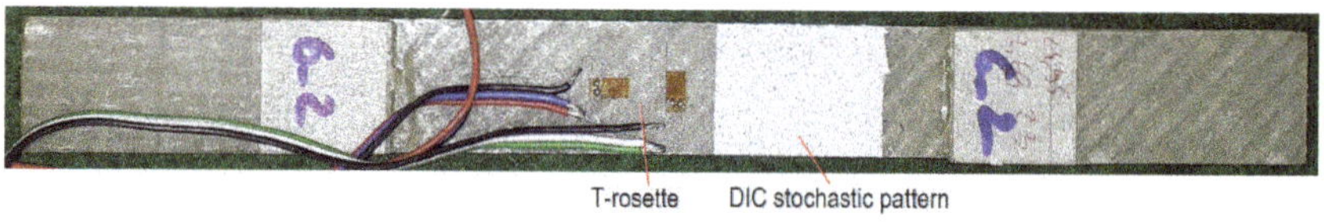

Figure 2. A sample with a T-rosette attached and a stochastic pattern applied.

For the purpose of DIC measurements, a stochastic black–white pattern was applied to a section of the sample used for measurements (see Figure 2). The pattern had dimensions of approximately 25 × 25 mm. The T-rosette strain gauges were located at the centre of a sample, while the DIC stochastic pattern was positioned close to this centre. It was considered essential to keep this pattern as far away as possible from the gripping area.

In the DIC measurements, the GOM Aramis system (producer: GOM mbH., Braunschweig, Germany), specifically designed for capturing deformations and strains on materials under load, was employed. The testing setup, along with cameras of the DIC system, is presented in Figure 3. The system utilises two cameras equipped with a CMOS sensor (Figure 3b) to capture a 3D image with resolution of 4 megapixels. The cameras, fitted with lenses with a fixed focal length of 50 mm, were positioned at a distance of 345 mm from the measurement object. The distance between the cameras in this configuration was 126 mm, while the cone angle was set to 25°.

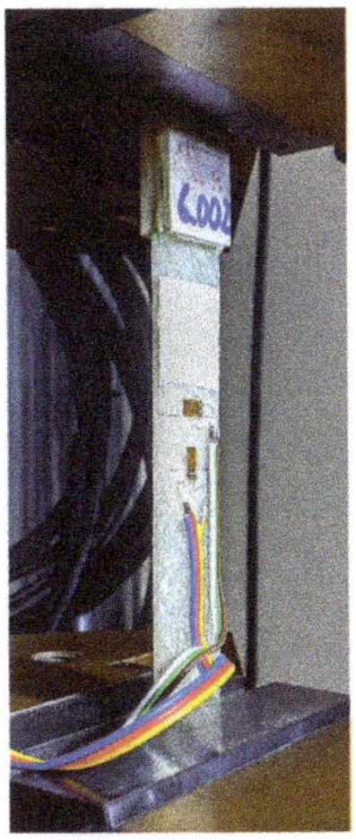

(a)

(b)

Figure 3. Testing setup (**a**) with DIC measurement cameras (**b**).

The calibration procedure was applied to determine parameters such as the relative position of cameras, lens distortion, and intensity and uniformity of lighting. The system was calibrated within a measurement volume of 85 × 65 × 45 mm. The calibration deviation was 0.028 pixels.

The utilized DIC system involves the application of a virtual regular grid of regions, known as facets, onto the recorded image sequence. These facets are then correlated with the corresponding areas on the image captured using the adjacent camera at the same stage, followed by correlation with the remaining pairs of images taken. The facet size was set at 18 × 18 pixels (approximately 0.4 × 0.4 mm) with separation of 13 pixels (resulting in 2-pixel overlap). The default overlap between neighbouring facets was 2 pixels. By utilising directional strain measurements, the shear curve and related strength properties were determined. The strain maps revealed an uneven strain distribution. The DIC strain measurements and the strain gauge measurements were compared.

Shear via tension testing of rectangular GFRP composite samples was conducted at an ambient temperature of 24 °C using the Instron 8802 (producer: Instron Illinois Tool Works Inc., Norwood, MA, USA) universal strength testing machine (Figure 3a). The strains were measured using both strain gauges and the DIC method. All data was synchronised using the force signal from the testing machine. Two loading rates were tested: 0.02 mm/min and 20.0 mm/min. The frequency of the images was set at 0.05 frames/s for the slower loading rate and 5 frames/s for the higher loading rate.

Gamma strains presented in the shear curves were calculated using the procedure outlined in standard [21]:

$$\gamma = |\varepsilon_x| + |\varepsilon_y|, \tag{1}$$

where ε_x represents strain in the axial direction measured with DIC or a strain gauge method, and ε_y is strain in the perpendicular direction measured with DIC or a strain gauge method.

Identical procedures were applied for both strain gauge and DIC measurements. Although the standard sets the limit for the shear strain at 5%, the tests were actually conducted until the sample fracture. Optical measurements allowed for determination of the gamma until fracture, while the strain gauges were damaged at a gamma strain level below 10%.

3. Results

The research presented in paper [4] aimed to investigate the influence of a loading rate on the shear stress–strain curve and material properties of the G/V composite. The determined values for material constants G_{12} and R_{12}, along with their standard deviations obtained during the research, are presented in Table 2. These values were derived following the procedures described in [5,6].

Table 2. Mean values and standard deviations for in-plane shear modulus G_{12} and in-plane shear strength R_{12} of the G/V composite [4].

Loading Rate	In-Plane Shear Modulus G_{12}	Standard Deviations G_{12}	In-Plane Shear Strength R_{12}	Standard Deviations R_{12}
[mm/min]	[mPa]	[mPa]	[mPa]	[mPa]
0.02	3395	74	38.4	0.6
2.0	3885	109	47.4	1.9
20.0	3952	194	49.3	1.6
2000	3502	100	58.2	3.1

It is worth noting that the shear strength is limited to a gamma strain of 0.5% on the shear curve. However, if the tests were analysed until failure, the in-plane shear strength would likely be higher. The reason for limiting the gamma strain to 5% was the assumption that the state of pure shear would be lost in such cases.

Based on the data presented in Table 2, a graph in Figure 4 was created to illustrate the relationship between the in-plane shear strength and the loading rate. The X-axis of the graph was scaled logarithmically to accurately represent the dependencies. The observed increase in strength is typical for polymer composites exhibiting viscoelastic properties [14]. In this representation, linear regression was applied, which resulted in a good fit to the data. The coefficient of determination for the regression is close to 1.

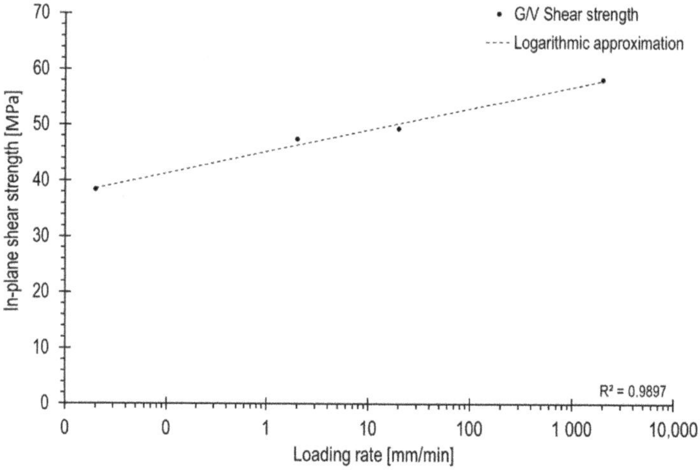

Figure 4. In-plane shear strength as a function of loading rate.

Directional strain maps calculated in the DIC system are presented in the subsequent figures. The maps illustrate the axial strains (direction X) corresponding to the measurements taken by the strain gauge installed parallel to the sample axis (Figure 2). In order to compare the results from both measurement methods, DIC strains in the perpendicular direction (Y direction) were also obtained. Due to the material and load symmetry, the directional strains are approximately equal to each other and are approximately half of the corresponding gamma strain, as described by Equation (1).

The directional maps illustrating strain rates of 0.02 and 20 mm/min are presented in Figure 5 and Figure 6, respectively. Four maps representing selected stages are provided for each loading rate. The corresponding stage numbers, displayed below the maps, indicate comparable levels of shear state advancement for both loading rates. The time interval between subsequent stages depends on the set frequency. The frequency of the images on the DIC system was set at 0.05 frames/s for the slower loading rate and 5 frames/s for the higher loading rate. The time interval between subsequent stages was 20 s and 0.02 s for the 0.02 and 20 mm/min speeds, respectively.

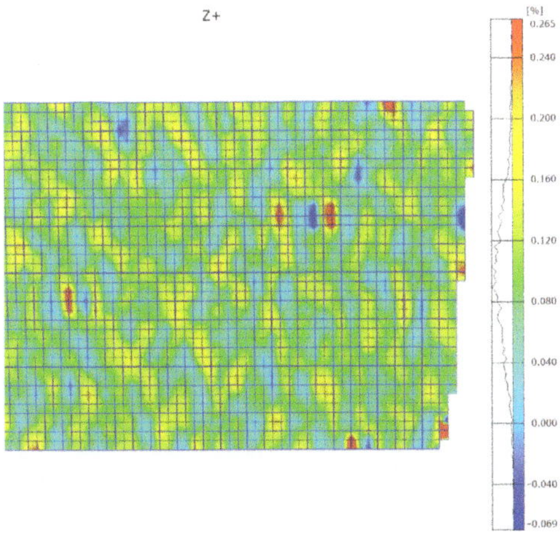

(**a**) stage 2

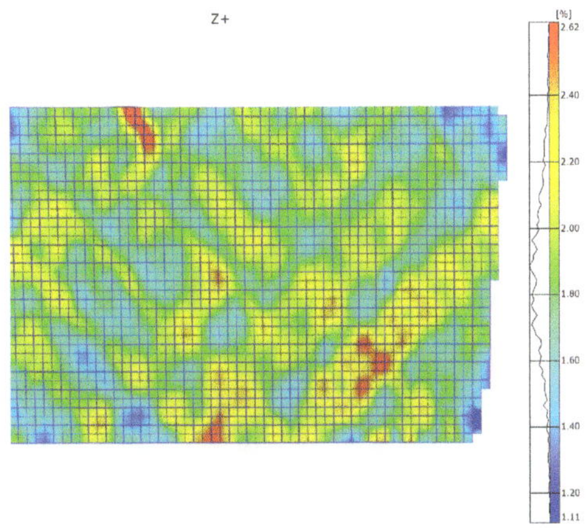

(**b**) stage 30

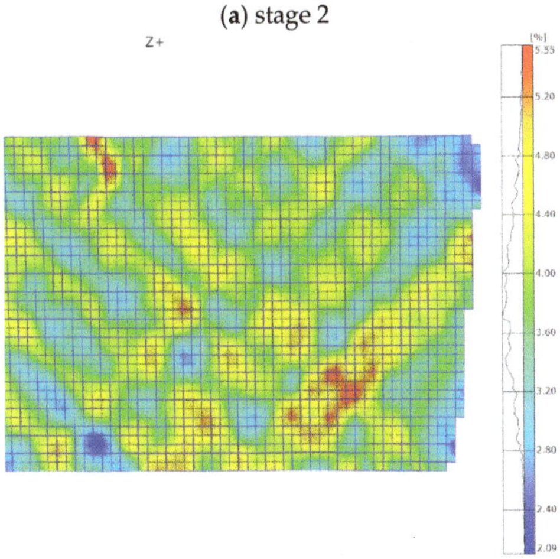

(**c**) stage 60

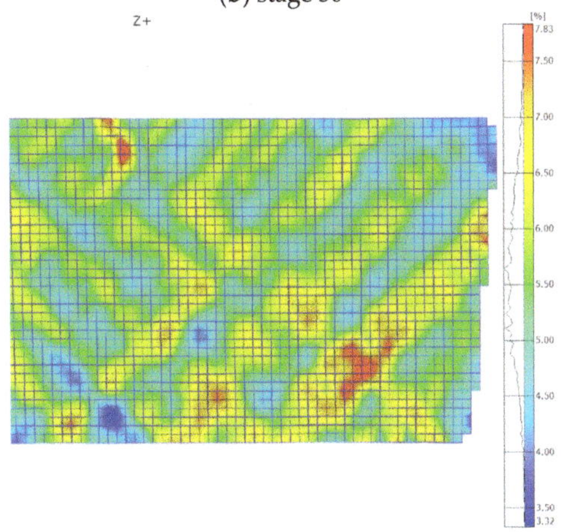

(**d**) stage 90

Figure 5. Facet maps representing strain distribution in the X direction for a sample tested at a loading rate of 0.02 mm/min: (**a**) stage 2; (**b**) stage 30; (**c**) stage 60; (**d**) stage 90.

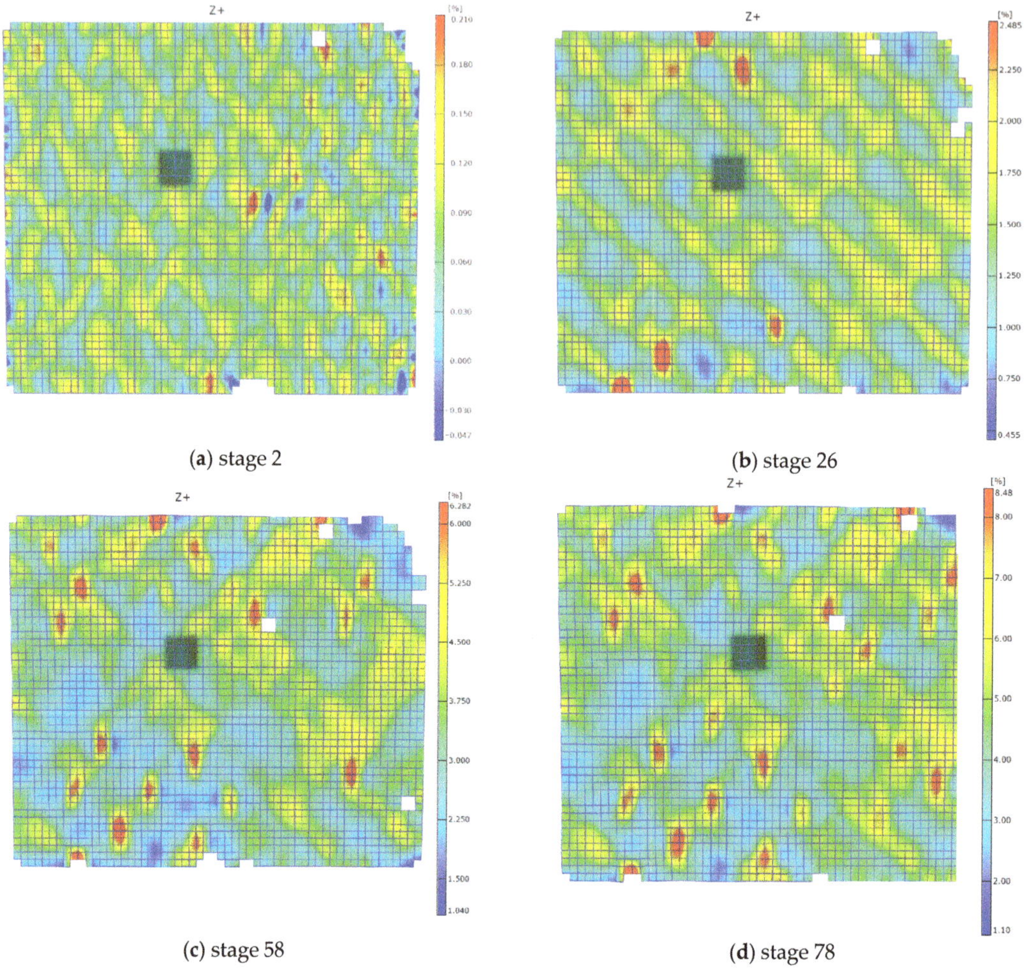

Figure 6. Facet maps representing strain distribution in the X direction for a sample tested at a loading rate of 20 mm/min: (**a**) stage 2; (**b**) stage 26; (**c**) stage 52; (**d**) stage 78.

The results reveal significant non-uniformities in the strain distribution across the measured area of the sample. The traces in the form of a net pattern, caused by the presence of the near-surface layers of glass fabric in the composite, are visible. In the first two stages, the net pattern is distinct; however, it becomes increasingly blurred in the subsequent maps.

It is also noticeable that as the damage progresses, an increasing number of areas with higher strain values, exceeding two percent, appear in the DIC maps. These areas become more pronounced in the third and fourth images, accompanied by an increase in local strain values. During the 20 mm/min shear test, a higher number of defects were observed (Figure 6c,d) compared to the slower test (Figure 5c,d). They are distributed across the entire area and have higher strain values. This effect can be associated with the occurrence of microcracks and defects within the composite structure. In the case of the slower test, the areas seem to grow less and appear less frequently. These defects may contribute to the non-linearity of the shear curve, as described in [4].

A quantitative analysis of these differences was conducted. Three different measurement areas of facets were selected for each tested loading rate. The areas selected are presented in Figures 7 and 8, for 0.02 and 20 mm/min, respectively. The facets' fields had dimensions of approximately 2 × 2 mm, which is comparable to the dimensions of the strain gauge used in the test (Figure 2). For each area, an average value was calculated using the DIC software. The first measurement area (labelled as "aver") included facets with both low and high strain values, resulting in a balanced average value. The second area (labelled as "high") was selected only from locations with higher strain facets. The third area (labelled as "low") consisted of facets, in which lower strain values were measured. In Figure 6, the shifting of the low-value facet area during testing is illustrated.

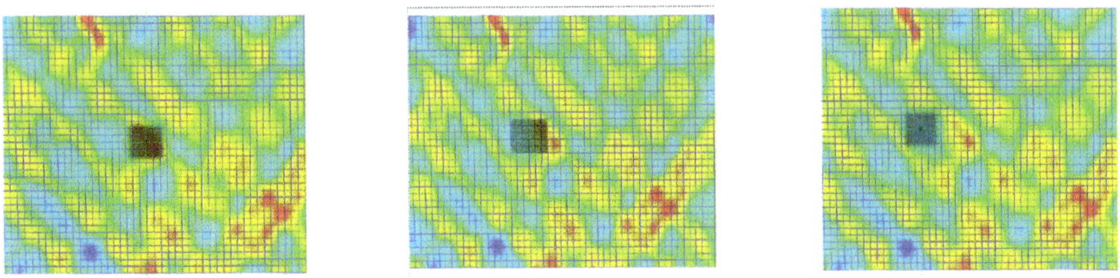

(a) (b) (c)

Figure 7. Facet fields: (**a**) area of high-value facets, (**b**) area of average-value facets, and (**c**) area of low-value facets for a sample investigated with 0.02 mm/min loading rate.

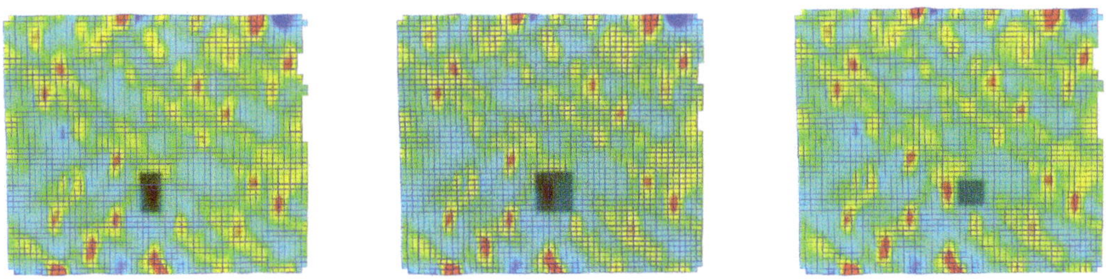

(a) (b) (c)

Figure 8. Facet fields: (**a**) area of high-value facets, (**b**) area of average-value facets, and (**c**) area of low-value facets for a sample investigated with 20 mm/min loading rate.

In order to quantitatively compare the strains obtained from all three groups of facets and validate them in relation to the strain gauge measurements, appropriate graphs illustrating directional strains were created. Figure 9 displays the results for a loading rate of 0.02 mm/min, while Figure 10 presents the results for a loading rate of 20 mm/min. The strain gauge measurements were presented using a black line, while the facet groups (average, high, and low) were represented by red, blue, and green curves, respectively. The XY coordinate system is assumed to lie within the plane of the composite plate, with the X-axis aligned along the sample axis and the Y-axis perpendicular to the sample axis.

The strain in the X direction is positive, while the strain in the Y direction is negative. Due to the load and symmetry material properties, the moduli of strain values are similar. A deformation range for strain gauges is limited, typically around 5%. As a result, the strain gauge curves terminate within that range. Optical methods, such as this DIC system, allow for less precise strain measurements compared to strain gauges but offer a wider strain range, exceeding even 100%. Strain gauges can accurately measure strains of a few μm/m, while the DIC system is limited to an accuracy of approximately 0.2%. However, the strain measurement range for strain gauges is narrower, about 5%. The presented curves were limited to approximately 8%. Within this range, DIC maps are also presented (Figures 5 and 6). Testing at a speed of 0.02 mm/min is 1000 times longer.

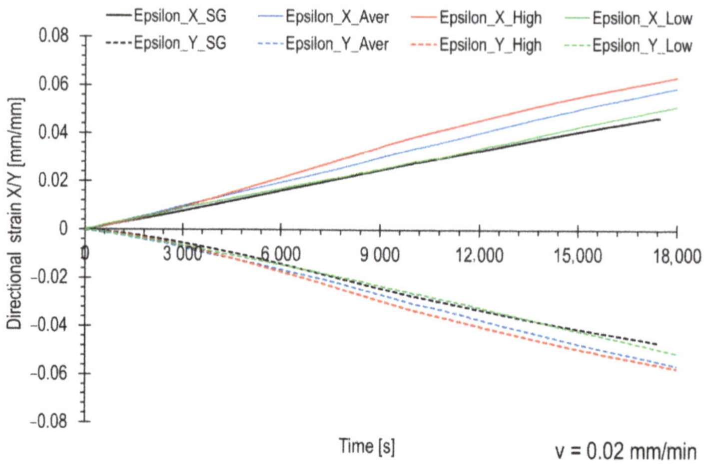

Figure 9. Directional strain comparison for different strain measuring methods. The loading rate is equal to 0.02 mm/min.

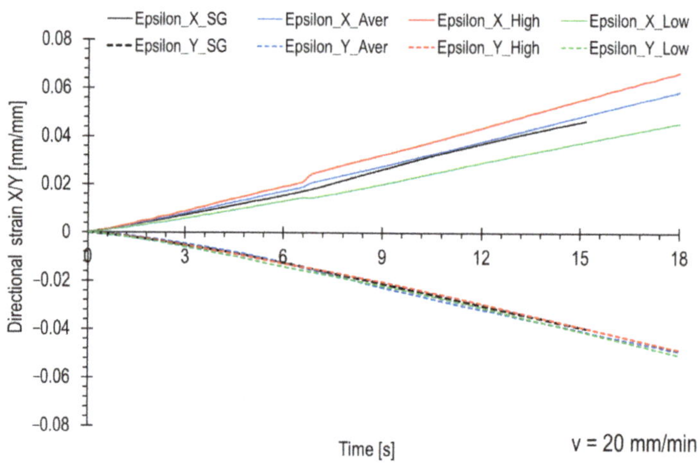

Figure 10. Directional strain comparison for different strain measuring methods. The loading rate is equal to 20 mm/min.

As expected, higher strain values were measured in the regions with high-value facets (in terms of the modulus), corresponding to the areas depicted in Figures 7b and 8b. The curves for the average-value facets (Figures 7a and 8a) fall in the middle, while the curves for the low-value areas (Figures 7c and 8c) are the lowest. The differences between the regions are more pronounced in the loading direction than in the transverse direction. These dependencies were observed for both tested loading rates.

The strain gauge curves in both figures lie between the high and low DIC measurements. However, it can be observed that for a velocity of 0.02 mm/min (Figure 9), the strain gauge curve closely aligns with the strains from the low-value region. This suggests that the strain gauge in this particular is located in the area dominated by lower strain values, randomly selected within the measurement area. In the case of the second test

(Figure 10), the curve runs more towards the middle, indicating that the measurement is more averaged.

There is a noticeable peculiarity in the strain curves (Epsilon X), clearly visible in Figure 10 at approximately the 7-s mark. At that specific moment, a significantly increased localised strain began to emerge in the analysed faced fields, indicating the occurrence of a microcrack within the sample. This crack is visibly depicted as a red area within the selected facet fields, as illustrated in Figure 8. It is worth noting that this red area is absent at the beginning of the test and starts to manifest around the 7 s mark. Additionally, the crack specifically manifests in the X direction and does not appear in the Epsilon Y curves. Furthermore, the strain gauge measurement does not capture this crack as the T-rosette is positioned in a different area of the sample.

The relative differences between the high-value and average-value areas, as well as between the low-value and average-value facets, were analysed. Table 3 presents the results for the test at a loading rate of 0.02 mm/min, with values determined at 5000, 10,000, and 15,000 s. Table 4 presents analogous results for 5, 10, and 15 s of testing at a loading rate of 20 mm/min.

Table 3. Analysis of relative differences in directional strain measurements with different measurement fields selected. The loading rate is set at 0.02 mm/min.

Time	Epsilon_X_{aver}	Epsilon_X_{max}	Epsilon_X_{min}	The Difference Relative to X_{aver} of	
[s]	[mm/mm]	[mm/mm]	[mm/mm]	X_{max}	X_{min}
5000	0.0163	0.0173	0.0143	6.2	−12.4
10,000	0.0338	0.0385	0.0283	13.9	−16.2
15,000	0.0502	0.0553	0.0430	10.3	−14.3
Time	**Epsilon_Y_{aver}**	**Epsilon_Y_{max}**	**Epsilon_Y_{min}**	**The Difference Relative to Y_{aver} of**	
[s]	[mm/mm]	[mm/mm]	[mm/mm]	Y_{max}	Y_{min}
5000	−0.0133	−0.0136	−0.0118	1.9	−12.9
10,000	−0.0304	−0.0333	−0.0261	9.6	−21.6
15,000	−0.0472	−0.0494	−0.0419	4.5	−15.1

Table 4. Analysis of relative differences in directional strain measurements with different measurement fields selected. The loading rate is set at 20 mm/min.

Time	Epsilon_X_{aver}	Epsilon_X_{max}	Epsilon_X_{min}	The Difference Relative to X_{aver}	
[s]	[mm/mm]	[mm/mm]	[mm/mm]	X_{max}	X_{min}
5	0.0141	0.0158	0.0106	12.0	−25.0
10	0.0311	0.0360	0.0232	15.9	−25.4
15	0.0485	0.0553	0.0376	14.1	−22.6
Time	**Epsilon_Y_{aver}**	**Epsilon_Y_{max}**	**Epsilon_Y_{min}**	**The Difference Relative to Y_{aver}**	
[s]	[mm/mm]	[mm/mm]	[mm/mm]	Y_{max}	Y_{min}
5	−0.0093	−0.0101	−0.0112	8.3	11.6
10	−0.0256	−0.0232	−0.0247	−9.4	6.4
15	−0.0402	−0.0388	−0.0406	−3.5	4.6

Differences in strain measurements depending on the selected facet groups can reach several percent. At the beginning of the test, with a loading rate of 0.02 mm/min, choosing facets from a high-value area resulted in approximately 6% higher strains in the X direction compared to a balanced selection (average-value area). On the other hand, selecting extremely low-strain fields led to measurements underestimating the values by over 12% compared to the averaged values. As the test progressed, the relative differences increased. After 10,000 s, a relative difference for the high-strain field reached nearly 14%, while the low-strain field underestimated the values by 16%. It is important to note that the difference between the highest and lowest strain fields also increased. During the subsequent reading at 15,000 s, the differences stabilised and slightly decreased.

The analysis of transverse strains confirms the aforementioned trend. The differences in the second reading reach up to 20% for the lowest strain field. However, it is worth noting that the differences are lower compared to the axial direction.

The results obtained from the strain gauge measurement method and various DIC facet areas were used to generate shear stress–strain curves, which are presented in Figures 11 and 12. In all cases, the shear strain values were calculated using Equation (1). These graphs illustrate how the location of the strain measurement area influences the shape of the shear stress curve. The electro-resistant curve (black colour) generally lies roughly between the high- and low-value DIC curves.

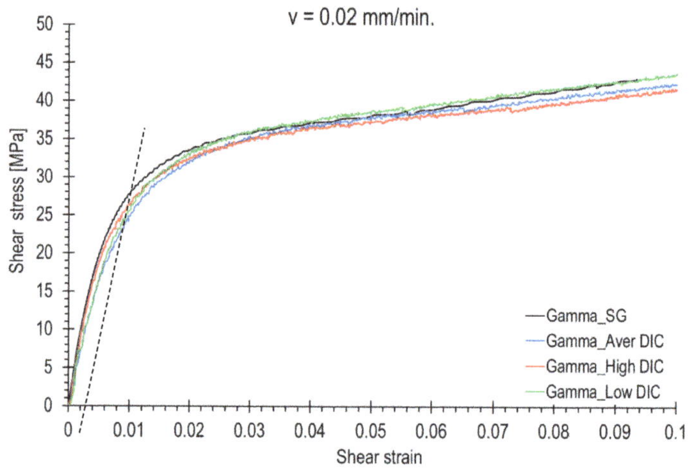

Figure 11. Shear stress–shear strain curve waveforms for different strain measuring methods and various facet areas. The loading rate is equal to 0.02 mm/min. The shear offset yield strength was determined using dashed lines.

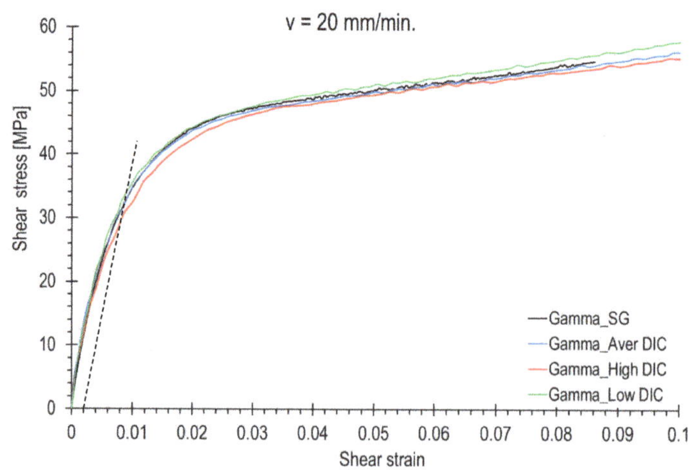

Figure 12. Shear stress–shear strain curve waveforms for different strain measuring methods and various facet areas. The loading rate is equal to 20 mm/min. The shear offset yield strength was determined using dashed lines.

Noteworthy differences can be observed, particularly in the transitional region where the shear curve shifts from linear to nonlinear. This can have a significant impact on the measurement of strength parameters, such as the shear offset yield strength. For a loading rate of 20 mm/min, readings can range from 27 to 32 MPa (indicated by the dotted line in Figure 12), depending on the selection of the measurement area or the positioning of the strain gauge. At a lower loading rate, the values can fluctuate between 21 and 27 MPa (indicated by the dotted line in Figure 11).

It can be concluded that choosing a location on the sample surface with a lower level of strain results in an increase in the shear stress curve because the given load value is recorded for a lower strain. The electro-resistant curves lie roughly between the upper and lower DIC curves. In the DIC analysis, it is possible to choose any area. The choice of the location for strain gauge mounting is random.

4. Discussion of the Results

The determined material properties of the G/V composite, as a function of strain rate, indicate that the shear strength increases with higher strain rates. This relationship can be represented by a straight line on logarithmic loading rate coordinates. Such a trend is commonly observed in polymer matrix composites that demonstrate viscoelastic behaviour and is consistent with findings reported in the existing literature.

The investigation of the shear stress–strain curve was extended with optical measurements, enabling the determination of strain distribution on the G/V composite. An uneven distribution of the strain field was observed, characterised by a net-like pattern caused by the presence of surface layers of glass fabric in the composite.

Based on measurements from various facet areas, it was observed that there are relative differences of several percent between the areas of the averaged facets and the areas of the facets with high or low values. These differences are more pronounced at higher loading rates, where they can reach up to 25%.

DIC measurement is well suited for detecting anomalies and damaged areas in FRP composites, as investigated in [32]. During the G/V test, it was observed on the DIC maps that as the damage progresses, an increasing number of areas exhibit significant levels of strain. This phenomenon can be attributed to the development of damage, formation of defects, and microcracks. This could potentially explain the non-linear shape of the shear stress–strain curve.

In the shear graphs (Figures 11 and 12), the gamma strain is limited to 10%. However, shear-by-tensile tests were carried out until failure. An example image of a DIC area from one of the samples is provided in Figure 13. Despite being observed at low magnification, numerous cracks, indicated by ellipses, are visible on the surface of the samples. Nevertheless, associating specific cracks with the corresponding DIC image can be challenging due to potential overlapping and alterations that may occur throughout a test.

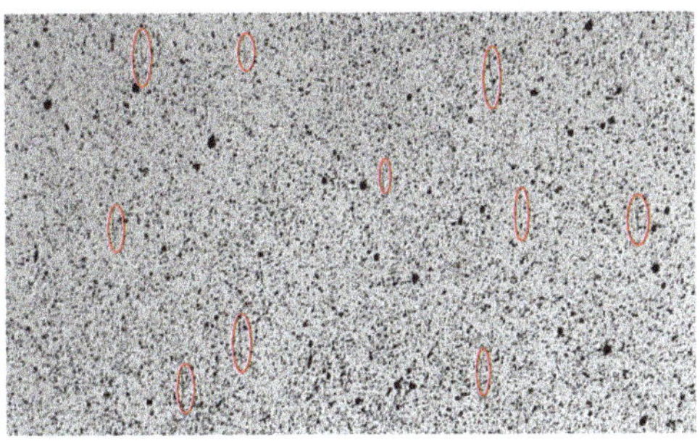

Figure 13. An example image of a DIC area from one of the samples tested with a loading rate of 20 mm/min reveals the presence of multiple visible cracks in the field, examples of which were indicated with the red ellipses.

Strain field optical measurement methods, such as Digital Image Correlation (DIC), appears to be the optimal choice for capturing non-uniform strain distribution on the surface of GFRP composites. When analysing DIC results, it is essential to pay attention to potential variations in the strain distribution. In the case of observing regions with higher strain values, it is important to verify whether they are a result of deformation or material damage.

The shear stress–strain curve obtained via tensile testing does not take into account the non-uniform strain distribution on the composite. Usually, the curve is determined based on strain gauge measurements. However, the installation of T-rosette strain gauges is random with respect to the described strain distribution. Consequently, the placement of the strain gauge or choosing a DIC facet area can influence the determined gamma strain and strength parameters. In particular, it may influence the measurement of the shear offset yield strength. When using the electrical resistance method, it is recommended to place multiple strain gauges on the measurement surface of the tested material in order to achieve more reliable results and understand the variability of values. This allows for the determination of strains at multiple measurement points and the identification of any areas with atypical responses.

The literature on FRP composite research is extensive, with numerous studies utilising the DIC method for strain field measurement. However, most of these studies concentrate on carbon FRP composites and cover a wide range of strength tests conducted under different conditions, making direct comparisons with this work challenging or even impossible. Nevertheless, the overall findings remain consistent. The strain field is observed to be non-uniform, as demonstrated in studies such as [26,27,33], which were dedicated to CFRP composite. During testing of the G/V, the occurrence of cracks, damage, and delamination becomes clearly visible on DIC maps.

When conducting strain measurements using both strain gauges and DIC, it is important to compare the results obtained from these methods to identify any differences and understand their causes.

5. Conclusions

This study aimed to investigate the non-uniformity of the strain field during the off-axis tensile test of the G/V composite and its influence on the shape of the shear curve. The following key conclusions can be drawn:

- Significant non-uniformities in strain distribution were observed on the tested samples' surface, manifested as a mesh-like pattern.
- The choice of the measurement area had an impact on the shape of the stress–strain curve. Differences in strain measurements depending on the selected areas could reach several percent and influence the obtained stress–strain curves and strength parameters.
- The DIC method proved to be valuable in detecting anomalies and damages in FRP composites. As the damage progressed, an increasing number of regions with higher strain values exceeding two percent, were appearing on the DIC maps. Microcracks and defects were identified within the composite structure. This could potentially explain the non-linear shape of the shear stress–strain curve.
- It is essential to consider potential variations in strain distribution when analysing DIC results. In the case of observing regions with higher strain values, it is important to verify whether they result from deformation or material damage.
- When utilising the electrical resistance method, it is recommended to place multiple strain gauges on the measurement surface of the tested material in order to achieve more reliable results and understand the variability of values. This allows for the determination of strains at multiple measurement points and the identification of any areas with unusual responses.

- When conducting strain measurements using both strain gauges and DIC, it is important to compare the results obtained from these methods to identify any differences and understand their causes.

Funding: Materials for this work were supported by the Institute of Mechanics and Computational Engineering, Faculty of Mechanical Engineering, Military University of Technology, Warsaw, Poland [Block Grant No. PBS/23-894/2019]. This support is gratefully acknowledged.

Institutional Review Board Statement: Not applicable.

Informed Consent Statement: Not applicable.

Data Availability Statement: The necessary data are contained within the article.

Acknowledgments: I would like to express my appreciation to Marian Klasztorny for his support and provision of materials throughout this research project.

Conflicts of Interest: I declare no conflict of interest.

References

1. Özkılıç, Y.O.; Gemi, L.; Madenci, E.; Aksoylu, C. Effects of stirrup spacing on shear performance of hybrid composite beams produced by pultruded GFRP profile infilled with reinforced concrete. *Archiv. Civ. Mech. Eng.* **2023**, *23*, 36. [CrossRef]
2. Madenci, E.; Özkılıç, Y.O.; Aksoylu, C.; Safonov, A. The Effects of Eccentric Web Openings on the Compressive Performance of Pultruded GFRP Boxes Wrapped with GFRP and CFRP Sheets. *Polymers* **2022**, *14*, 4567. [CrossRef] [PubMed]
3. Gemi, L.; Madenci, E.; Özkılıç, Y.O. Experimental, analytical and numerical investigation of pultruded GFRP composite beams infilled with hybrid FRP reinforced concrete. *Eng. Struct.* **2021**, *244*, 112790. [CrossRef]
4. Klasztorny, M.; Nycz, D.B.; Bogusz, P. Rheological effects in in-plane shear test and in-plane shear creep test on glass-vinyl-ester lamina. *Compos. Theory Pract.* **2020**, *20*, 35–42.
5. *ASTM D3518/D3518M-94(2001)*; Standard Test Method for In-Plane Shear Response of Polymer Matrix Composite Materials by Tensile Test of a ±45° Laminate. ASTM International: West Conshohocken, PA, USA, 2001.
6. *EN ISO 14129:1997*; Fibre-Reinforced Plastic Composites—Determination of the In-Plane Shear Stress/Shear Strain Response, Including the In-Plane Shear Modulus and Strength, by the ±45° Tension Test Method. ISO: Geneva, Switzerland, 1997.
7. Zhou, G.; Davies, G.A.O. Characterization of thick glass woven roving/polyester laminates: 1. Tension, compression and shear. *Composites* **1995**, *26*, 579–586. [CrossRef]
8. Dickson, T.; Munro, M. Selection of an in-plane shear test method based on the shear sensitivity of laminate tensile modulus. *Composites* **1995**, *26*, 17–24. [CrossRef]
9. Liang, Y.; Wang, H.; Gu, X. In-plane shear response of unidirectional fiber reinforced and fabric reinforced carbon/epoxy composites. *Polym. Test.* **2013**, *32*, 594–601. [CrossRef]
10. Cui, H.; Thomson, D.; Pellegrino, A.; Wiegand, J.; Petrinic, N. Effect of strain rate and fibre rotation on the in-plane shear response of ±45° laminates in tension and compression tests. *Compos. Sci. Technol.* **2016**, *135*, 106–115. [CrossRef]
11. Herakovich, C.T.; Schroedter, R.D.; Gasser, A.; Guitard, L. Damage evolution in [±45]s laminates with fiber rotation. *Compos. Sci. Technol.* **2000**, *60*, 2781–2789. [CrossRef]
12. Van Paepegem, W.; De Baere, I.; Degrieck, J. Modelling the nonlinear shear stress-strain response of glass fibrereinforced composites. Part I: Experimental results. *Compos. Sci. Technol.* **2006**, *66*, 1455–1464. [CrossRef]
13. Van Paepegem, W.; De Baere, I.; Degrieck, J. Modelling the nonlinear shear stress–strain response of glass fibrereinforced composites. Part II: Model development and finite element simulations. *Compos. Sci. Technol.* **2006**, *66*, 1465–1478. [CrossRef]
14. Ng, W.H.; Salvi, A.G.; Waas, A.M. Characterization of the in-situ non-linear shear response of laminated fiberreinforced composites. *Compos. Sci. Technol.* **2010**, *70*, 1126–1134. [CrossRef]
15. Klasztorny, M.; Nycz, D.B.; Romanowski, R.K.; Gotowicki, P.; Kiczko, A.; Rudnik, D. Effects of operating temperature and accelerated environmental ageing on glass-vinylester composite mechanical properties. *Mech. Compos. Mater.* **2017**, *53*, 335–350. [CrossRef]
16. Wilczynski, A.; Klasztorny, M. Determination of complex compliances of fibrous polymeric composites. *J. Compos. Mater.* **2000**, *34*, 2–26. [CrossRef]
17. Klasztorny, M.; Wilczynski, A. Constitutive equations of viscoelasticity and estimation of viscoelastic parameters of unidirectional fibrous polymeric composites. *J. Compos. Mater.* **2000**, *34*, 1624–1639. [CrossRef]
18. Dan, X.; Li, J.; Zhao, Q.; Sun, F.; Wang, Y.; Yang, L. A Cross-Dichroic-Prism-Based Multi-Perspective Digital Image Correlation System. *Appl. Sci.* **2019**, *9*, 673. [CrossRef]
19. Chu, T.C.; Ranson, W.F.; Sutton, M.A. Applications of digital-image-correlation techniques to experimental mechanics. *Exp. Mech.* **1985**, *25*, 232–244. [CrossRef]
20. Tang, Z.; Liang, J.; Xiao, Z.; Guo, C.; Hu, H. Three-dimensional digital image correlation system for deformation measurement in experimental mechanics. *Opt. Eng.* **2010**, *49*, 103601. [CrossRef]

21. Li, J.; Xin, X.; Yang, G.; Zhang, G.; Siebert, T.; Yang, L. Whole-field thickness strain measurement using multiple camera digital image correlation system. *Opt. Lasers Eng.* **2017**, *90*, 19–25. [CrossRef]
22. Grytten, F.; Daiyan, H.; Polanco-Loria, M.; Dumoulin, S. Use of digital image correlation to measure large strain tensile properties of ductile thermoplastics. *Polym. Test.* **2009**, *28*, 653–660. [CrossRef]
23. Seon, G.; Makeev, A.; Schaefer, J.D.; Justusson, B. Measurement of Interlaminar Tensile Strength and Elastic Properties of Composites Using Open-Hole Compression Testing and Digital Image Correlation. *Appl. Sci.* **2019**, *9*, 2647. [CrossRef]
24. Meng, F.; Zhang, X.; Wang, J.; Li, C.; Chen, J.; Sun, C. 3D Strain and Elasticity Measurement of Layered Biomaterials by Optical Coherence Elastography based on Digital Volume Correlation and Virtual Fields Method. *Appl. Sci.* **2019**, *9*, 1349. [CrossRef]
25. Sutton, M.A.; Ke, X.; Lessner, S.M.; Goldbach, M.; Yost, M.; Zao, F.; Schreier, H.W. Strain field measurements on mouse carotid arteries using microscopic three-dimensional digital image correlation. *J. Biomed. Mater. Res. Part A* **2008**, *84*, 178–190. [CrossRef] [PubMed]
26. Gliesche, K.; Hübner, T.; Orawetz, H. Investigations of in-plane shear properties of ±45°-carbon/epoxy composites using tensile testing and optical deformation analysis. *Compos. Sci. Technol.* **2005**, *65*, 163–171. [CrossRef]
27. Ma, Y.; Li, Y.; Liu, L. Off-Axis Tension Behaviour of Unidirectional PEEK/AS4 Thermoplastic Composites. *Appl. Sci.* **2023**, *13*, 3476. [CrossRef]
28. Tekieli, M.; De Santis, S.; de Felice, G.; Kwiecień, A.; Roscini, F. Application of Digital Image Correlation to composite reinforcements testing. *Compos. Struct.* **2017**, *160*, 670–688. [CrossRef]
29. Gupta, S.; Pal, S.; Ray, B.C. An overview of mechanical properties and failure mechanism of FRP laminates with hole/cutout. *J. Appl. Polym. Sci.* **2023**, *140*, 53862. [CrossRef]
30. Thomson, D.; Quino, G.; Cui, H.; Pellegrino, A.; Erice, B.; Petrinic, N. Strain-rate and off-axis loading effects on the fibre compression strength of CFRP laminates: Experiments and constitutive modelling, Composites Science and Technology. *Compos. Sci. Technol.* **2020**, *195*, 108210. [CrossRef]
31. Thomson, D.; Cui, H.; Erice, B.; Petrinic, N. A study on the longitudinal compression strength of fibre reinforced composites under uniaxial and off-axis loads using cross-ply laminate specimens. *Compos. Part A Appl. Sci. Manuf.* **2019**, *121*, 213–222. [CrossRef]
32. del Rey Castillo, E.; Allen, T.; Henry, R.; Griffith, M.; Ingham, J. Digital image correlation (DIC) for measurement of strains and displacements in coarse, low volume-fraction FRP composites used in civil infrastructure. *Compos. Struct.* **2019**, *212*, 43–57. [CrossRef]
33. Merzkirch, M.; Foecke, T. 10° off-axis testing of CFRP using DIC: A study on strength, strain and modulus. *Compos. Part B Eng.* **2020**, *196*, 108062. [CrossRef]
34. Polimal VE-11M Resin. *Technical Information*; CIECH Sarzyna: Nowa Sarzyna, Poland, 2017.

polymers

MDPI

Article

Simulation-Based Identification of Operating Point Range for a Novel Laser-Sintering Machine for Additive Manufacturing of Continuous Carbon-Fibre-Reinforced Polymer Parts

Michael Baranowski [1,2,*], Zijin Shao [1], Alexander Spintzyk [1], Florian Kößler [1] and Jürgen Fleischer [1]

1 Institute of Production Science, Faculty of Mechanical Engineering, Karlsruhe Institute of Technology (KIT), Kaiserstaße 12, 76131 Karlsruhe, Germany; taopiangearen@gmail.com (Z.S.)
2 Karlsruhe Research Factory, Karlsruhe Institute of Technology (KIT), Rintheimer Querallee 2, 76131 Karlsruhe, Germany
* Correspondence: michael.baranowski@kit.edu; Tel.: +49-1523-9502642

Abstract: Additive manufacturing using continuous carbon-fibre-reinforced polymer (CCFRP) presents an opportunity to create high-strength parts suitable for aerospace, engineering, and other industries. Continuous fibres reinforce the load-bearing path, enhancing the mechanical properties of these parts. However, the existing additive manufacturing processes for CCFRP parts have numerous disadvantages. Resin- and extrusion-based processes require time-consuming and costly post-processing to remove the support structures, severely restricting the design flexibility. Additionally, the production of small batches demands considerable effort. In contrast, laser sintering has emerged as a promising alternative in industry. It enables the creation of robust parts without needing support structures, offering efficiency and cost-effectiveness in producing single units or small batches. Utilising an innovative laser-sintering machine equipped with automated continuous fibre integration, this study aims to merge the benefits of laser-sintering technology with the advantages of continuous fibres. The paper provides an outline, using a finite element model in COMSOL Multiphysics, for simulating and identifying an optimised operating point range for the automated integration of continuous fibres. The results demonstrate a remarkable reduction in processing time of 233% for the fibre integration and a reduction of 56% for the width and 44% for the depth of the heat-affected zone compared to the initial setup.

Keywords: laser sintering (LS); continuous carbon-fibre-reinforced polymer parts (CCFRP); fibre integration unit; heat-affected zone; finite element model; central composite design (CCD)

1. Introduction

The utilisation of continuous carbon-fibre-reinforced polymer (CCFRP) parts in industrial applications presents a substantial opportunity for achieving significant reductions in future product consumption and CO2 emissions while maintaining economic viability [1]. CCFRP parts are notable for their favourable weight-to-strength ratio and impressive mechanical tensile properties along the fibre orientation. Continuous fibres play a crucial role in enhancing the mechanical characteristics of fibre-reinforced parts along the load-bearing pathway [2].

Additive manufacturing processes offer a promising avenue for the tool-less and time-efficient production of CCFRP parts, allowing for high levels of customisation and shape complexity. Material extrusion (MEX), which includes techniques like fused layer modelling (FLM) and ARBURG plastic freeforming (APF), has gained prominence in the literature as a viable method for the additive manufacturing of CCFRP parts [3–9]. Another category of processes employed for CCFRP parts is vat photopolymerisation (VPP) [10–12]. However, it is essential to note that CCFRP parts produced using these processes (MEX and VPP) have decisive disadvantages. The inherent nature of these processes necessitates using support structures, which must be removed and managed after production. This results in additional time and cost implications for the disposal and post-processing steps. Furthermore, the reliance on support structures restricts the ability to create features such as overhangs, cavities, and undercuts, consequently limiting the complexity of the produced parts. Additionally, removing support structures can potentially introduce surface defects to the remaining part surfaces, leading to a less uniform appearance of the parts. Moreover, MEX and VPP processes do not facilitate cost-effective small-batch production.

In contrast, the laser-sintering (LS) process is a compelling alternative for CCFRP parts. In a comparative analysis of the mechanical and thermal properties, as well as the long-term stability of polymer parts produced via material extrusion (MEX), vat photo polymerisation (VPP), and LS, the LS process exhibits notable advantages, notably in the creation of durable, functional parts akin to those produced through injection moulding [13,14]. As reported in [13], laser-sintered specimens demonstrate a higher Young's modulus compared to their injection-moulded counterparts, attributed to the increased crystallinity in the molecular structure of the semi-crystalline thermoplastics. The tensile strength in the LS-produced specimens closely approximates that of their injection-moulded counterparts. Additional merits of the LS process include the absence of support structures, offering greater design flexibility. In LS, unprocessed powder is a natural support structure, eliminating the need for labour-intensive and costly post-processing steps [15]. Furthermore, LS can generate near-net-shape functional parts with intricate features like undercuts, cavities, and overhangs in a single manufacturing step. The LS process's ability to efficiently position parts vertically and horizontally within the powder bed facilitates cost-effective small-batch production [13,16]. Compared to parts produced through fused layer modelling (FLM), LS-produced parts exhibit significantly reduced anisotropy, enhanced dimensional accuracy, and reduced surface roughness [13,17]. Consequently, the LS process offers promising foundational properties (matrix) for CCFRP parts. However, it is worth noting that there are currently no commercially available LS machines that fully integrate the advantages of the LS process with those of continuous fibres. Challenges associated with integrating continuous fibres into the LS process primarily revolve around intricate temperature control and the repetitive recoater movement for each layer.

To harness the distinct advantages of the LS process and leverage the promising attributes of continuous carbon fibres, a technical demonstration of feasibility was conducted [18–20]. This involved the layer-wise integration of continuous rovings (fibre strands) into laser-sintered parts composed of PA12, facilitated by a newly developed LS machine. Central to this prototype LS machine is a dedicated fibre integration unit designed to seamlessly incorporate the rovings into the pre-existing layers of the parts. A heated fibre nozzle is employed to liquefy the polymer locally, creating a heat-affected zone (HAZ) characterised by specific width and depth parameters. Subsequently, a synchronised series of movements involving the roving feed rate and the nozzle feed rate is used to introduce the roving into the liquefied polymer melt. In a prior study [19], an in-depth examination of the extended LS process was conducted, focusing on temperature control during the integration of fibres to determine the optimal operational conditions for achieving curl-free roving integration. Furthermore, an investigation into the impact of the process parameters and their interactions on the dimensions of the HAZ was carried out [18] using a split-plot design (SPD). The findings from these previous studies [18,19] served as the foundational knowledge for this research. However, the insights gained from [18] led to an inefficient production process for continuous carbon-fibre-reinforced polymer CCFRP parts. The production speed was relatively slow, necessitating a low feed rate for roving integration. This low feed rate resulted in an economically disadvantageous and time-consuming production of CCFRP parts. Additionally, the dimensions of the HAZ, particularly its width and depth, were more significant than desired, requiring the placement of rovings at a significant distance from the part edges. This oversized HAZ resulted in a limitation in achieving a higher fibre volume content (FVC) and the corresponding mechanical properties. As a result, the full potential of the new LS process with continuous fibre integration remained untapped. Only by identifying an optimised operational point for roving integration is it possible to reduce the dimensions of the HAZ and, crucially, the processing time in a targeted manner. A systematic approach to determining an operating point that ensures a secure process and a suitably compact HAZ while enabling swift roving integration is imperative for the cost-effective production of LS components with elevated FVC and enhanced mechanical properties.

Therefore, the objective of this paper is the simulation-based identification of an optimal operating point range within which the smallest possible HAZs can be generated, and the rovings can be integrated into the part as quickly as possible but simultaneously in a process-safe manner. Using COMSOL Multiphysics (Version 6.1), the process zone is first modelled with the help of a macroscopic modelling approach. Using this FE model, the creation process of the HAZ, due to the heat input of the heated fibre nozzle, is to be simulated, and an optimised operating point range is to be identified in a subsequent simulation study. Section 2.1 presents the principle of roving integration. The influencing and target variables on the roving integration caused by the fibre integration unit are discussed in the same section. The procedure for deriving the FE model in COMSOL (Version 6.1) is described in Section 2.2. The same section presents the evaluation procedure for evaluating the FE model's accuracy using a convergence analysis and a plausibility check. Section 2.3 describes the approach for the simulation-based identification of an optimised operating point range with the help of a central composite design (CCD) with initially assumed factor levels within which an optimised operating point range is sought. Finally, the identified operating point range derived by the FE model is experimentally validated, and the roving integration is analysed in the same chapter using an adjusted, more detailed CCD. Section 3 presents and discusses the results concerning the initial state.

This study successfully showcased a substantial 233% reduction in the processing time required for roving integration. Consequently, this achievement paves the way for the more cost-effective production of CCFRP parts using the developed LS machine. Furthermore, 56% and 44% reductions in the width and depth of the heat-affected zone (HAZ) were attained. This advancement enables the integration of rovings closer to the edges of the parts, thereby permitting a higher fibre volume content (FVC). As a result, this study offers optimised operation points that can guide future research efforts in systematically enhancing the FVC and the accompanying mechanical properties of CCFRP parts.

2. Materials and Methods

2.1. Principle of Roving Integration

This section outlines the process flow to provide a fundamental understanding of roving integration within the developed LS machine—see Figure 1a.

It is important to note that this paper does not offer an exhaustive description of the machine itself or the achievable properties of the produced parts, as those details can be found in [18,20]. The process begins within a heated process chamber of the LS machine, maintained at approximately 110 °C. Subsequently, a fresh layer of powder is evenly applied by the recoater, and the powder bed's surface temperature is homogenised using infrared (IR) emitters. Black PA12 powder (Sintratec AG, Brugg, Switzerland) with a melting temperature of approx. 185 °C is used as the matrix material. The laser beam then liquefies the applied powder layer. Following ISO 6983 G-code instructions, the layer-specific (2D) integration of one or more rovings is carried out sequentially. In order to be able to process parts with the developed LS machine, a MATLAB app was developed in xy to slice the parts (.stl) and generate the G-code for roving integration [21]. The developed LS machine uses 1K rovings (HTA40) with 67 tex (Teijin Limited, Tokyo, Japan), a width of about 365 μm and a thickness of about 110 μm. The 1K rovings have an elliptical shape in their delivery condition. To facilitate this process, the entire fibre integration unit, with an additional heat source and a heated fibre nozzle, is moved rapidly in the x and y directions to reach the initial point of the first roving path located in the stable sintering region. Detailed descriptions of all symbols used in this process can be found in Tables 1 and 2.

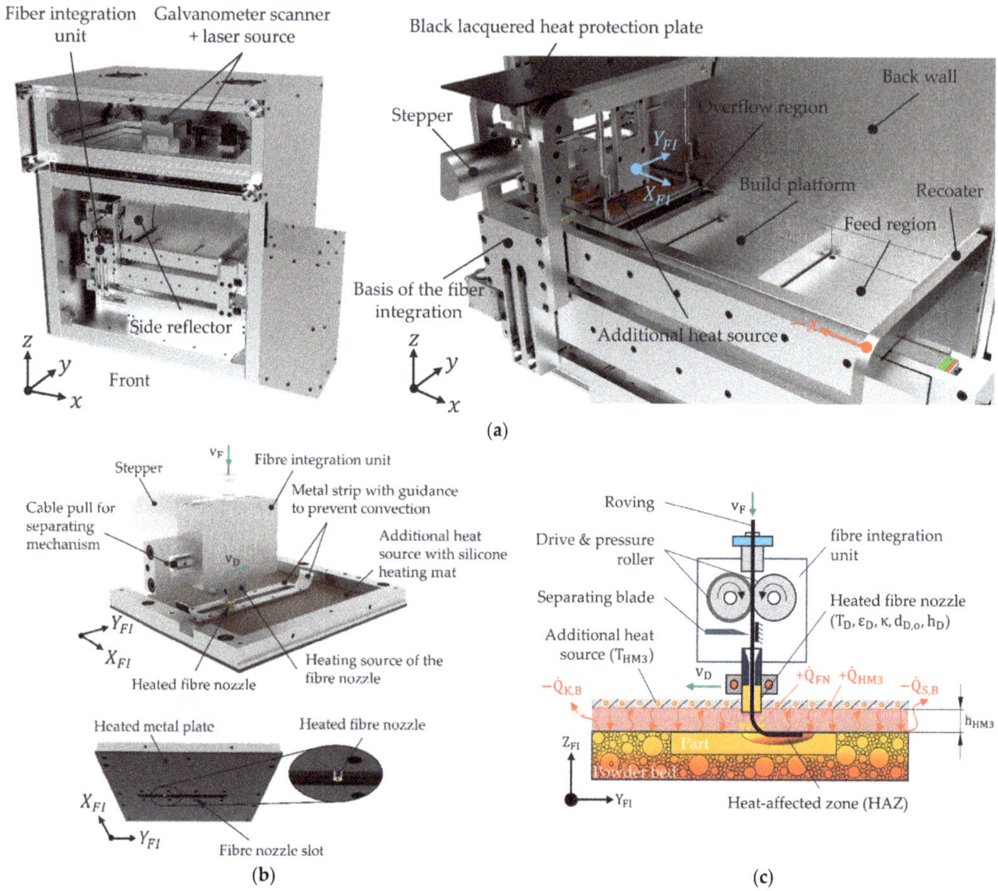

Figure 1. View of the process chamber of the developed LS machine (**a**) with a detailed view of the fibre integration unit (**b**). Schematic representation of the heat fluxes and influencing factors (**c**).

Detailed illustrations of the fibre integration unit and its constituent components are provided in Figure 1b,c. Throughout the entire roving integration process, the structure of the fibre integration unit moves into the path of radiation from the infrared (IR) emitters installed within the LS machine. This movement results in the shadowing of IR radiation. To maintain the powder bed's part and surface within the sintering window, a metal plate fitted with an adhered silicone heating mat (functioning as an additional heat source) is positioned parallel to the powder surface. It is situated at a distance of h_{HM3} beneath the bottom side of the fibre integration unit. The term "sintering window" refers to a temperature range encompassing the interval between the onset of crystallisation and the melting of the semi-crystalline thermoplastic material used. The PA12 material employed in this study corresponds to an approximate crystallisation temperature of 154 °C and a melting temperature of around 184 °C, resulting in a sintering window of 30 °C [19,22]. The interaction between the heated fibre nozzle, operating at temperature T_{HM3}, and the part's surface induces the formation of a localised melt zone or heat-affected zone (HAZ). The dimensions of this HAZ, namely, its width b_{HAZ} and depth t_{HAZ}, describe the extent to which the polymer's viscosity is locally reduced. The successful embedding of rovings within the part relies on creating a sufficiently deep HAZ through the action of the heated fibre nozzle. The resultant molten material adheres to the roving, firmly anchoring it to the underlying layers. Positioned above the fibre nozzle, a cutting blade trims the continuous roving to a length specified by the programmed instructions in the G-code. It is important to note that the built-in diode laser, operating at 450 nm with a power output of 1.6 W, remains inactive during the roving integration process. Following the successful integration of the rovings, the fibre integration unit returns to its designated home position, and the recoater applies a fresh layer of powder. Once the infrared (IR) emitters have sufficiently heated and homogenised the powder bed surface to reach the sintering temperature of approximately 175 °C, the laser is employed to melt the new powder layer, thereby fully incorporating the roving within the polymer matrix. This sequence is repeated until all rovings have been integrated per the instructions specified in the G-code. Subsequently, after the printing process concludes, a controlled cooling process is initiated for the powder bed housing the CCFRP parts.

Table 1. Influencing variables for keeping the part/powder bed surface temperature within the sintering window.

Symbol	Description	Unit	Setting	Value
d_R	Thickness of the 1K roving	µm	State of delivery	≈ 365
h_D	Distance between fibre nozzle and powder bed surface	mm	PLC	0–2
$d_{D,o} \mid d_{D,i}$	Outer $d_{D,o}$ and inner diameter $d_{D,i}$ of the fibre nozzle	mm	Lathe	$d_{D,o} \geq 2$ $d_{D,i} = 0.6$
κ	The curvature of the fibre nozzle	-	Lathe	Planar/concave
A_{HM3}	Area of the heat source (metal plate)	mm^2	Laser cutting	≈ 22,500
h_{HM3}	Air gap width (additional heat source to powder surface)	mm	Feeler gauge tape	0–2
ε_D	The emissivity of the fibre nozzle (black-oxidised)	-	Varnished	≈ 0.9
ε_P	The emissivity of the already molten part (Sintratec PA12)	-	State of delivery	≈ 0.9
ε_{HM3}	The emissivity of the matte-black-painted metal plate	-	Varnished	≈ 0.9
T_D	Fibre nozzle temperature	°C	PLC	...400
T_O	Powder bed surface temperature	°C	PLC	...200
T_{HM3}	Heat source temperature (metal plate)	°C	PLC	...200
$-\dot{Q}_{S,B}$	Heat losses due to radiation	W	Disturbance variables	
$-\dot{Q}_{K,B}$	Heat losses due to radiation	W		

Table 2. Target variables for successfully heating the part/powder bed surface within the sintering window during roving integration.

Symbol	Description	Unit
v_D	Process time expressed by the nozzle feed rate	s
b_{HAZ}	Width of the HAZ	mm
h_R	Overlap of the roving as a measure of process reliability	mm
t_{HAZ}	Depth of the HAZ. For a process – safe roving integration, $t_{HAZ} \geq 350$ µm is assumed.	mm

2.1.1. Transferred Heat with Influencing and Target Variables during Roving Integration

An analytical examination of the heat transfer process is employed to determine the variables that have an affect and those are aimed for in the context of roving integration. One-dimensional heat flows were utilised to simplify this analysis to characterise the influencing and target variables outlined in this study. The heat transfer during the roving integration process can be mathematically described in Equation (1).

$$Q_{FI}[J] = Q_{HM3} + Q_{FN} \tag{1}$$

The first term in Equation (1) accounts for the heat delivered to the surface of the powder bed via the additional heat source. The part and the powder bed surface are kept warm within the sintering window, with the assistance of the additional heat source denoted as Q_{HM3}, located on the bottom side of the fibre integration unit to ensure the integration of rovings in a secure and replicable manner. For an extensive analysis of the additional heat source and Q_{HM3}, please refer to [19].

The second term in Equation (1) represents the heat the heated fibre nozzle conveys. This heat flow primarily plays a pivotal role in directing heat energy transfer, thus facilitating the formation of the HAZ. The heat transferred through the fibre nozzle is quantified according to Equation (2).

$$Q_{FN} = \pi \cdot \left(r_{D,o}^2 - r_{D,i}^2\right) \cdot \left(\lambda_L \cdot \frac{T_D - T_O}{h_D} + \frac{\sigma}{\frac{1}{\varepsilon_D} + \frac{1}{\varepsilon_P} - 1} \cdot \left(T_D^4 - T_O^4\right)\right) \cdot \Delta t_{FN} \quad (2)$$

The first term within the parentheses in Equation (2) pertains to Fourier's law, describing the heat exchange between the ring-shaped fibre nozzle characterised by inner diameter $d_{D,i}$ (responsible for guiding and transmitting the roving inside the fibre nozzle), outer diameter $d_{D,o}$, nozzle curvature κ, and the surface of the powder bed. It is assumed that heat is exclusively transferred through heat conduction across the air gap h_{HM3}, which constitutes a scenario of free convection involving internal flow and heat radiation [18,20]. Under these conditions, the Rayleigh number (Ra) is less than 1, indicating a stable stratified fluid layer with no induced flow (Nusselt number, Nu, equals 1) [23]. The second term within the parentheses in Equation (2) corresponds to the Stefan–Boltzmann law, incorporating an additional radiation exchange between the front surface of the fibre nozzle and the surface of the powder bed. From the standpoint of a stationary point on the part's surface, Δt_{FN} signifies the temporal duration during which the heated fibre nozzle with a surface area $A_D(d_{D,i}, d_{D,o})$ imparts heat to the part at a feed rate of v_D. The duration Δt_{FN} for heat transfer can be influenced by the feed rate v_D of the fibre nozzle. It is important to note that potential convection currents between the fibre nozzle and the supplementary heat source and heat dissipation at the peripheries of the additional heat source into the surroundings were assumed but not considered in this analysis.

The parameters outlined in Equation (2) summarise the factors influencing the roving integration process within the developed laser-sintering machine, and their definitions are provided in Table 1. It is imperative to mention that this analysis did not incorporate influences arising from the laser-sintering process, such as interactions between the laser and the part, material composition, or ageing effects of the powder. These factors were maintained as constant as possible throughout the studies, as detailed in Section 2.3.

The target variables for successfully integrating rovings in the developed LS machine are depicted in Figure 2.

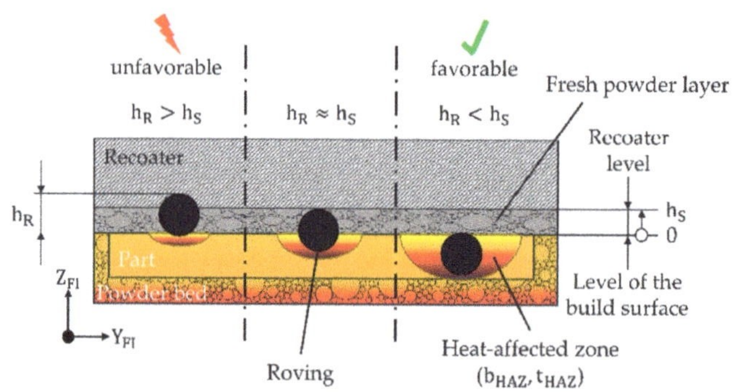

Figure 2. Schematic representation of the influence of the HAZ on the roving overlap h_R.

It is crucial to create a HAZ to position the roving below the recoater's movement level, denoted as h_S (the predetermined layer thickness during the printing process), to ensure the successful integration of rovings. In practical terms, t_{HAZ} needs to be adjusted to sufficiently immerse the roving in the molten material, meaning h_R should be less than h_S. This configuration avoids the formation of any disruptive contours for the recoater, ensuring process reliability.

Consequently, the finite element (FE) model must yield a value for t_{HAZ} corresponding to a roving thickness of approximately 365 µm. If the roving extends too far beyond the part's edges $h_R \geq h_S$, it could lead to a collision between the recoater and the roving during the subsequent recoating process. Such an occurrence may result in the part being displaced by the recoater, necessitating the stop of the printing process. Additionally, minimising b_{HAZ} is essential for enabling rovings to be placed as close to the edges of the part as possible without causing the unsintered and loose powder to melt beyond its edges. According to the results of the SPD from [21], b_{HAZ} and t_{HAZ} of the HAZ constantly change with a constant aspect ratio b_{HAZ} / t_{HAZ} when varying the process parameters (e.g., fibre nozzle feed rate and nozzle temperature). In other words, if, for example, t_{HAZ} of the HAZ is reduced by increasing the nozzle feed rate, b_{HAZ} is reduced simultaneously. For this reason, only t_{HAZ} is considered in this paper, as this is primarily responsible for the process reliability of the roving integration. Regarding cost-effective production, the duration required for roving integration plays a crucial role. This processing time is determined by the feed rate v_D of the fibre nozzle. The specific objectives for successful fibre integration are outlined in Table 2.

Initial studies were conducted in [18,19] for the influencing and target variables listed in Tables 1 and 2. The experimental analysis and quantification of the effects and interactions of v_D, h_D, $d_{D,o}$, h_{HM3}, κ, and T_D on the width and depth of the HAZ (without the influence of roving integration) were analysed using an SPD in [18,24]. According to [18], v_D, h_D, $d_{D,o}$, and T_D have the most significant influence on the shape of the HAZ. h_D cannot be set smaller than 0.6 mm from the powder bed surface; otherwise, the powder can adhere to the fibre nozzle and, thus, reduce the effective fibre nozzle distance. This leads to entrainment effects in the powder bed or the part. Furthermore, an outer diameter of the fibre nozzle of 2 mm has been established. $d_{D,o}$ directly influences the b_{WEZ}. Due to $d_{D,i} = 0.6$ mm, the outer diameter cannot be reduced further, as this would weaken the fibre nozzle too much. According to [19], an operating point was found for T_{HM3} and h_{HM3} at $T_{HM3} = 190$ °C and $h_{HM3} = 0.8$ mm, where rovings can be reliably integrated into the parts. This leaves T_D and v_D for an optimisation of t_{WEZ}. In other words, the two factors, T_D and v_D, can be varied to optimise t_{WEZ}. According to the SPD in [19] and Equation (2), a further reduction of t_{WEZ} is achieved by increasing T_D and v_D as optimisation direction.

This paper aims to determine an operating point range in which the condition $t_{HAZ} \geq 365$ µm is achieved with the highest possible fibre nozzle feed rate v_D using an FE model in COMSOL Multiphysics (Version 6.1). The findings from the SPD in [19] are used to evaluate the FE model's plausibility and accuracy with the help of a CCD with initially assumed factor levels within which an optimised operating point range ($t_{HAZ} \geq 365$ µm) is sought. Finally, the adjusted operating point range derived by the FE model is validated with the help of an adapted and experimentally performed CCD with more detailed factor levels. In addition, the influence of roving integration on the depth of the HAZ is analysed, and an optimised operating point for roving integration is experimentally derived. The results are described and discussed in Section 3.

2.2. Numerical Modelling

In this section, a systematic derivation of an FE model for modelling the formation process of the HAZ is carried out based on physical assumptions and simplifications. The results generated from the FE model are first compared with the results of the SPD from [18] to evaluate the plausibility and accuracy of the FE model. The FE model was then used to identify an optimal operating point range for T_D and v_D.

2.2.1. Modelling Approach

A macroscopic modelling approach is employed in which individual particles within the part are not individually considered. Instead, the molten part is treated as a continuous medium with homogenised properties. The fluctuations in temperature within the molten part due to the heat source Q_{FI} are described using a nonlinear heat transfer equation as defined in Equation (3) [25].

$$\rho_P c_P \frac{\partial T}{\partial t} = -\nabla q + Q \tag{3}$$

The quantities ρ_P $\left[kg/m^3\right]$ and C_P $[J/(kg{\cdot}K)]$ represent the part's density and heat capacity. Equation (3) delineates the alterations in the part's temperature resulting from the heat fluxes q $\left[W/m^2\right]$ and external heat sources Q $\left[W/m^3\right]$, predominantly involving conduction, convection, and radiation.

2.2.2. Model Assumptions and Simplifications

For the derivation of the FE model, some physical assumptions and simplifications are made based on the findings from [18,19] to keep the computing time and the required storage space low while still gaining maximum knowledge. The most important assumptions and simplifications are listed in the following points.

- The supplied heat flow Q_{HM3} compensates for heat losses due to radiation and convection according to Equation (1), consisting of heat conduction through the air gap and the radiation exchange between the installed additional heat source and the part surface. Consequently, the heat losses are not considered in the FE model but only the supplied heat Q_{HM3} of the additional heat source.
- According to Equation (2), the heat input is based on heat conduction through the air gap and the radiation exchange between the ring-shaped fibre nozzle and the part's surface. The FE model does not consider possible convection flows between the fibre nozzle and the additional heat source.
- The surrounding and loose powder bed is not considered in the FE model, but only the already manufactured part. The part edges are isolated in the FE model.
- Modelling approaches from the current state of research and technology for analysing the laser-sintering process consider the phase transformation from powdery to molten states [26,27]. However, in the developed LS machine, the HAZ is generated in the molten state of the part, which the laser has transformed. The phase transition has, therefore, already taken place when the HAZ was created. Therefore, only the material properties of the molten state of the part are used. Isotropic part properties are assumed. This means that constant values are used for the thermal conductivity of the part λ_P for the part density ρ_P, and, thus, for the porosity of the part Φ_P and the specific heat capacity c_P (constant pressure). The thermal conductivity of the air in the air gap is the only variable with a temperature dependence $\lambda_A(T)$.
- To determine the target value t_{HAZ} in the FE model, only the part area with a temperature value equal to or higher than the melting temperature of the PA12 part T_M is evaluated.
- According to Figure 1b, there is symmetry in the centre of the fibre nozzle and along the y-axis (axis of movement of the fibre nozzle). Due to this symmetry property, only half of the process zone is modelled in the FE model.
- To simplify the FE model, not the entire additional heat source is modelled, but only the immediate vicinity of the process zone, consisting of the PA12 part, fibre nozzle, air gap, feeler gauge tape, metal plate of the additional heat source, and guidance.
- Due to the roving's current scattering position/orientation in part [18], only the formation process of the HAZ is modelled. The influence of the roving on the HAZ is, therefore, not an object of investigation of the FE model. For a simulation-based identification of an operating point range with the highest possible fibre nozzle feed rate, a value of $t_{HAZ} \approx 365$ μm is assumed. According to Table 1, this corresponds to the approximate thickness of a 1K roving.

2.2.3. Geometric Model Structure

To build an FE model in COMSOL Multiphysics (Version 6.1), a geometric representation of the process zone to be investigated must first be created. According to Section 2.1.1, the process zone consists of the PA12 part in which the HAZ is generated, along with the air gap, the feeler gauge tape with guidance, and the fibre nozzle, which is mandatory for creating the HAZ. These components are modelled within COMSOL so that the settings to be investigated can be parameterised. A parameterisation of the geometry parameters (h_D, h_{HM3}, $d_{D,o}$, and κ) enables a simple and automated variation of the influencing variables using a MATLAB script linked to the FE model when carrying out parameter studies. Figure 3a shows the CAD model, including the installed components of the process zone. For comparison, Figure 3b shows the simplified geometry of the process zone realised in COMSOL.

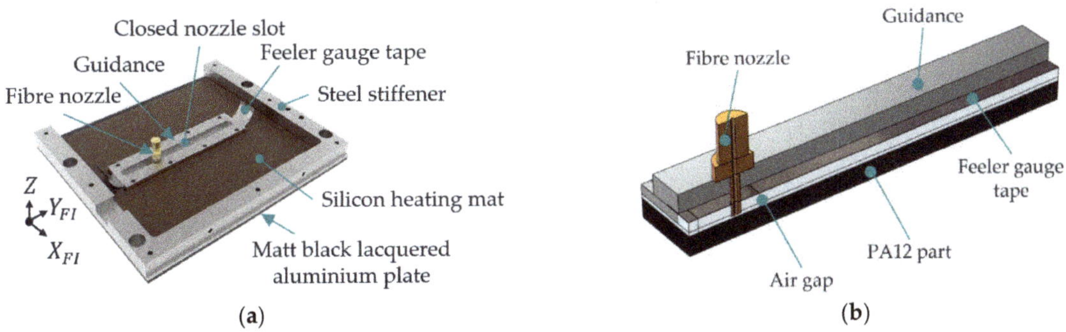

Figure 3. Three-dimensional view of the additional heat source with fibre nozzle, feeler gauge tape, and guide (**a**), and geometry implemented in COMSOL for process zone (**b**).

2.2.4. Material Properties with Initial Settings

Table 3 shows the material parameters relevant to the FE model with numerical values and sources. In addition, Table 3 shows the initial values for all materials that form the starting point for the simulation.

Table 3. Material parameters and initial values are defined in COMSOL.

Symbol	Description	Value	Unit	Source
ρ	Density of the molten part	1040	$\frac{kg}{m^3}$	[28]
Φ	Porosity of the part	8.5	%	[29]
T_M	Melting temperature of PA12	184	°C	[30]
T_O	Temperature of the sintered part, as well as the part surface (initial value)	179	°C	[19]
T_A	The initial temperature of the air in the air gap	179	°C	[31]
λ_P	Thermal conductivity of the PA12 part	0.26	$\frac{W}{(m \cdot K)}$	[28]
c_P	Specific heat capacity of the PA12 part	2.66	$\frac{kJ}{(kg \cdot K)}$	[26]
ε_P	The emissivity of the PA12 part	0.90	-	[19]
ε_{HM3}	The emissivity of the black lacquered metal plate of the additional heat source	0.97	-	[32]
ε_{FN}	The emissivity of the copper fibre nozzle (oxidised)	0.76	-	[32]
ε_{FGT}	The emissivity of the feeler gauge tape	0.85	-	[32]

As described in Section 2.2.2, isotropic material properties are assumed for the PA12 part. For modelling heat conduction through the air gap, the heat conduction coefficient through the air is relevant. The temperature-dependent thermal conductivity can be approximated according to [23] using Equation (4).

$$\lambda(T) = 0.0651T + 24.881 \quad (4)$$

2.2.5. Meshing Zones and Moving Mesh

To reduce the calculation time, the individual meshing zones of the assembly are meshed with different degrees of resolution according to their expected influence on the simulation results. For this reason, the geometry from Figure 3 is partitioned into sections. The following Figure 4 shows the meshing zones used.

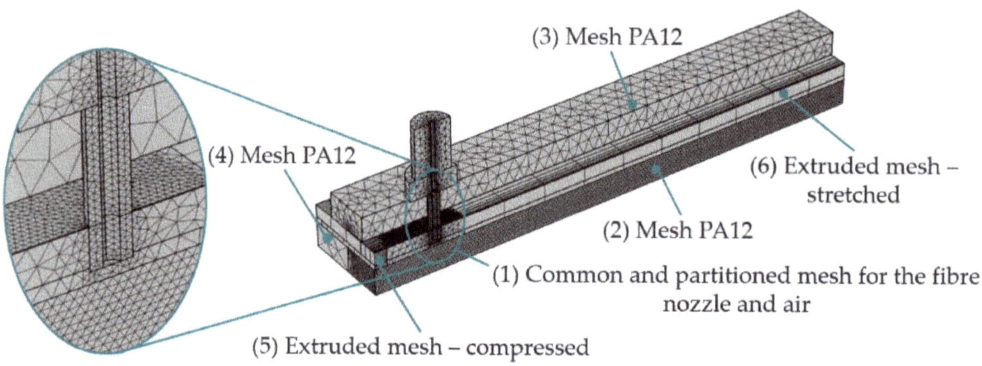

Figure 4. Process zone with the occurring mesh zones and initially assumed mesh resolution.

Table 4 lists the individual meshing zones with the initially assumed mesh resolution. The meshing parameters for the individual mesh resolutions are in the COMSOL documentation [33].

Table 4. Individual meshing zones with initially assumed mesh resolution.

Meshing Zone	Description	Mesh Resolution
1	Common mesh for the fibre nozzle and air.	Fine
2	PA12	Fine
3	PA12	Coarse
4	PA12	Coarse
5	Extruded mesh—compressed	Coarse
6	Extruded mesh—stretched	Coarse

For a detailed description of the mesh specifications, please refer to [33]. The PA12, the front surface of the nozzle, and the air near the nozzle are significantly involved in the heat transfer and the formation of the HAZ. Consequently, a fine mesh is initially assumed there. The nozzle and the air section mesh in a standard mesh using a parameterisable selection cylinder along the rotation axis of the fibre nozzle in COMSOL. The PA12 part is also partitioned. In this way, a correspondingly fine mesh can be selected in the area where the HAZ is formed, and the low-influence areas of the PA12 part can be more coarsely meshed. The guidance, the feeler gauge tape, and the black lacquered metal plate are provided with a coarser mesh. By default, geometries in COMSOL are meshed tetrahedrally. To replicate the one-dimensional movement of fibre nozzle feed along the y-direction, the moving mesh knot is used in COMSOL. The moving mesh knot allows the deformation of the extruded meshes (zones 5 and 6), which realises a relative movement between zones 1 and 2. The extruded mesh in zones 5 and 6, represented by a rectangular mesh, includes the edge zones of the feeler gauge tape, the black lacquered plate of the additional heating mat, and the air. It is assumed that these zones along the extrusion axis have a minor influence on the target variables. The following Figure 5 shows the displacement of the mesh along the y-axis.

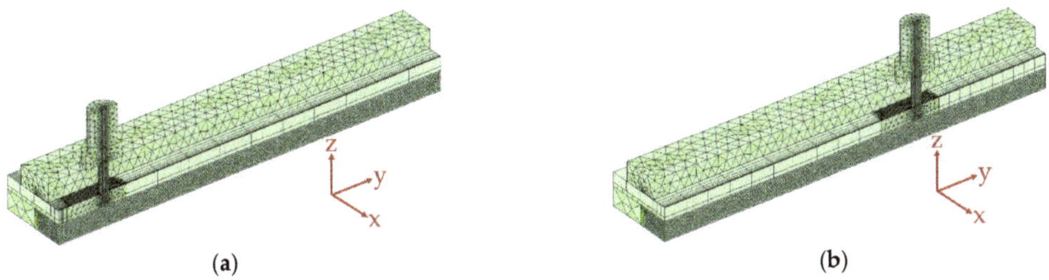

Figure 5. The initial state of the moving mesh at time t = 0 s (**a**) and the extruded mesh at a relative fibre nozzle offset of 40 mm (**b**).

Figure 5a shows the initial state of the mesh at time t = 0 s. Figure 5b shows the extruded mesh at a relative fibre nozzle offset of 40 mm. Zone 1 moves continuously, whereas the extruded meshes in zones 5 and 6 deform. Equation (5) is used to simulate the nozzle feed rate v_D.

$$v_D = \frac{\Delta s}{\Delta t}\left[\frac{mm}{min}\right] \quad (5)$$

Here, Δs represents the travelled path of the fibre nozzle within a discrete time step Δt. To ensure relative movement between zones 1, 5, and 6, sliding conditions are defined on the surfaces of the deformable zones so that no fixed nodes to neighbouring areas are generated during meshing.

The mesh resolutions initially assumed in Table 4 are optimised in Section 2.3.1 with the help of a convergence analysis. The model's accuracy can be increased with more account points. However, this happens at the expense of computing time and memory requirements. Therefore, a compromise must always be found between accuracy and computing time or storage capacity.

2.2.6. Physics with Initial and Boundary Conditions

The meshing zones for the PA12 part and the air are selected in the physics module heat conduction—see Figure 6a.

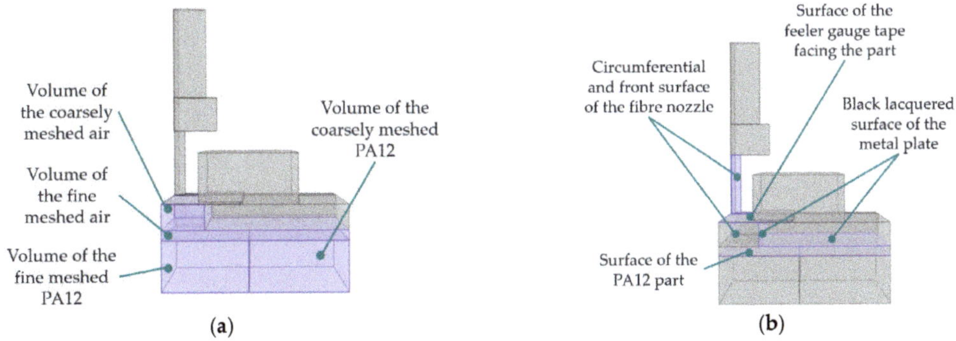

Figure 6. The volumes (purple) involved in the FE model in which heat conduction occurs (**a**) and the surfaces (purple) involved in the FE model which are involved in radiation exchange (**b**).

The heat conduction within the heating mat is insignificant for the FE model. Therefore, only the additional heat source's black lacquered metal plate surfaces are given a temperature boundary condition (T_{HM3}), which can be adjusted via the model's parameter list. This conducts heat through the air gap between the black lacquered metal plate and the PA12 part. The heat conduction from the fibre nozzle is also ensured by a temperature boundary condition (T_D). The air and the PA12 part are given an initial value (T_O) at the beginning of the simulation, corresponding to the LS machine's preheating. Furthermore, the circumferential surfaces of the FE model receive a thermal insulation condition. On the one hand, the surfaces in the symmetry plane of the FE model must be insulated, as there are no heat flows here. On the other hand, the outer surfaces of the air region and the PA12 part are isolated. Since only a section of the black lacquered metal plate is shown in the FE model, no heat transfer from the heating mat to the surrounding process chamber occurs there.

The thermal radiation physics module in COMSOL is applied to all surfaces involved in radiation exchange—see Figure 6b. These include the surface of the black lacquered metal plate, the bottom of the feeler gauge tape, and the surfaces of the PA12 part and the fibre nozzle. The direction of the emitted radiation is determined by the opacity, which is why the radiation exchange takes place through the air areas. The total heat Q_{FN} to be transferred according to Equation (2), the multiphysics function heat transport with surface-to-surface radiation, is applied in COMSOL.

2.2.7. Determination of the Depth of the HAZ

To determine the width and depth of the HAZ, the General Projection function in COMSOL is used [34]. For each discrete time step Δt of the moving mesh, the heat propagation in the PA12 part is calculated. Using the General Projection operator, starting from the centre of the HAZ, the distance along the Cartesian axes is integrated until the condition $T_M \geq 184$ °C is fulfilled. A detailed description of how the General Projection Operator functions is given in [34]. The result is a value for the width and depth of the HAZ. The result of the simulation, the HAZ, is shown in Figure 7.

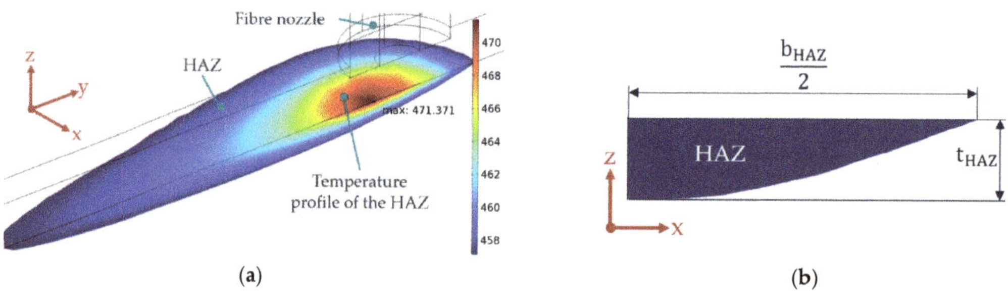

Figure 7. Isometric view of the evolution of the HAZ within the part when moving mesh is executed (**a**), as well as a front view of the HAZ with half width and total depth of the HAZ (**b**).

By default, COMSOL Multiphysics defines the time steps $\Delta t = 3$ s used to solve a time-dependent problem. The predefined settings, however, lead to irregular HAZs that do not occur in reality. The greater the feed rate of the nozzle, the worse the HAZ reproduced. To counteract this, time steps are given to the temporal solver of the FE model. A further convergence analysis examines the effect of the selected time step on the HAZ—see Section 2.3.1.

2.2.8. Evaluation of Model Quality

As described in Section 2.2, a compromise must be found between model accuracy and computing time or storage capacity. The meshing zones introduced in Section 2.2.5 are increasingly refined during a convergence analysis, starting from the initially assumed mesh resolution. The resulting changes in the target value t_{HAZ} as a function of the number of elements/mesh resolution are presented with the help of tables and a diagram indicating the number of elements and calculation time. In addition to the standardised mesh resolutions in COMSOL (fine, finer, extra fine, and extreme fine) [33], two additional customised mesh resolutions, according to Table 5, are also used.

Table 5. Additional and customised meshing parameters of the convergence analysis.

Mesh Properties	Max. Element Size [mm]	Min. Element Size [mm]	Max. Element Growth Rate	Curvature Factor	Resolution of Narrow Areas
Custom 1	1	0.017	1.3	0.2	1
Custom 2	0.5	0.017	1.3	0.2	1

To determine a suitable time step Δt of the moving mesh function, the procedure is analogous to the convergence analysis described above. The only difference is using a time step Δt instead of the mesh's number of elements/mesh resolution. In addition to the time step of three seconds automatically defined by COMSOL, the time steps 2 s, 1 s, 0.75 s, and 0.5 s are used as time steps.

After completion of the convergence analysis, it is qualitatively checked whether the FE model correctly determines the calculated results for the depth of the HAZ. In other words, it is checked whether the physics and material parameters in the FE model correspond to the experimental results up to that point. Using a plausibility check, the results of the FE model are compared with the results of the SPD from [18]. The results of this SPD are the main effect and interaction diagrams for the investigated parameters (T_D, h_{HM3}, T_{HM3}, h_D, κ, and $d_{D,o}$). To check the plausibility of the developed FE model, the exact repetition of the SPD from [18], including the factor-level combinations, is carried out with the help of the FE model and Minitab (2022 Cloud App). A detailed description of the SPD is not given here. For this, please refer to [18]. The plausibility is checked using arrows by comparing the main effect and interaction diagrams concerning the slope. If, for example, the effect has a positive slope when changing from the first to the second level value of a factor, an ascending arrow symbol ↑ is assigned to this diagram. A descending arrow symbol ↓ is assigned in the case of a negative slope. The procedure for the other factors is analogous. Finally, the agreement of the arrow symbols, i.e., the agreement between simulation and experiment, is checked (✓). For the interactions, two arrows are used instead of one arrow.

To determine the model accuracy, the quantitative results (mean values) of the SPD are compared with the simulated results of the FE model for the same factor-level combinations. The degree of agreement, i.e., the model accuracy, e_{FE}, between simulation and experiment is expressed in %.

2.3. *Simulation-Based Identification of an Optimal Operating Point Range*

The starting point for the simulation-based optimisation of v_D and t_{HAZ} is the knowledge of the SPD from [18]. Section 2.1.1 describes process parameters T_D and v_D, which must be increased to reduce the target value t_{HAZ} (optimisation direction). The correlation between the dependent target variable t_{HAZ} and the independent influencing variables T_D and v_D is required to determine an optimal operating point range.

2.3.1. Determination of an Optimised Operating Point Range

To identify an optimised operating point range for T_D and v_D, in which $t_{HAZ} \approx 365$ µm can be achieved, the relationship between $t_{HAZ}(T_D, v_D)$ is first derived with the help of a central composite design (CCD) with initially widely spaced factor levels. A CCD consists of a full factorial or partial factorial ground plan and a central star. CCD designs have an orthogonal design with a two-stage structure. The two-stage basic design can be evaluated in advance. The star can be used to create both square and cubic models. Such experimental designs are preferably used for the optimisation of target variables. Due to two independent influencing variables and a dependent target variable, the cause–effect relationship of these variables can be described in three-dimensional space. Following the Stefan–Boltzmann law as outlined in Equation (2), it is observed that alterations in temperature exhibit a nonlinear impact on the transferred heat Q_{FN}. Therefore, Equation (6) serves as a fundamental framework for comprehensively depicting the linear and nonlinear influences of T_D and v_D on t_{HAZ}.

$$f(x, y) = a_0 + a_1x + a_2y + a_3xy + a_4x^2 + a_5y^2 + a_6x^2y^2 \quad (6)$$

To determine the coefficients from Equation (6), assumed factor-level combinations of a central composite design (CCD) are first set and varied using the developed FE model and a linked MATLAB script, and the target value t_{HAZ} is determined. The initial factor levels are selected so that the results of the CCD include $t_{HAZ} \approx 365$ µm. Information on the construction and derivation of a CCD can be found in [35]. Since the results of the FE model are not subject to scatter, the factor-level combinations are not repeated, nor are randomisation and block formation.

The following considerations determine the initial factor levels for T_D and v_D of the CCD within which the value $t_{HAZ} \approx 365$ is expected.

- The basis for optimisation is the best operating point of the production process so far [35]. According to the results of the SPD, the best operating point range so far is based on the process understanding gained in [18], where the process parameters have a constant value of $T_{HM3} = 190$ °C and $d_{D,o} = 2$ mm and the fibre nozzle has a planar front surface. This factor-level combination achieves an average value of t_{HAZ}(TD = 280 °C, $h_D = 0.8$ mm, and $v_D = 60$ mm/min) = 52 µm. According to Equation (2), a heat quantity of approx. $Q_{FN} = 28.7$ joules is transferred to the PA12 part. This heat quantity of 28 joules, thus, forms the lower edge of the operating point range, which must be applied to generate a HAZ. The largest heat quantity of approx. Q_{FN} =152.8 joules is reached at a factor-level combination of t_{HAZ} ($T_D = 310$ °C, $h_D = 0.4$ mm, and $v_D = 30$ mm/min). A mean value t_{HAZ} ($T_D = 310$ °C, $h_D = 0.4$ mm, and $v_D = 30$ mm/min) = 783 µm is achieved. Thus, the heat quantity of 152.8 joules represents the upper edge of the operating point range. The heat quantities were calculated using Equation (2), along with an averaged heat conduction coefficient for the air between the heated fibre nozzle T_D and powder bed surface T_O.
- It is assumed that, for higher values of T_D and, especially, v_D, a $t_{HAZ} \approx 365$ µm can be identified as long as the amount of heat transferred to the part is within the range of 28.7 joules $\leq Q_{FN} \leq$ 152.8 joules.
- As the first upper factor level (1), a value of $T_D = 360$ °C and $v_D = 150$ mm/min is initially set. The amount of heat Q_{FN} ($T_D = 360$ °C, $v_D = 150$ mm/min) thus transferred is 29.3 joules. The setting of the SPD with $T_D = 310$ °C and $v_D = 60$ mm/min is determined as the CCD's lower factor level (−1). The factor-level combinations of the initial CCD are shown in Table 6.

Table 6. Factor levels used for the initial CCD with abbreviations and heat quantity transferred.

T_D		v_D		Q_{FN}
310	(−1)	60	(−1)	50.9
360	(+1)	60	(−1)	73.2
310	(−1)	150	(+1)	20.4
360	(+1)	150	(+1)	29.3
299.7	$\left(-\sqrt{2}\right)$	105	(0)	26.6
370.4	$\left(+\sqrt{2}\right)$	105	(0)	44.6
335	(0)	41.4	$\left(-\sqrt{2}\right)$	89.7
335	(0)	168.6	$\left(+\sqrt{2}\right)$	22
335	(0)	105	(0)	35.3

The first four columns contain the factor-level combinations of the CCD. The last column contains the amount of heat transferred to the part. Three values are below 28 joules. Since this initial CCD intends to find a parameter constellation for T_D and v_D at which a value of $t_{HAZ} \approx 365$ µm can be achieved, the CCD assumed in Table 6 is used as a first approximation. According to [18], and for reasons of process safety (Section 2.1.1), h_D has a constant value of $h_D = 0.6$ mm for all factor-level combinations. Using the FE model and based on the factor-level combinations given in Table 6, an optimised operating point can be found for which a $t_{HAZ}(T_D, v_D) \approx 365$ µm can be achieved. Based on this identified operating point, a more detailed CCD is derived, based on which the operating point determined by the FE model is to be validated experimentally.

2.3.2. Experimental Validation of the Adjusted Operating Point Range

The starting point for the experimental validation is the more detailed CCD determined by the FE model and adjusted concerning the factor levels. The samples with integrated HAZ are produced with the developed LS machine's help to validate the operating point range according to the adapted CCD's factor-level combinations. The generated HAZ is then measured in its depth. The specimens used in this study were rectangular parts with dimensions of $60 \times 15 \times 3$ mm^3 ($l \times w \times h$) and were constructed using PA12 material (specifically, Sintratec black). The design of these specimens was adapted from a previously established geometry described in [19]. Notably, the specimens' thickness and width were significantly larger than the HAZ. Consequently, the heat input occurred exclusively within the generated part structure, preventing any melting of the unsintered powder. The positioning of these specimens within the powder bed of the developed LS machine is depicted in Figure 8 below.

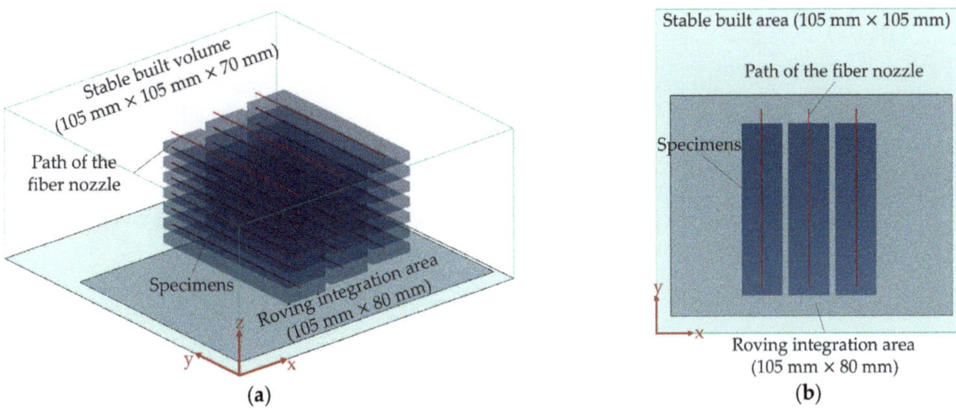

Figure 8. 3D view of the arrangement of the specimens in the built volume of the LS machine (**a**) and a top view (**b**) with the movement path of the fibre nozzle (red lines).

The specimens were placed at a separation of 2 mm from each other along the x-axis. Furthermore, adjacent specimens were positioned with a 0.5 mm offset along the z-axis, corresponding to the build direction. The HAZ insertion occurred after 2.5 mm, corresponding to the 25th layer of each part. The fibre integration unit executed a rapid traverse to reach the initial position for the path of motion, indicated by the red line in Figure 8b. Subsequently; it moved at the feed rate specified in the experimental design, following which it returned to its home position via rapid traverse. Five central points were selected for each block to generate orthogonal blocks [35]. Each combination of factor levels, including two replications and the corresponding central point, was produced within a single print run, forming one block. The sequencing of experiments within two blocks was randomised using Minitab software (2022 Cloud App). The settings employed for producing the samples with the developed LS machine aligned with those described in [18]. Notably, the travel distance of the fibre nozzle, and, thus, the area subjected to melting, exceeded the dimensions of the part itself, resulting in a protrusion at the end faces of the specimens. The specimens were prepared in a manner that allowed the HAZ to be observable for evaluation to ascertain the depth of the HAZ. This was achieved by trimming off the projecting portion (a "slug") of the melted area at the front of the specimens using a scalpel. The depth of the HAZ was then measured using a microscope (specifically, the Keyence VHM 7000), following the procedures outlined in [18].

2.3.3. Influence of Roving Integration on the Depth of the HAZ

In addition to the previously outlined experimental design, an analysis was conducted to assess the impact of roving integration on the HAZ. The modified CCD was replicated while considering the influence of roving integration to accomplish this. The results were compared with those from the CCD experiments conducted without roving integration. However, given that it is challenging to visually distinguish the roving from the surrounding matrix when examining the specimen from the front (as illustrated in Figure 9a), a scalpel was employed to incise the specimen, as demonstrated in Figure 9b. This incision allowed for the measurement of the overlap of the roving [26].

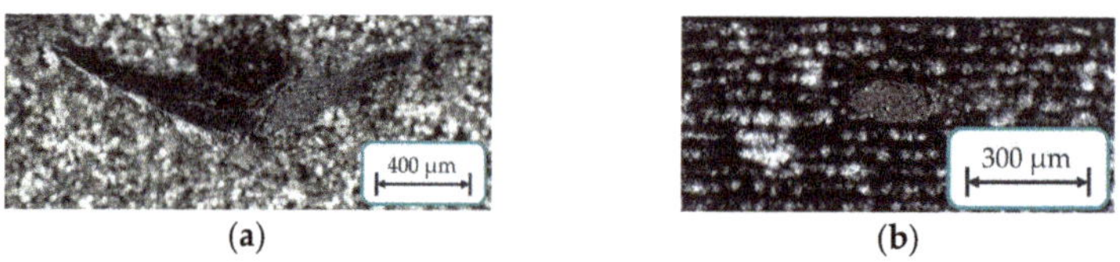

Figure 9. The surface of a sample with embedded 1K roving uncut (**a**) and cut open with a scalpel (**b**).

In addition to the depth of the HAZ, the roving overlap is an important target value for roving integration. If the roving overlap is too high, a dragging effect of the part and the roving can occur. To determine the roving overlap, the total part thickness is first determined. Based on the known number of powder layers per part n_{tot} = 30, the thickness per layer h_S^* (≈0.095 mm) can, thus, be determined—see Figure 10. This layer thickness h_S^* is necessary since the part will shrink during cooling. In addition, the roving overlap h_R can be determined from the known integration layer of the roving at n_{FI} = 25. The roving overlap is compared to the set powder layer thickness during printing and/or the movement level of the recoater at h_S (= 0.1 mm) and should ideally be $h_R < h_S$.

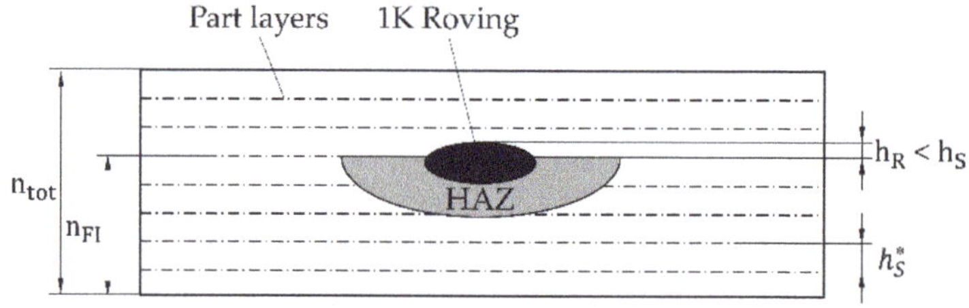

Figure 10. Schematic representation for calculating the roving overlap within a specimen.

The machine settings employed in this study mirrored those utilised in [18]. The roving selected was a coated 1K roving (67 tex, HTA40) sourced from Teijin Limited. The chosen roving featured a coating material comprising a thermoplastic-compatible polymer dispersion known as PERICOAT AC250 [36] to facilitate adequate bonding between the fibre and matrix. The coating content was maintained at 5% [35]. Finally, the process understanding gained in this paper is validated using a demonstrator part. The demonstrator part is a battery tab gripper for handling battery electrodes within an agile battery cell production line [37].

3. Results and Discussion

3.1. Model Quality

During the convergence analysis, a change in the mesh resolution for the initially assumed coarse meshing zones 3 to 6 does not lead to a significant difference in the target value t_{HAZ}. For meshing zones 1 and 2 and time step Δt of the moving mesh, changes in t_{HAZ} could be recorded based on the convergence analysis.

Figure 11 below shows the result of the convergence analysis for meshing zone 1.

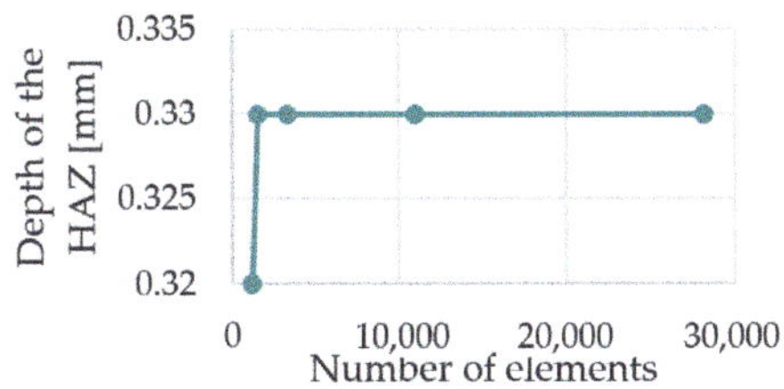

(**a**)

Number of elements	Time
1201	54 min
1504	1h 4 min
3273	1h 11 min
11022	1h 25 min
28165	2h 11 min

(**b**)

Figure 11. Depth of the HAZ as a function of the number of elements (**a**) and required calculation time (**b**) for meshing zone 1.

It can be seen in Figure 11a that the target t_{HAZ} converges relatively early. According to Figure 11b, the required calculation time is 54 min for the smallest number of elements and 2 h 11 min for the highest number of elements. A number of elements of 110,022 are selected for this mesh (Custom 1).

Figure 12 shows the result of the convergence analysis for meshing zone 2 (PA12).

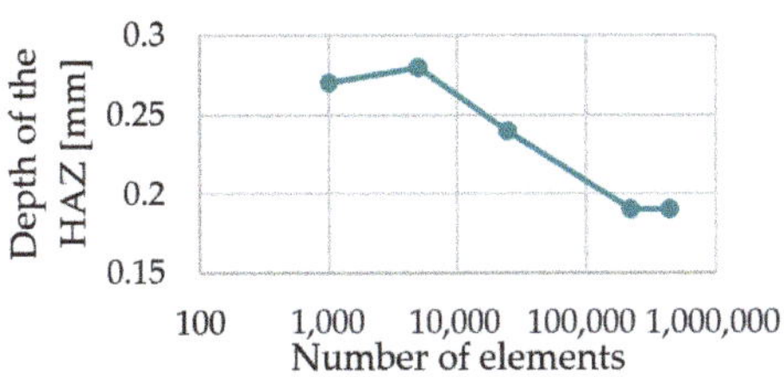

(**a**)

Number of elements	Time
996	21 min
4930	25 min
24591	38 min
220384	2h 25 min
440891	5h 10 min

(**b**)

Figure 12. Depth of the HAZ as a function of the number of elements (**a**) and required calculation time (**b**) for meshing zone 2.

It can be seen in Figure 12a that the target value t_{HAZ} appears to change extremely late, i.e., at a high number of elements. According to Figure 12b, the required calculation time is 21 min for the smallest number of elements and 5 h 10 min for the highest number of elements. Due to the significant jump between the extremely fine mesh and the Custom 1 mesh concerning t_{HAZ} of about 0.05 mm, a number of elements of 220,384 are chosen for meshing zone 2 (Custom 1).

Figure 13 shows the result of the convergence analysis for the time step Δt.

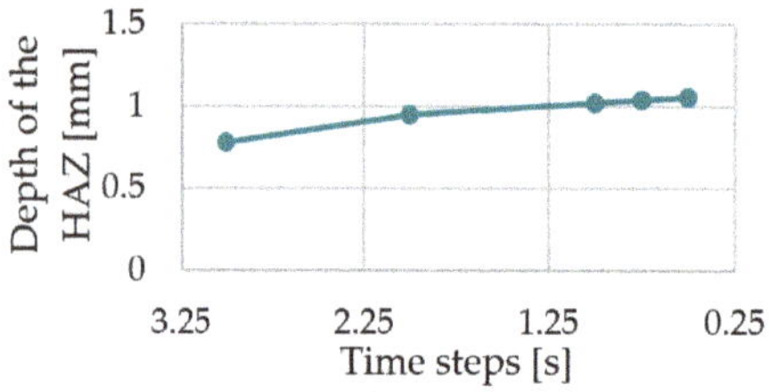

(a)

Time step [s]	Time
3	34 min
2	33 min
1	52 min
0.75	1h
0.5	1h 22 min

(b)

Figure 13. Depth of the HAZ as a function of the time step Δt (**a**) and required calculation time (**b**) for the moving mesh.

It can be seen in Figure 13a that the target value t_{HAZ} does not converge even at a very high mesh resolution, i.e., at an increased number of elements. However, the changes in t_{HAZ} are still approximately 0.02 mm for the tiniest time steps. According to Figure 13b, the required calculation time is 34 min for the smallest number of elements and 1 h 22 min for the largest number of elements. Due to the relatively small change in the target value with small time steps, a time step of 0.5 s is, nevertheless, selected since it was shown during the CCD that the high values of the fibre nozzle feed rate could be better reproduced with this time step.

Table 7 summarises the mesh resolutions for the meshing zones 1–6.

Table 7. Selected mesh resolutions for meshing zones 1 to 6.

Meshing Zone	Description	Mesh Resolution
1	Common mesh for the fibre nozzle and air	Custom 1
2	PA12	Custom 1
3	PA12	Coarse
4	PA12	Coarse
5	Extruded mesh—compressed	Coarse
6	Extruded mesh—stretched	Coarse
Time step	Moving mesh	0.5 s

In the following, the results of the plausibility check are presented. Table 8 contains the results of the plausibility check comparing the slopes of the main effects of the SPD from [18] and the slopes of the main effects of the FE model. For this purpose, the results of the FE model were evaluated using Minitab (2022 Cloud App).

Table 8. Comparison of the slopes of the main effect diagrams between the FE model and the SPD.

Factor	Factor Level 1→Factor Level 2	Experiment Slope	Simulation Slope	Concordance
κ	(planar→concave)	↓	↓	✓
$d_{D,o}$	(2 mm→4 mm)	↑	↑	✓
h_D	(0.4 mm →0.8 mm)	↓	↓	✓
T_D	(280 °C→310 °C)	↑	↑	✓
v_D	(30 mm/min→60 mm/min)	↓	↓	✓
T_{HM3}	(190 °C→200 °C)	↑	↑	✓

It can be seen from Table 8 that the slope characteristics simulated by the FE model correspond to the slope characteristics of the main effects of the experimentally conducted SPD from [18]. Thus, it can be assumed that, at least for the main (linear) effects, the FE model provides a correct estimation for an optimised operating point range.

In Table 9, the slopes identified from the SPD for the significant interactions are compared with those from the FE model for the same interactions.

Table 9. Comparison of the slopes of the interaction diagrams between the FE model and the SPD.

Interaction Factor 1/Factor 2	Experiment Slope	Simulation Slope	Concordance
κ / $d_{D,o}$	↑↑	↑↑	✓
κ / h_D	↓↓	↓↓	✓
$d_{D,o}$ / T_D	↑↑	↑↑	✓
$d_{D,o}$ / v_D	↓↓	↓↓	✓
v_D / T_D	↓↓	↓↓	✓

As in Table 8, it can be seen in Table 9 that the slope curves of the interactions simulated by the FE model match the slope curves of the experimentally performed SPD from [18]. Thus, in addition to the main effects, it can be assumed that the FE model provides a correct estimate for an optimised operating point range when performing the CCD in Section 3.2.

To determine the model accuracy e_{FE}, the results of each factor-level combination of the FE model were compared with the measured values for t_{HAZ} from the SPD. The maximum model deviation is 22%. This results in a model accuracy of $e_r = 78\%$. The most probable cause for the model deviation of 22% is the unfavourable position of the factor-level combination. For this factor-level combination, the transferred heat quantity Q_{FN} is a value of 28.7 joules. This heat quantity corresponds to an average value of t_{HAZ} ($T_D = 280$ °C, $h_D = 0.8$ mm, $v_D = 60$ mm/min, κ =Concave, $T_{HM3} = 190$ °C, and $d_{D,o} = 2$ mm) = 52 µm. When evaluating this factor-level combination, it was found that the generated HAZ was barely visible and, thus, could not be reliably measured. This factor-level combination was repeated several times to obtain a statistically reliable result. With another factor-level combination, an average value of t_{HAZ} ($T_D = 310$ °C, $h_D = 0.4$ mm, $v_D = 30$ mm/min, κ =planar, $T_{HM3} = 190$ °C, and $d_{D,o} = 2$ mm) = 783 µm was obtained. At this setting, a heat quantity of 152.8 joules is transferred to the part. For this factor-level combination, a model accuracy of 82% could be achieved in comparison. Further causes for reduced model accuracy are listed in the following points:

- Depending on the position of the parts in the powder bed, the porosity of the parts can vary significantly due to temperature differences caused by the installed IR emitters or the additional heat source on the powder bed surface [13]. This has a direct influence on how the heat propagates within the part. The result is scattered values for the width and depth of the HAZ.
- Furthermore, the mixing ratio of the PA12 powder used has a decisive influence on the part properties. Although a mixing ratio of 60% new powder and 40% old powder was used for the SPD, according to [38], the proportion of old powder contributes to the scattering.
- Another cause of the reduced model accuracy is the possible convection flows in the air gap between the heated fibre nozzle and the part surface due to the nozzle velocity. These convection currents can influence the transferred heat quantity Q_{FN}.
- The values in the FE model are partly based on literature values. Depending on the powder composition of the manufacturer, the material characteristics assumed in Table 3 may differ from the real material characteristics. Furthermore, the mesh resolutions have an additional influence.

In summary, it can be said that, according to the results of the plausibility check, the FE model can reflect the experiments well concerning the slope curves. A model accuracy of $e_r = 78\%$ is classified as acceptable. Due to the proven plausibility and considering the FE model's accuracy of $e_r = 78\%$, the FE model is used to identify an operating point range for T_D and v_D, in which $t_{HAZ} \approx 365$ µm occurs with high probability.

3.2. Operating Point Range for Roving Integration

In this section, the simulation-based derivation of an operating point range is carried out within which the rovings can be integrated into the part as reliably as possible, i.e., sufficiently deeply and with the highest possible fibre nozzle feed rate. Since the FE model does not consider the roving, an operating point range is sought with the help of the FE model by achieving a value of $t_{HAZ} \approx 365$ µm. $t_{HAZ} \approx 365$ µm is the average roving thickness. It is assumed that, at $t_{HAZ} \approx 365$ µm, the roving is integrated sufficiently deeply into the part without forming an interfering contour for the recoater.

3.2.1. Adjustment of the Operating Point Range

The starting point for the simulation-based identification of an operating point range for v_D and T_D are the factor-level combinations of the CCD listed in Table 6 and initially assumed. With the help of the FE model and the MATLAB link, these factor-level combinations were set, and the regression equation and the coefficient of determination R^2 were determined. In the following Figure 14a, the generated surface response diagram and the simulated points of the FE model can be seen in red.

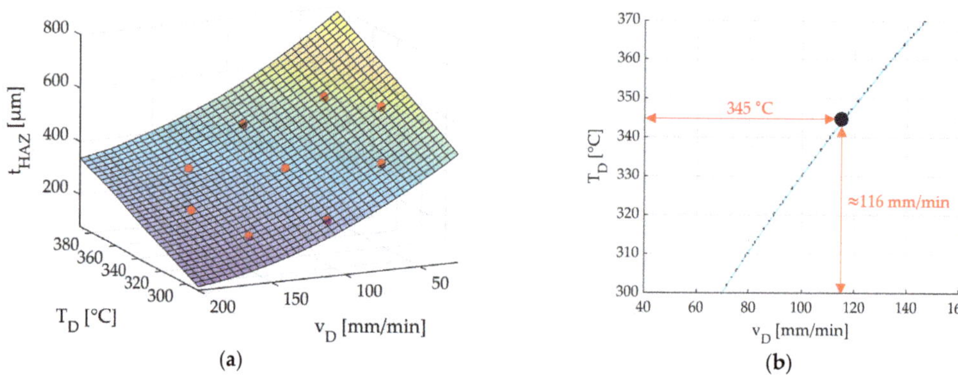

Figure 14. The results (red dots) of the initial CCD simulated with the FE model and the surface response diagram (**a**) derived with Minitab. Possible parameter constellations (blue line) for v_D and T_D for which the condition $t_{HAZ} \approx 365$ µm applies (**b**). Selected operating point at $T_D = 345$ °C (black dot).

It can be seen that the determined factor-level combinations from Table 6 include and can map the assumed condition $t_{HAZ}(T_D, v_D) \approx 365$ µm. The regression equation calculated from Minitab (2022 Cloud App) is shown in Equation (7).

$$t_{HAZ}(T_D, v_D) = -400 + 3.7v_D - 3.4T_D - 0.001v_D^2 + 0.009T_D^2 - 0.003v_D T_D \tag{7}$$

With a coefficient of determination of $R^2 = 0.98$, this value is close to 1. Thus, with the help of the determined regression equation, the simulated target value $t_{HAZ}(T_D, v_D)$ is very well reproduced. Figure 14b shows the possible parameter constellations for T_D and v_D, where a value of $t_{HAZ} \approx 365$ µm is achieved. The fibre nozzle feed rate in mm/min is plotted on the x-axis, and the nozzle temperature in °C is plotted on the y-axis. The graph in Figure 14b represents the function values calculated by the regression model at constant $t_{HAZ}(T_D, v_D) \approx 365$ µm.

To achieve a high fibre nozzle feed rate for $t_{HAZ}(T_D, v_D) \approx 365$ µm and, thus, a reduced process time, the nozzle temperature T_D must be increased at the same time according to Figure 14b so that the same amount of heat is transferred to the part. According to Figure 14b, the desired operating point range for T_D and v_D, thus, moves to the upper right corner. Since a high fibre nozzle feed rate is sought, an initial setting at a temperature of $T_D = 345$ °C is defined as the first starting point of the CCD to be adjusted, which will be used for the experimental validation. This temperature value, while maintaining the requirement $t_{HAZ}(T_D, v_D) \approx 365$ µm and according to Figure 14b, corresponds to a feed rate value of approximately $v_D = 116$ mm/min. This factor-level combination $T_D = 345$ °C and $v_D = 116$ mm/min is, thus, determined as the central point of the new, adjusted CCD. According to [35], starting from this starting point (central point), a factor-level combination is selected in the assumed optimisation direction, i.e., in the direction of a higher fibre nozzle feed rate v_D (reduced process time for roving integration). Table 10 shows the factor-level combinations selected for the adjusted CCD with the new, more detailed operating point range.

Table 10. Factor levels used for the adjusted CCD with abbreviations.

Factor	$-\sqrt{2}$	-1	0	1	$\sqrt{2}$
T_D(°C)	335	338	345	352	355
v_D(mm/min)	92	99	116	133	140

3.2.2. Experimental Validation of the FE Model in the Adjusted Operating Point Range without the Influence of Roving Integration

Based on the factor levels for the adjusted CCD in Table 10, PA12 parts were produced using the developed LS machine and the generated HAZ (without roving) was measured. The measured and FE-model-simulated values for the depth of the HAZ are listed in Table 11, together with the prevailing model deviation.

Table 11. Comparison of simulated and experimental results for t_{HAZ} with deviation and an indication of model accuracy.

T_D		v_D		t_{HAZ}[µm] FE Model	$\bar{t}_{HAZ}$[µm] Experiment	Δt_{HAZ}[µm]	Deviation [%]
338	(−1)	99	(−1)	388	391	4	1
352	(+1)	99	(−1)	426	400	−26	6
338	(−1)	133	(+1)	308	275	−33	12
352	(+1)	133	(+1)	345	292	−53	18
335	$\left(-\sqrt{2}\right)$	116	(0)	337	295	−42	14
355	$\left(+\sqrt{2}\right)$	116	(0)	391	336	−55	16
345	(0)	92	$\left(-\sqrt{2}\right)$	426	447	21	5
345	(0)	140	$\left(+\sqrt{2}\right)$	313	284	−29	10
345	(0)	116	(0)	365	354	−11	3

The experimental results for the depth of the HAZ are close to the simulated results. The last column shows that the relative maximum model deviation between the experiment and the FE model is max. 18%. Compared to the initial model accuracy of 22%, which was determined with the help of the SPD, the model accuracy could be reduced with the CCD. A possible reason for this could be the reduced number of factors. Compared to the SPD, where six factors are varied, with the CCD, only two factors are varied, and the rest are kept constant. This reduces the scattering influence of the factors that are held constant. Figure 15 compares the surface response diagrams of the CCD derived using Minitab for the FE model (top) and the experiments carried out (bottom).

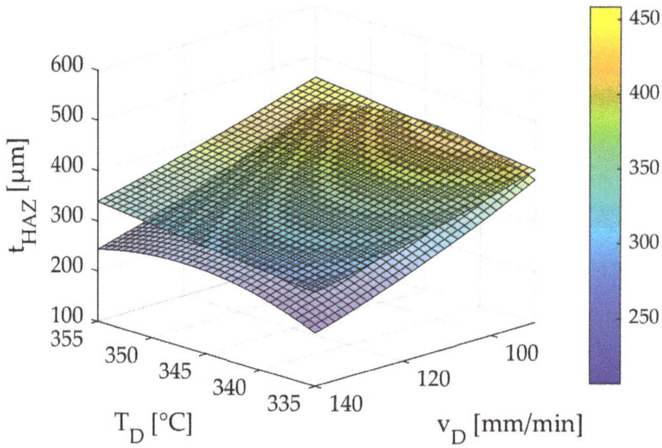

Figure 15. Surface response diagram for the results determined with the FE model (**top**) and the experimental results (**bottom**).

Equation (8) shows the regression function generated by Minitab (2022 Cloud App) to describe the depth of the HAZ for the experimentally performed CCD without the influence of the roving integration (lower surface response diagram).

$$t_{HAZ} = 46,214 - 274.1T_D - 13.56v_D + 0.3980T_D^2 + 0.01903v_D^2 + 0.0168T_Dv_D \quad (8)$$

The standard error of the regression, S, for the depth of the HAZ was low at 32.41 µm. Therefore, the predicted deviation from the actual value was only 36.68 µm. The coefficient of determination is $R^2 = 0.78$ and is rated as acceptable. Furthermore, Equation (9) shows the regression function describing the depth of the HAZ for the simulation-based CCD (upper surface response diagram).

$$t_{HAZ} = -400 + 3.7T_D - 3.4v_D - 0.001T_D^2 + 0.009v_D^2 + 0.003T_Dv_D \quad (9)$$

The standard error of the regression, S, for the depth of the HAZ was also low at 32.41 µm. Therefore, the predicted deviation from the actual value was only 36.68 µm. The coefficient of determination is $R^2 = 0.98$, close to 1 and rated as very good.

It can be said that the two surface response diagrams are relatively close. The surface response diagrams are further apart for high feed rate values v_D than for lower feed values. Furthermore, it can be seen in Figure 15 that the surface response diagram for Equation (9) drops for both high and lower temperatures. The causes for deviations between the surface response diagrams could be the same as in Section 3.1. In addition, the longer sample geometry could be another reason for a reduced coefficient of determination for Equation (9). Due to the longer length of the samples compared to the samples from the SPD, the samples from the CCD may lie in areas on the powder bed where the heat distribution is no longer constant. The result is an increased pore content of the parts and a reduced value for the depth of the HAZ.

3.2.3. Influence of Roving Integration on the Depth of the HAZ

The identical sets of factor levels from the modified CCD were selected to examine the impact of roving integration on the depth of the HAZ, as outlined in Table 10 [21]. Equation (10) represents the regression function derived from the CCD outcomes, used to model the HAZ depth with consideration of the roving's influence.

$$t_{HAZ} = 39,969 - 224.5T_D - 18.14v_D + 0.33T_D^2 + 0.0651v_D^2 \quad (10)$$

The standard error of the regression (S) for the HAZ depth was notably low, measuring 35.69 μm. Although the coefficient of determination R^2 of 0.72 was lower than that of the model presented in Equation (10), it is considered acceptable. In Figure 16a, a surface response plot based on Equation (9) of the FE model (excluding the influence of the roving) was compared with Equation (10) (considering the influence of the roving).

With the integration of rovings, it is evident that the depth of the HAZ becomes more substantial compared to scenarios without roving integration. Furthermore, roving integration primarily influences the depth of the HAZ, increasing it by approximately 100 μm. In comparison, the width of the HAZ experiences a marginal increment of about 32 μm, consistent with findings in [18]—this increase in depth primarily results from the heat transferred from the heated fibre nozzle to the roving. The stored heat of the roving contributes to additional melting, resulting in an average increase of 110 μm in the HAZ depth. Figure 16b presents the contour diagram for Equation (10) (CCD with roving influence). According to the contour lines in Figure 16b, the minimum depth is observed in the top-left region. In this case, it corresponds to a nozzle temperature between 335 °C and 345 °C and a feed rate between 130 mm/min and 140 mm/min. The predicted HAZ depth within this range falls below 540 μm and above 520 μm.

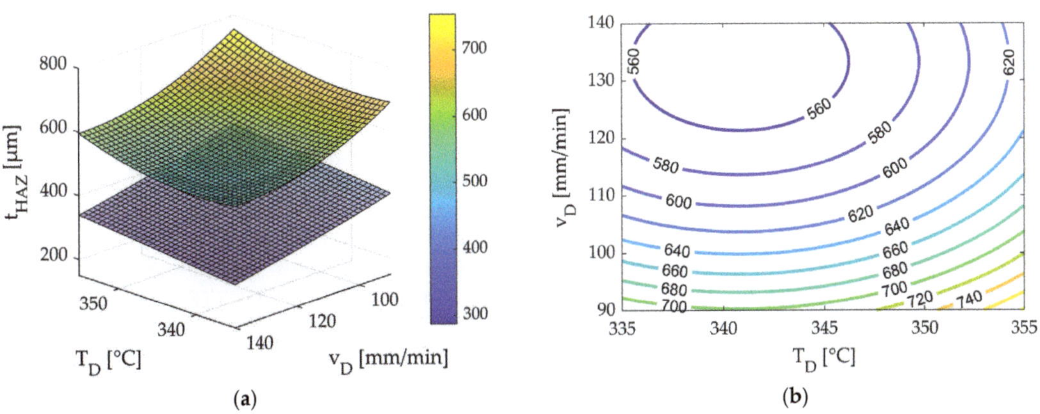

Figure 16. Surface response diagram of the FE model (bottom) and surface response diagram of the experimentally performed CCD with the influence of roving integration (top) (**a**). Contour diagram of the upper surface response diagram from (**a**) with contour lines (**b**).

To assess the reliability of the roving overlap above the integrated part level, it was observed that there was no significant influence on the nozzle feed rate within the range of 92 mm/min to 140 mm/min. Consequently, additional specimens were employed to explore the limitations of the nozzle feed rate. A nozzle temperature of 345 °C was set for these particular specimens, while the nozzle feed rate was systematically adjusted. The supplementary feed rates used were 160, 180, 200, 220, 240, and 300 mm/min, each repeated twice for reliability. Furthermore, only those derived from the modified CCD using a nozzle temperature of 345 °C were used for consistency among the specimens. Figure 17a presents the results concerning the measured roving overlaps (blue dots), the recoater's movement level (red), and a linear trend line (black) to represent the estimated measurement points. Figure 17b illustrates the rovings' orientation in the selected specimens' parts.

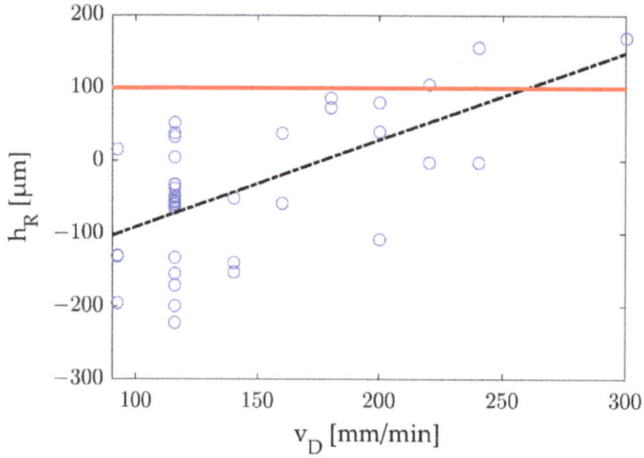

(a)

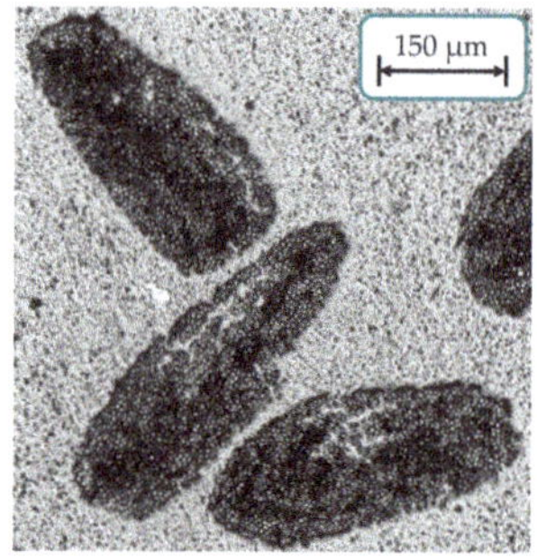

(b)

Figure 17. Influence of nozzle feed rate on roving overlap (**a**) and uncontrolled roving orientation within the specimen (**b**). The blue dots indicate the roving overlap and the red line in (**a**) indicates the layer thickness during printing.

The recoater's movement level depicted in Figure 17a remained constant at 100 µm, aligning with the predetermined layer thickness during printing. Figure 17a provides a visual representation of the measured values' variability. However, an increase in the nozzle feed rate resulted in the roving being less deeply embedded in the heat-affected zone (HAZ), which raised the risk of the recoater encountering the roving or potentially becoming entangled with it. Notably, while there were fewer data points for specimens with feed rates exceeding 140 mm/min, three values (representing 25% of these specimens) surpassed the 100 µm limit. Examining Figure 17b reveals significant variations in the shape and orientation of the rovings among the specimens. Some specimens exhibited flat and wide rovings, while others had tall and narrow ones. This underscores the substantial impact of orientation as a significant source of variation affecting the successful integration of rovings. The arbitrary orientation of the rovings has a significant effect on the overlap of rovings and, consequently, the reliability of the process. Process limits are established based on the conducted tests and the previously established relationships to address this issue. Increasing the nozzle feed rate while reducing the nozzle temperature led to the shallower embedding of the roving within the part. This heightened the risk of the recoater encountering and entangling the roving. Furthermore, this condition increased the likelihood of powdered material adhering to the nozzle, potentially disrupting the printing process. The powder adhering to the nozzle was observed at 335 °C with a feed rate of 160 mm/min and 345 °C with a 200 mm/min feed rate. In the experimental CCD study, a detailed examination was conducted within the temperature range of 335 °C to 355 °C for the nozzle and within the feed rate range of 92 mm/min to 140 mm/min. Notably, no disturbances were encountered during the production of 52 specimens within this range, leading to the conclusion that a process-reliable integration of fibres can be achieved within these parameters.

The following factors are potential explanations for the lower coefficients of determination and the lack of control over roving orientation:

- The entire structure of the fibre integration unit remains within the process chamber of the developed LS machine during the printing process, leading to thermal expansion and potential changes in manually set values (e.g., nozzle distance) compared to the system's cold state. Additionally, thermal expansion of the feed spindles that position the fibre integration unit can introduce errors and alter nozzle feed rate values.

- The PLC temperature setting accuracy is approximately ±1 °C, which can contribute to variations in the measurement results.
- The inner diameter of the fibre nozzle exceeds the thickness of the roving, resulting in an increased play of the roving within the fibre nozzle. This play can lead to uncontrolled roving placement within the part.
- Other factors, such as the placement of specimens in the powder bed, the condition of the powder's ageing, and the condition of roving delivery, may also contribute to result deviations.

3.2.4. Determination of an Optimal Operating Point

All target parameters must be considered to determine the optimal operating point for the developed LS machine. Notably, the settings for nozzle temperature and nozzle feed rate exert opposing effects on the target variables. A higher nozzle feed rate proves advantageous for the width and depth of the HAZ (heat-affected zone) and processing speed. Conversely, lowering the nozzle temperature results in reduced HAZ dimensions. Conversely, a higher nozzle temperature and a lower nozzle feed rate enhance process reliability. The adjusted CCD analysis identified an optimal range between 335 °C and 345 °C at a nozzle feed rate of 130 mm to 140 mm per minute for the HAZ depth. Further reductions in either nozzle temperature or increases in nozzle feed rate should be avoided for process safety. Considering these factors, a nozzle feed rate increase enhances all three remaining target values. Therefore, a process limit of 140 mm per minute was selected for optimisation. At the same time, nozzle temperature variations have a relatively minor impact within the range of 335 °C to 345 °C; a limit of 340 °C was set to ensure greater process reliability. Considering process reliability, speed, and HAZ size, the optimised operating point corresponds to a nozzle temperature of 340 °C and a nozzle feed rate of 140 mm per minute. According to the FE model, this setting is expected to yield a HAZ width of 2638.72 μm and a HAZ depth of 523.36 μm. An overview of all settings for the optimal operating point is provided in Table 12.

Table 12. Identified operating points evaluated as optimal for reproducible and process-reliable roving integration.

Operating Points	
T_D (°C)	340
v_D (mm/min)	140

In the future, several approaches can be considered to enhance process reliability and reduce processing time:

- Modifying the fibre nozzle's inner diameter to match the roving's shape or making it smaller can provide better control over the orientation of the rovings within the part.
- Implementing additional twisting of the rovings before coating could result in a rounder shape. This, combined with adjustments to the inner diameter of the fibre nozzle, may improve deposition accuracy within the part.

3.2.5. Manufacturing of a Battery Tab Suction Gripper

For the experimental validation of the identified and optimised operating points from Table 12, this section involves the production of a battery tab suction gripper with a function-integrated spring, i.e., without subsequent assembly, for agile fuel cell production. Figure 18a shows the 3D model of the suction grip with integrated roving paths (red) inside the struts to be reinforced, which was created using generative design.

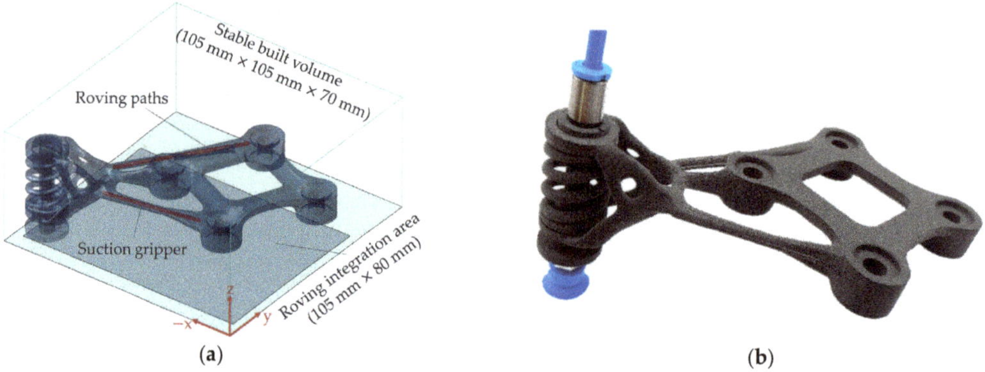

Figure 18. Three-dimensional view of the 3D model in the slicer app with roving paths in red (**a**) and manufactured suction gripper with integrated spring (**b**).

In the struts, within which the rovings are to be integrated, the part has a minimum part diameter of 3.5 mm. The risk of the HAZ protruding from the part, thus affecting the struts' overall appearance and reducing the struts' surface quality, is, therefore, relatively high. Using the optimised operating point from Table 12, three rovings could be integrated into each strut according to Figure 18b, and, thus, the part could be successfully manufactured and reinforced. The surface of the struts differs only marginally from the rest of the part.

4. Conclusions

The LS machine developed for the additive manufacturing of CCFRP parts aims to combine the specific advantages of the LS process with continuous roving reinforcement. This technology enables the future production of intricate, near-net-shape functional parts with favourable matrix properties. The approach allows for load-path-oriented reinforcement with continuous rovings, offering economic benefits by eliminating the need for support structures and reducing post-processing efforts.

This study pursued the systematic optimisation of roving integration within a newly developed LS machine. This involved utilising an FE model in COMSOL and conducting experiments using a CCD. The optimisation focused on crucial aspects such as processing time, process reliability, and the HAZ's shape. The critical findings of this study are summarised as follows:

- Using a convergence analysis and plausibility check, the developed FE model could be verified concerning model plausibility. The FE model shows the same physical behaviour as the split-plot design (SPD) in [18]. When comparing the results from the SPD and the FE model, an initial model accuracy of the FE model of 78% is achieved.
- A large percentage of the deviations between the developed FE model and the conducted experiments most likely originate in the pure LS process and the course of roving integration. Depending on the position of the parts in the powder bed, the porosity of the parts can vary significantly on the powder bed surface due to temperature differences caused by the installed IR emitters in the LS machine or the additional heat source of the fibre integration unit [20]. This has a direct influence on how the heat propagates within the part. The result is scattered values for the width and depth of the HAZ. Furthermore, the mixing ratio of the PA12 powder used has a considerable influence on the part properties. Another cause for reduced model accuracy is the occurrence of possible convection flows in the air gap between the heated fibre nozzle and the part surface due to the nozzle feed rate. These convection currents can influence the transferred heat quantity Q_{FN}. The material parameters or the mesh

fineness used in the FE model in COMSOL can also affect the target values and, thus, the model accuracy.
- With the help of the derived FE model and a CCD with initially widely spaced factor-level combinations, an operating point range could be identified in which $t_{HAZ}(T_D, v_D) \approx 365$ µm occurs. Based on a selected operating point at $T_D = 345$ °C and $v_D = 116$ mm/min, a new, more detailed CCD was derived as the basis for experimental validation of the FE model.
- The adapted CCD was carried out experimentally and simulatively. In a result comparison, the model accuracy of the FE model could be reduced to 18%. The reasons for this are the reduced number of varying factors and, thus, a reduced scattering effect. The experimental regression model for the detailed CCD has a coefficient of determination of 0.78 and is close to 1. The target variable $t_{HAZ}(T_D, v_D)$ can thus be described relatively well by the influencing factors.
- Additional factors contributing to the lower coefficients of determination include the thermal expansion of the fibre integration unit, potential deviations in temperature settings controlled by the PLC, the interaction between the roving and the inner diameter of the fibre nozzle, and scattering effects linked to the laser-sintering process (such as material properties and specimen placement within the powder bed).
- Integrating rovings into the part results in an expanded HAZ. Specifically, the depth of the HAZ is notably increased compared to scenarios without roving integration. Roving integration primarily influences the depth, which sees an average increase of 100 µm. This heightened depth can be attributed to the heat transferred from the heated fibre nozzle to the roving. Moreover, the roving's intrinsic heat contributes to additional melting, increasing the average HAZ depth of 110 µm.
- This study successfully demonstrated a substantial 233% increase in the nozzle feed rate, achieving a 140 mm/min rate for roving integration. Consequently, more cost-effective production of CCFRP parts in the developed LS machine becomes feasible. Furthermore, the width and depth of the HAZ were effectively reduced to 2638.72 µm −56%) and 523.36 µm (−44%), respectively. This reduction enables the integration of rovings closer to the part edges, facilitating higher fibre volume content (FVC) settings. The study, thus, provides optimised operational parameters for future research endeavours.
- However, certain limitations persist, particularly in terms of processing time. Although a 233% increase in processing time may seem substantial, the manufacturing duration for CCFRP parts with a high FVC can still be considerable. The increase in manufacturing time is only marginal for CCFRP parts, necessitating localised reinforcement in highly stressed areas with a lower FVC requirement. Additionally, it should be noted that, when rovings are placed near part edges, this may lead to protrusions of melted material or the HAZ from the part surface. Achieving a uniform surface requires the removal of these protruding materials.

Subsequent research efforts will focus on systematically augmenting the FVC and its correlated mechanical properties. In tandem with positioning rovings close to part edges, careful consideration will be given to the relative distances between rovings in both the vertical (build-up direction within the LS process) and horizontal (across the powder bed surface) axes. Tensile specimens will be fabricated to elucidate the mechanical attributes and, ultimately, reveal the potential of this LS process integrated with continuous rovings. Moreover, an investigation into the influence of nozzle temperature on roving properties is slated for exploration.

Furthermore, the established FE model can be leveraged to determine an operational point conducive to successful roving integration with alternative materials in the LS process (such as PA11, PA6, TPU, etc.). The potential applications of this LS process for CCFRP parts extend into the domain of production engineering. Notably, the cost-effective and time-efficient production of lightweight tools, purposefully tailored for industrial robot applications (e.g., gripper fingers featuring internal air channels for parallel jaw grippers or suction grippers equipped with integrated springs), stands as a promising avenue for reducing both moving masses and energy demands.

Author Contributions: Conceptualisation, M.B.; data curation, Z.S. and A.S.; formal analysis, M.B., Z.S. and A.S.; investigation, M.B., Z.S. and A.S.; methodology, M.B.; project administration, M.B. and F.K.; supervision, J.F.; validation, M.B., Z.S. and A.S.; visualisation, M.B.; writing—original draft, M.B.; writing—review and editing, M.B., F.K. and J.F. All authors have read and agreed to the published version of the manuscript.

Funding: This research received no external funding.

Institutional Review Board Statement: Not applicable.

Informed Consent Statement: Not applicable.

Data Availability Statement: Not applicable.

Acknowledgments: The authors would like to thank the Karlsruhe Research Factory of the Karlsruhe Institute of Technology for the opportunity to carry out the results of this paper. Furthermore, we acknowledge the support by the KIT-Publication Fund of the Karlsruhe Institute of Technology.

Conflicts of Interest: The authors declare no conflict of interest.

References

1. Schlüsseltechnologie Leichtbau. Available online: https://www.bmwi.de/Redaktion/DE/Schlaglichter-der-Wirtschaftspolitik/2019/05/kapitel-1-6-schluesseltechnologie-leichtbau.html (accessed on 2 May 2020).
2. Schürmann, H. *Konstruieren mit Faser-Kunststoff-Verbunden*, 2nd ed.; Springer: Berlin/Heidelberg, Germany, 2007. [CrossRef]
3. Baumann, F.; Sielaff, L.; Fleischer, J. Process Analysis and Development of a Module for Implementing Continuous Fibres in an Additive Manufacturing Process. In Proceedings of the SAMPE Europe Symposium 2017, Stuttgart, Germany, 14–16 November 2017.
4. Vaneker, T. Material extrusion of continuous fiber reinforced plastics using commingled yarn. *Procedia CIRP* **2017**, *66*, 317–322. [CrossRef]
5. Matsuzaki, R.; Ueda, M.; Namiki, M.; Jeong, T.-K.; Asahara, H.; Horiguchi, K.; Nakamura, T.; Todoroki, A.; Hirano, Y. Three-dimensional printing of continuous-fiber composites by in-nozzle impregnation. *Sci. Rep.* **2016**, *6*, 23058. [CrossRef] [PubMed]
6. Yang, C.; Tian, X.; Liu, T.; Cao, Y.; Li, D. 3D printing for continuous fiber reinforced thermoplastic composites: Mechanism and performance. *Rapid Prototyp. J.* **2017**, *23*, 209–215. [CrossRef]
7. Akhoundi, B.; Behravesh, A.; Saed, A.B. Improving mechanical properties of continuous fiber-reinforced thermoplastic composites produced by FDM 3D printer. *J. Reinf. Plast. Compos.* **2019**, *38*, 99–116. [CrossRef]
8. Bettini, P.; Alitta, G.; Sala, G.; Di Landro, L. Fused deposition technique for continuous fiber reinforced thermoplastic. *J. Mater. Eng. Perform.* **2017**, *26*, 843–848. [CrossRef]
9. Dickson, A.; Barry, J.; McDonnell, K.A.; Dowling, D.P. Fabrication of continuous carbon, glass and Kevlar fibre reinforced polymer composites using additive manufacturing. *Addit. Manuf.* **2017**, *16*, 146–152. [CrossRef]
10. Karalekas, D.; Antoniou, K. Composite rapid prototyping: Overcoming the drawback of poor mechanical properties. *J. Mater. Process. Technol.* **2004**, *153–154*, 526–530. [CrossRef]
11. Goh, G.; Yap, Y.; Agarwala, S.; Yeong, W.Y. Recent Progress in Additive Manufacturing of Fiber Reinforced Polymer Composite. *Adv. Mater. Technol.* **2019**, *4*, 1800271. [CrossRef]
12. Lu, Y.; Han, X.; Gleadall, A.; Chen, F.; Zhu, W.; Zhao, L. Continuous fibre reinforced Vat photopolymerisation (CONFIB-VAT). *Addit. Manuf.* **2022**, *60*, 103233. [CrossRef]
13. Schmid, M. *Laser Sintering with Plastics: Technology, Processes, and Materials*; Hanser: München, Germany, 2018. [CrossRef]
14. Fischer, M.; Josupeit, S. Material Properties of Additive Manufactured Polymer Parts. In Proceedings of the Inside 3D Printing Conference and Expo, Berlin, Germany, 10–11 March 2014.
15. Gebhardt, A. *Additive Fertigungsverfahren: Additive Manufacturing und 3D-Drucken für Prototyping—Tooling—Produktion*, 5th ed.; Hanser: München, Germany, 2016.
16. Lasersintern—Optimal Für Hohen Durchsatz. Available online: https://www.industrial-production.de/additive-fertigung/lasersintern-fuer-hohen-durchsatz-898725.htm (accessed on 23 November 2021).
17. Breuninger, J.; Becker, R.; Wolf, A.; Rommel, S. *Verl, Alexander Generative Fertigung mit Kunststoffen: Konzeption und Konstruktion für Selektives Lasersintern*; Springer: Berlin, Germany, 2013. [CrossRef]
18. Baranowski, M.; Völger, L.; Friedmann, M.; Fleischer, J. Experimental Analysis and Optimisation of a Novel Laser-Sintering Process for Additive Manufacturing of Continuous Carbon Fibre-Reinforced Polymer Parts. *Appl. Sci.* **2023**, *13*, 5351. [CrossRef]
19. Baranowski, M.; Basalla, F.; Kößler, F.; Fleischer, J. Investigation of the Thermal Characteristics of a Novel Laser Sintering Machine for Additive Manufacturing of Continuous Carbon Fibre-Reinforced Polymer Parts. *Polymers* **2023**, *15*, 3406. [CrossRef] [PubMed]
20. Baranowski, M.; Basalla, F.; Friedmann, M.; Fleischer, J. Thermal Analysis of a Novel Laser Sintering Machine for Additive Manufacturing of Continuous Carbon Fiber Reinforced Polymer Parts. *SSRN J.* **2023**. [CrossRef]

21. Knispel, H. Further Development of a Software for the Generation of Control Code for Integration of Continuous Fibres in the Selective Sintering Process. Bachelor Thesis, Karlsruhe Institute of Technology, Karlsruhe, Germany, 24 February 2023.
22. Wegner, A.; Witt, G. Correlation of Process Parameters and Part Properties in Laser Sintering using Response Surface Modeling. *Phys. Procedia* **2012**, *39*, 480–490. [CrossRef]
23. Stephan, P.; Kabelac, S.; Kind, M.; Mewes, D.; Schaber, K.; Wetzel, T. (Eds.) *VDI-Wärmeatlas*, 12th ed.; Springer: Berlin, Germany, 2019. [CrossRef]
24. Völger, L. Design of Experiments (DoE) to Optimize an SLS-System for Additive Manufacturing of Plastic Components with Continuous Fibers. Bachelor Thesis, Karlsruhe Institute of Technology, Karlsruhe, Germany, 2022.
25. The Heat Transfer Module User's Guide. Available online: https://doc.comsol.com/5.4/doc/com.comsol.help.heat/HeatTransferModuleUsersGuide.pdf (accessed on 4 September 2023).
26. Riedlbauer, D.; Drexler, M.; Drummer, D.; Steinmann, P.; Mergheim, J. Modelling, simulation and experimental validation of heat transfer in selective laser melting of the polymeric material PA12. *Comput. Mater. Sci.* **2014**, *93*, 239–248. [CrossRef]
27. Soldner, D.; Greiner, S.; Burkhardt, C.; Drummer, D.; Steinmann, P.; Mergheim, J. Numerical and experimental investigation of the isothermal assumption in selective laser sintering of PA12. *Addit. Manuf.* **2021**, *37*, 101676. [CrossRef]
28. VESTOSINT®Polyamide 12 (Nylon 12) Powders for Demanding Coatings Applications. Available online: https://products.evonik.com/assets/78/55/VESTOSINT_polyamide_12_powders_for_demanding_coating_applications_EN_Asset_1807855.pdf (accessed on 4 September 2023).
29. Rabenseifner, V. Experimental Analysis of the Process Limits of a Novel Laser Sintering Process with Automated Integration of Continuous Fibres. Master Thesis, Karlsruhe Institute of Technology, Karlsruhe, Germany, 4 April 2023.
30. Sintratec PA12 High Performance Powder. Available online: https://sintratec.com/3d-printing-materials/sintratec-pa12-powder/ (accessed on 2 October 2023).
31. Spintzyk, A. Simulative and Experimental Investigation of the Heat-Affected Zone for Continuous Fiber Integration in the Laser Sintering Process. Bachelor Thesis, Karlsruhe Institute of Technology, Karlsruhe, Germany, 2022.
32. Emissivity Data Sheet. Available online: https://static-int.testo.com/media/b5/14/6481a245e502/Emissivity-table-DE.pdf (accessed on 2 November 2022).
33. Mesh Element Quality and Size. Available online: https://doc.comsol.com/5.5/doc/com.comsol.help.comsol/comsol_ref_mesh.15.18.html (accessed on 4 September 2023).
34. General Projection. Available online: https://doc.comsol.com/5.5/doc/com.comsol.help.comsol/comsol_ref_definitions.12.077.html (accessed on 4 September 2023).
35. Kleppmann, W. *Versuchsplanung: Produkte und Prozesse Optimieren*, 10th ed.; Hanser: München, Germany, 2020. [CrossRef]
36. Hoesch, F. Development of an Experimental Setup for Twisting Carbon Fiber for Additive Fabrication of Fiber Reinforced Plastics. Bachelor Thesis, Karlsruhe Institute of Technology, Karlsruhe, Germany, 11 April 2022.
37. Baranowski, M.; Kößler, F.; Fleischer, J. 2023 Laser-Sintern und Gleichzeitig Endlosfasern in Kunststoffteile Integrieren. Additive, 14 June 2023. Available online: https://additive.industrie.de/news/laser-sinteranlage-endlosfasern-additiv-gefertigte-kunststoffteile-integrieren/ (accessed on 20 September 2023).
38. Pham, D.; Dotchev, K.; Yusoff, W.A.Y. Deterioration of polyamide powder properties in the laser sintering process. *Proc. Inst. Mech. Eng. Part C J. Mech. Eng. Sci.* **2008**, *222*, 2163–2176. [CrossRef]

polymers

MDPI

Article

Spatial 3D Printing of Continuous Fiber-Reinforced Composite Multilayer Truss Structures with Controllable Structural Performance

Daokang Zhang, Xiaoyong Tian *, Yanli Zhou, Qingrui Wang, Wanquan Yan, Ali Akmal Zia, Lingling Wu and Dichen Li

State Key Laboratory for Manufacturing Systems Engineering, Xi'an Jiaotong University, 28 Xian Ning West Road, Xi'an 710049, China; zhangdk0821@stu.xjtu.edu.cn (D.Z.); zyl1631118746@163.com (Y.Z.); wqr@qzc.edu.cn (Q.W.); 15332080756@163.com (W.Y.); ziaaliakmal@stu.xjtu.edu.cn (A.A.Z.); lingling.wu@mail.xjtu.edu.cn (L.W.); dcli@mail.xjtu.edu.cn (D.L.)

* Correspondence: leoxyt@mail.xjtu.edu.cn

Abstract: Continuous fiber-reinforced composite truss structures have broad application prospects in aerospace engineering owing to their high structural bearing efficiency and multifunctional applications. This paper presents the design and fabrication of multilayer truss structures with controlled mechanical properties based on continuous fiber-reinforced thermoplastic composite 3D printing. Continuous fiber composite pyramid trusses fabricated by 3D printing have high specific stiffness and strength, with maximum equivalent compression modulus and strength of 401.91 MPa and 30.26 MPa, respectively. Moreover, the relative density of a truss structure can be as low as 1.45%. Additionally, structural units can be extended in any direction to form a multilayer truss structure. Structural performance can be controlled by designing the parameters of each layer. This study offers a novel approach for designing a multifunctional multilayer truss structure, a structure with low-density needs and unique load-bearing effects.

Keywords: spatial 3D printing; continuous fiber; lightweight structure; mechanical properties

1. Introduction

Continuous fiber-reinforced composites are used for a wide range of applications in the automotive and aerospace industries owing to their excellent mechanical properties, which include low density, high specific strength, and high modulus [1–4]. Because 3D printing is flexible and customizable, it has improved the use of continuous fiber-reinforced materials in industrial applications and made it possible to fabricate complex structures [5–7]. Studies on 3D printing of continuous fiber composites have primarily focused on ways of increasing fiber content and achieving a denser microstructure for better mechanical characteristics [8–10]. However, the interest in the design and manufacture of lighter structures with superior mechanics has been increasing among researchers.

Topological optimization [11] and lattice structure [12,13] designs are extensively utilized in 3D printing to produce lightweight designs, and the optimization method is used to produce a variety of composite structures. The topological optimization of composite structures has been studied extensively. The difficulties in the topological optimization of continuous fibers, such as the fiber discontinuity, length scale separation, decreased design freedom, and fiber orientation of CFRSs with complicated forms, have recently been resolved by a number of studies. Li et al. [14], for instance, provided a brilliant path-designed 3D printing method based on the composite's evaluated stress states, taking into account the load transmission path and anisotropic property of the continuous fiber filament. Based on this methodology, Wang et al. [15] developed a load-dependent path planning method under the stress-vector-tracing algorithm for 3D-printed CFRPCs, where fiber trajectories are generated along the load transmission path. In order to produce these, Tian et al. [16] created a multiscale design and production approach that included the concurrent optimization of fiber orientation and macro structural topology, taking into account hatch space and printed fiber radius. Continuous fiber composite structures with simultaneously optimized fiber orientations and topology structures were produced by 3D printing. At present, regardless of the kind of topology optimization method, the path planning is carried out within a plane, and compared with the spatial structure, there is still a certain gap in the density and specific intensity. For the lattice structure design, 2D CFRSs with various filling forms [17], including rectangular, circular, honeycomb, rhombus, trapezoidal, and corrugation, can be created using the conventional layer-by-layer 3D printing technique. Unique cell shapes, such as negative Poisson's ratio units [18] and re-entrant units [19], are also used to create 2D cellular architectures. Due to the limitations of the additive manufacturing process, such as requiring layer-by-layer stacking, it is difficult to manufacture spatial lattice structures.

Some researchers are starting to manufacture 3D lattice structures using new processes. First the continuous carbon fiber-reinforced thermosetting composite grid was fabricated with the winding printing approach [20]. Similarly, He et al. [21] printed spring and grid structures using UV-curable thermosets. Eichenhofer et al. [22] printed an ultra-lightweight sandwich structure with a pyramidal truss core using a continuous lattice fabrication process. In addition, Li et al. [23] and Luan et al. [24] manufactured continuous fiber-reinforced thermoplastic lattice structures using a free-hanging 3D printing method. However, research into space composite structures still needs to be further improved from the aspects of technology and structure. Owing to the unique properties of continuous fibers and the requirement for a continuous path, it is challenging to create continuous fiber composites with a general structure [25,26]. However, because most partial composite truss structures need to be fabricated and assembled, the overall fabrication of truss structures remains a challenge [27]. The integrated fabrication of continuous fiber composite truss structures still requires further research, and the mechanical properties of continuous fiber truss structures are still lacking, especially the control of the mechanical properties of multilayer truss structures.

In this study, multilayer pyramid truss structures with variable structural parameters are designed and fabricated using continuous fiber-reinforced thermoplastic composite additive manufacturing technology for pyramid truss structures with different structural parameters and by employing spatial path planning, as opposed to the conventional additive manufacturing process of layer-by-layer manufacturing. Pyramid trusses with different structural parameters were tested for compressive strength, and the effects of different structural parameters on the mechanical properties of the pyramid truss structure were theoretically analyzed. Based on the study of the pyramidal truss structure, horizontal and vertical extensions of the pyramidal truss were carried out, allowing for the extension of the truss unit in all directions. Each layer of the multilayer truss was individually designed to obtain a multilayer truss structure with special mechanical properties. In the face of the shortcomings of existing research, the design and manufacture of multilayer truss structures with different mechanical properties were achieved through the design of parametric truss structures on the basis of continuous fiber additive manufacturing. The manufacture of structures with better mechanical properties was completed at as low a cost as possible while still being able to cope with a variety of different load environments.

2. Design and Manufacturing

2.1. Pyramid Truss Unit Design

Truss structures are widely used in various industrial scenarios, and the concise and efficient design of a structure can be achieved by programming its mechanical properties [28]. We needed to choose a truss for parametric design, and in the end, we chose the pyramid truss because the pyramid truss has the following advantages: the structural form is simple and there are fewer structural parameters, which is conducive to using the structural parameters to control the mechanical properties; the structure is stable and uniform, with a strong bearing capacity; and the shape of the structure makes it easy to plan a continuous print path, with no overlap in the paths in the important bearing structure part, which is the most important reason. Maintaining a continuous print path can achieve better mechanical properties. As shown in Figure 1, the main parameters included in the unit are truss length (L), truss inclination angle (θ), truss width (d), truss thickness (h), and overlap thickness at the fold (Δx). A pyramid truss structure was formed by overlapping two continuous fiber composite filaments; furthermore, the four fulcrums at the bottom were connected and fixed with composite filaments. Here, the pyramidal cells are extended in two ways: (1) along the direction of the pyramidal cell composite wire, which allows for a simple array in both the horizontal and vertical directions but has the disadvantage of having many layer-to-layer individual nodes, which prevents the formation of tight connections and force transfer; (2) along the 45° direction, which doubles the relative density compared with the first method and allows for connections between the nodes and a tighter bond between the layers in the vertical direction.

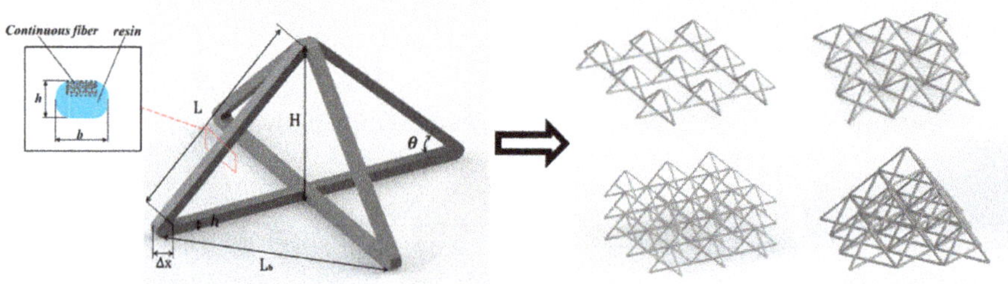

Figure 1. Pyramid composite truss units and their different types of arrays stacked in 0/90°and 45° directions.

The relevant parameters of the pyramid truss structural unit can be obtained using the following calculation:

The base length of the pyramid truss structural element is expressed as follows:

$$L_b = \sqrt{2} \times (L \times cos\theta + \Delta x) \quad (1)$$

The height of the pyramid truss structural element is expressed as follows:

$$\mathrm{H} = L \times \sin\theta + h \quad (2)$$

The relative density $\overline{\rho}$ of the pyramid truss structural element is expressed as follows:

$$\overline{\rho} = \frac{(4 \times (L + \Delta x) + 4 \times L \times cos\theta) \times b \times h + 2 \times b^2 \times h}{{L_b}^2 \times H} \quad (3)$$

2.2. *The Manufacture of Pyramid Truss Structure*

2.2.1. Material and Equipment

The pyramid truss structure was fabricated using a continuous fiber-reinforced thermoplastic composite 3D printer (Shanxi Fibertech Technology Development Co., Ltd., Xi'an, China), as shown in Figure 2. In this study, continuous fiber-reinforced thermoplastic composites were composed of continuous aramid fibers for reinforcement and polylactic acid (PLA) polymer material as the matrix. Combining the two materials can provide a composite truss with a better bearing capacity and a lighter relative density. In the present study, a continuous aramid-fiber material (Kevlar fiber with linear density of 670 dtex, density of 1440 kg/m^3, breaking strength of 152 N, specific strength of 23.0 cN/dtex, modulus of 700 cN/dtex, from DuPont Corp) with excellent stability in 3D printing for CFRCs was used as the reinforcement material, and polylactide (PLA/1.75 mm, density of 1240 kg/m^3) from Polymaker in China was used as the thermoplastic material.

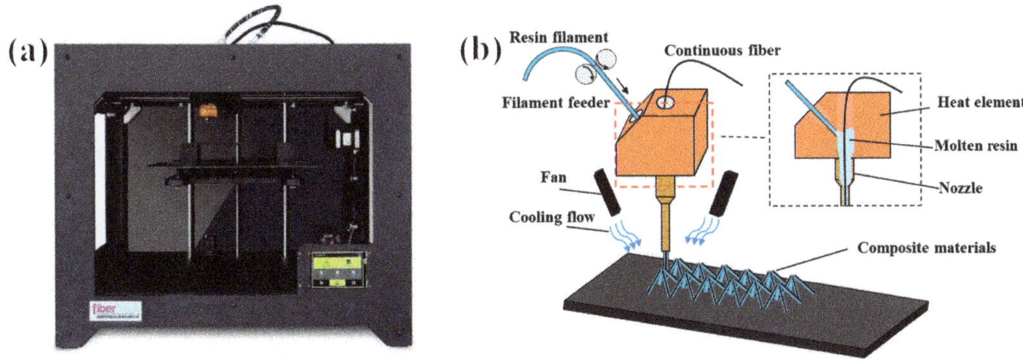

Figure 2. (**a**) Continuous fiber composite 3D printer physical image, and (**b**) continuous fiber composite space structure printing schematic.

To prevent interference between the nozzle and forming structure, a custom extended nozzle was used to finish the printing process. The outer layer of the nozzle was pasted onto a thermal insulation layer to make the temperature inside the nozzle as uniform and as constant as possible. The wire-feeding motor fed the resin wire into the heating chamber. The continuous fiber and melt resin were completely compounded inside the heating chamber. The composite monofilament was extruded from the nozzle while being pulled by the continuous fiber and rapidly cooled by an air cooler.

2.2.2. Pyramid Truss Manufacturing Process

The printing of a continuous-fiber thermoplastic composite space truss differs from that of the standard 3D printing layer-by-layer manufacturing procedure. Although the printing path and forming structure can be freely designed in space, spatial printing results in the deformation of the printed portion owing to the nozzle movement, which changes the printing results. Therefore, it is necessary to replicate the print path.

The nozzle temperature analysis produced the following results: The mechanical properties of thermoplastic resin changed with the temperature; specifically, its stiffness decreased with increasing temperature. As a result, after cooling, the section that was farther away had a higher rigidity, whereas the part closer to the nozzle always maintained a lower rigidity. The material in this portion was softened and easily distorted when the nozzle moved because the temperature of the thermoplastic resin near the nozzle was lower than its glass transition temperature (Tg). The length of the softened part depended on the nozzle temperature and cooling rate and remained a fixed value during the printing process. As a result, when the nozzle moved, certain continuous traction force was generated in the fiber, which deformed the printed structure. To avoid this deformation, it was necessary to re-plan the printing path to ensure that the final printing result was consistent with the designed structure.

In the truss 3D printing process, the main errors arose from the following three factors: First, the deformation of the structure was caused by the traction force of the nozzle on the printed structure during the printing process, as shown in Figure 3. For this part, we changed the initial print path to print a larger tilt angle and then printed the designed tilt angle by moving it horizontally. We denoted the angle deviation value and nozzle-generated traction force as $\Delta\theta$ and F, respectively. The traction force perpendicular to the truss structure is F_x. The deflection of the truss is regarded as the process of bending the single rod of the truss, and the maximum deflection μ_{max} of the beam deformation under a simple load gives the formula for the angle deviation value $\Delta\theta$, as follows:

$$\Delta\theta = \frac{F * \sin\theta * (l + \Delta l)^2}{3EI} \quad (4)$$

For instance, we selected a pyramid structure with a truss length of 12 mm and a tilt angle of 50° for theoretical calculation. The process parameters of the 3D printer included a nozzle diameter of 1.6 mm and a nozzle temperature of 230 °C to ensure the quality of the unsupported printing, and the adopted printing speed was 50 mm/min. The tractive force F of the printing nozzle under this process parameter was measured with a tension meter. The modulus of elasticity of the material was E (the elastic modulus of the material here refers to the mechanical properties of continuous fiber-reinforced thermoplastic composites [8]), and we could easily calculate the angular deviation.

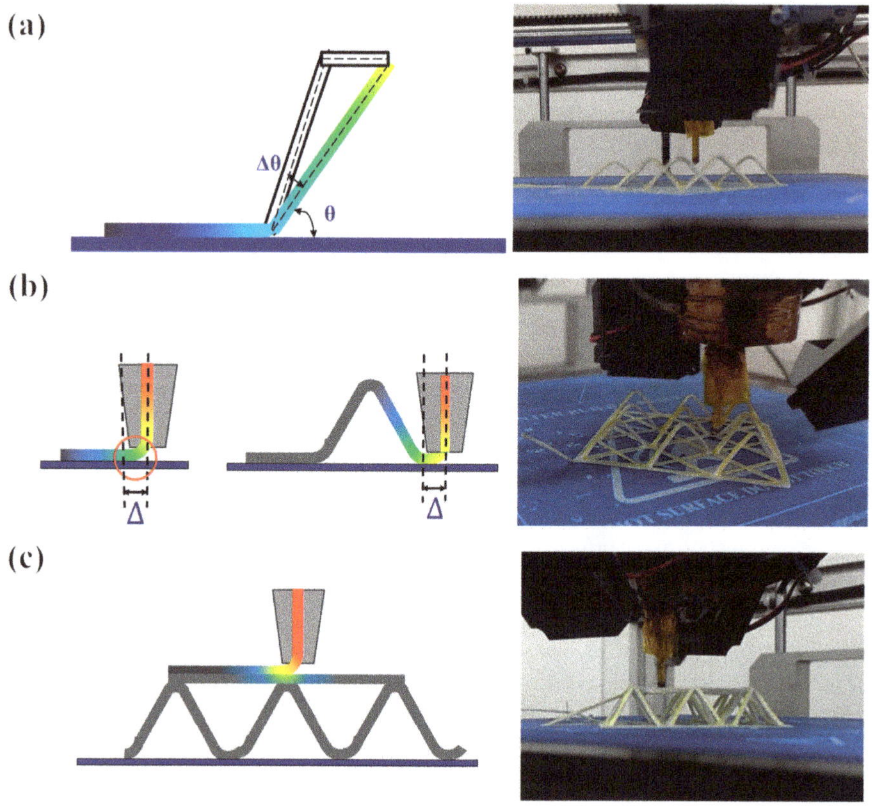

Figure 3. Main errors of continuous fiber-composite-printed space truss structure: (**a**) inclination angle error, (**b**) lateral error of starting and landing point, and (**c**) inter-layer error.

Second, in the nozzle lifting and falling process, because the nozzle diameter d was not negligible, the position of the truss node was shifted to the left by $\Delta = d/2$ via upward lifting. Similarly, when the nozzle fell, the position of the truss node shifted to the right by Δ. The third factor focused on the effects of pressure and heat on the lower layer when printing the upper structure when connecting trusses on each floor, as well as the bending deformation of the printed structure caused by gravity when printing horizontally without support. It is necessary to ensure that the fibers are always tensioned during the printing process. At this time, the printed structure maintains a certain straightness owing to the traction force and achieves a certain stiffness requirement through rapid cooling.

This study adopted measures such as error compensation and maintaining continuous fiber tension to achieve high-precision printing. To achieve this, the print path of the structure was designed, and the corresponding difference compensation was calculated. Furthermore, the specific coordinate points printed for each path and the resin wire feed rate were calculated, which could be used to convert these coordinates into the G-code for the 3D printer using MATLAB.

3. Result and Discussion

This section investigates the mechanical compression performance of continuous fiber-reinforced thermoplastic composite truss structures. Further design and manufacturing of the multilayer pyramid structures can be achieved by studying the mechanical properties of truss structures with different structural parameters. A compression rate of 1 mm/min was used to test the sample, Mechanical tests of the CFRTCTT samples were performed using a universal testing machine (MTS 850/25t, MTS Corp., New York, NY, USA).

3.1. Single-Layer Pyramid Structure

The pyramidal structural unit was expanded into a 2 × 2 array structure as a performance sample for compression testing, as shown in Figure 4. Furthermore, two main parameters of the continuous fiber composite pyramid truss structure, the inclination angle θ and truss size L, were studied. Here, the other parameters were constant, for example: truss width (d = 2 mm), truss thickness (h = 0.4 mm), and volumetric fiber content (8.75%) (compared with normal printing, the volume content of fiber in this sample was lower because the resin extrusion volume of the space truss was higher). In this study, we designed two groups of experiments. The first group, Group 1, had a fixed inclination angle of 50° and truss length L of the pyramid structure ranging from 6 mm to 18 mm, and each group consisted of three samples, with the same parameters for truss lengths differing by 2 mm. The second group was named Group 2, in which the length of the trusses was fixed at 12 mm and the truss inclination angle of each group of samples varied from 30° to 65° every 5°, and the number of samples in each group was also 3. The relative density of the pyramid truss with different structural parameters was calculated based on the design of the pyramid truss structure in Section 2, as shown in Tables 1 and 2.

Table 1. Structural parameters (variable angle) and relative density.

Type	Inclination Angle θ	Truss Size L/mm	Relative Density $\bar{\rho}$
A1	30°		2.98%
B1	35°		2.81%
C1	40°		2.74%
D1	45°	12	2.76%
E1	50°		2.88%
F1	55°		3.11%
G1	60°		3.48%
H1	65°		4.06%

Table 2. Structural parameters (variable size) and relative density.

Type	Inclination Angle θ	Truss Size L/mm	Relative Density $\bar{\rho}$
A2		6	8.30%
B2		8	5.46%
C2		10	3.86%
D2	50°	12	2.88%
E2		14	2.23%
F2		16	1.78%
G2		18	1.45%

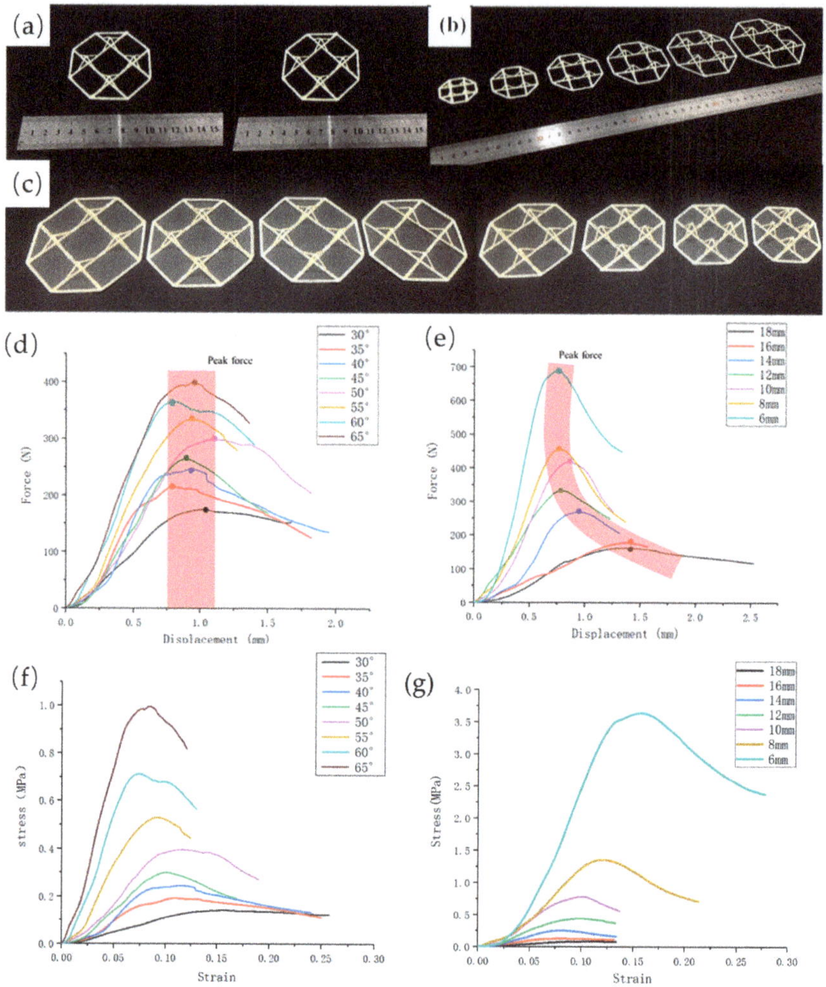

Figure 4. (**a**) The 2 × 2 pyramidal truss unit structure samples, (**b**) pyramidal truss structure samples of different scales, (**c**) pyramidal truss structure samples of different tilt angles, (**d**,**f**) compressive curves with different truss inclination angles, and (**g**,**e**) compressive curves with different truss scales.

The compression performance of a pyramidal truss unit was directly influenced by two crucial structural variables: truss length and truss angle. Here, we plot the force–displacement curve of the pyramidal truss structure in compression under various parameters using the data from a batch of samples that best reflect this structural parameter, as shown in Figure 4. Evidently, when the tilt angle was gradually increased from 30° to 65°, the load capacity of the pyramidal truss unit increased with the increasing tilt angle, and the maximum load capacity increased from 173 N to 485 N when the length of the truss was fixed at 12 mm. Therefore, from the perspective of energy absorption, an increase in the angle can further increase the energy absorption. However, for another structural parameter, the length of the truss, which gradually increased from 6 to 18 mm, the load capacity of the pyramidal unit decreased as the length of the truss increased; furthermore, when the length of the truss was increased, the energy absorption decreased. At a fixed tilt angle of 50°, the maximum load capacity dropped from 689 N to 130 N. From the force–displacement curves, it can be concluded that these two structural parameters are important for the pyramidal truss. This result can be easily explained because as the tilt angle increased, the direction of the pressure was closer to the direction of the continuous fibers. For composite structures, the more the direction of the force agreed with the direction of the material, the better the mechanical properties that the structure could exhibit. For lattice structures, the smaller the lattice size, the better the mechanical properties, so the structure showed that the smaller the truss size, the better the load-bearing performance. Therefore, by quantitatively analyzing the influence of the two parameters on the mechanical properties, pyramidal trusses can be designed with different structural parameters for different demand scenarios.

It is not comprehensive to judge structural parameters based only on their maximum load-carrying capacity. Further analyses should be conducted by comparing their equivalent strength and equivalent stiffness and the effect of different structural parameters on the mechanical properties of the pyramid truss. The equivalent bearing capacity, equivalent strength, and equivalent stiffness of the pyramid truss structure were calculated under different parameters. Their equivalent bearing capacity, equivalent strength, and equivalent stiffness, are the results obtained from the tests divided by the relative densities of the different structures. As shown in the Tables 3 and 4, with an initial increase in the angle, the equivalent bearing capacity of the pyramid truss with variable angles increased rapidly. When the angle was close to 60°, the specific bearing capacity had little difference, while the equivalent strength and equivalent stiffness increased significantly with the increase in the angle; furthermore, the maximum equivalent strength was 28.52 MPa, and the equivalent stiffness was up to 401.91 MPa. For a pyramid truss with a variable length, its equivalent bearing capacity exhibited a relatively complex law, and the equivalent strength and stiffness increased with a reduction in the truss length. When the truss length was L = 6 mm, the maximum equivalent strength was 30.26 MPa and the equivalent stiffness was 251.96 MPa. The mechanical behaviors of the 3D-printed pyramid truss structure also should be compared with those of the conventional truss structure, such as the pyramidal fiber composite lattice by Xiong [29] and the tetrahedral fiber composite lattice by Zhang [30]. The strength and stiffness of the pyramid truss structure created in this paper are not as good as those of the standard process truss structure in terms of mechanical parameters. In terms of equivalent strength, the truss structure in this study had a lower relative density and performed similarly to the standard process truss structure.

Table 3. Structural parameters (variable angle) and mechanical behaviors.

Type	The Equivalent Maximum force $F/\bar{\rho}$ (N)	Equivalent Strength $\sigma/\bar{\rho}$ (MPa)	Equivalent Elastic Modulus $E/\bar{\rho}$ (MPa)
A1	1451.34	4.73	37.82
B1	2117.43	7.57	68.89
C1	2454.38	9.80	99.39
D1	2463.77	11.21	124.48
E1	2586.80	13.71	164.42
F1	2813.50	17.83	228.03
G1	2765.80	21.61	291.59
H1	2850.98	28.52	401.91

Table 4. Structural parameters (variable size) and mechanical behaviors.

Type	The Equivalent Maximum Force $F/\bar{\rho}$ (N)	Equivalent Strength $\sigma/\bar{\rho}$ (MPa)	Equivalent Elastic Modulus $E/\bar{\rho}$ (MPa)
A2	2075.30	30.26	251.96
B2	2138.28	20.97	210.62
C2	2610.10	18.38	185.19
D2	2586.81	13.71	164.42
E2	2296.14	9.49	117.34
F2	1853.93	6.14	55.55
G2	2603.45	7.10	67.11

From the compression test results of the single-layer pyramid structure sample, we obtained the rule of the influence of these two parameters on the mechanical properties of the period. However, further analyses enabled us to design a pyramid-structured sample with controllable mechanical properties.

3.2. Multilayer Pyramid Structure

After systematically analyzing the relationship between the properties of the pyramidal unit structure and structural parameters, we further verified that the mechanical properties of the pyramidal unit were controllable in a multilayer structure. In the previous section, we accomplished the design and fabrication of a single layer of 2 × 2 pyramidal cells, in which it was necessary to fabricate a multilayer pyramidal structure; therefore, we designed several types of multilevel pyramid truss structures, as shown in Figure 5.

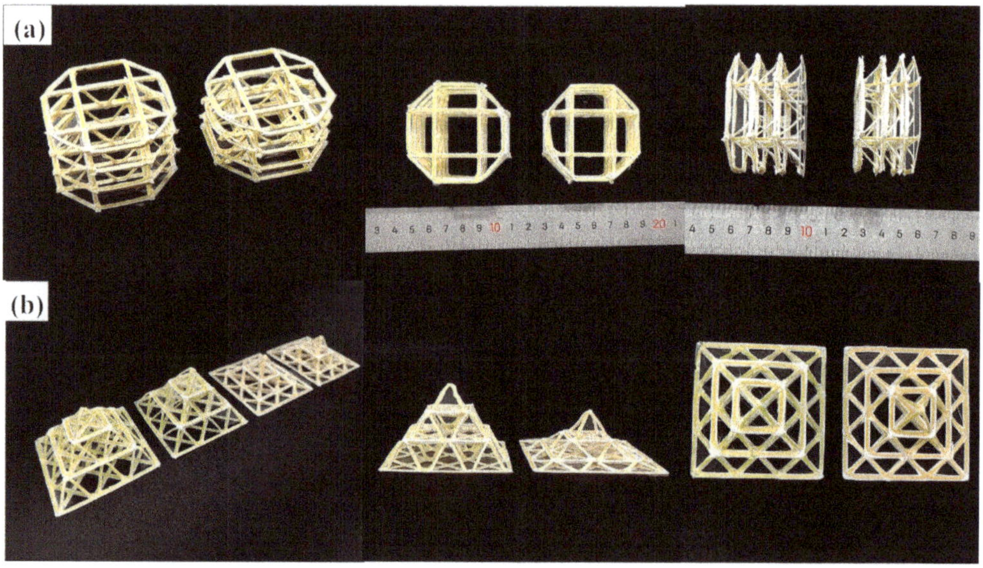

Figure 5. Samples of multilayer truss structure: (**a**) vertically stacked three-layer truss sample (uniform structure design and non-uniform structure design per layer), and (**b**) 45° connected stacked three-layer truss sample (uniform structure design and non-uniform structure design per layer).

The first type of pyramid truss adopted a vertical linear array. The apex of each pyramid structure was at the center of the upper pyramid structure. The advantage of this stacking method is that it maintains a low relative density, although this involves high requirements for the printing process and path planning. For the pyramid truss of this vertical array, samples with the same and different structural parameters for each floor were designed as three-floor truss structures. The height of each floor of the former was 10 mm with a 45° truss inclination. The heights of each floor of the latter from top to bottom were 10, 8, and 6 mm, respectively. The inclination also changed with height. In the previous section, based on the influence of the structural parameters on the mechanical properties, we speculated that, for a uniform single-layer truss structure, the stress of each layer is consistent when it is compressed. Under theoretical conditions, the three-layer structure undergoes uniform deformation and eventually fails simultaneously. For a non-uniform structure, owing to the difference in the bearing capacity and compression strength of structures with different parameters, the structure fails from top to bottom.

The second type of multilayer pyramid truss structure was designed by combining the pivot of the upper pyramid element with the vertex of the lower pyramid element;

thus, the number of elements in each floor is different. The number of pyramid elements on each floor from top to bottom was 1, 4, and 9, respectively. The advantage of this design is that the overall structure with overlapping nodes is more stable and reliable, and the force transmission is more continuous during the compression process. The overall structure remains a pyramid, reflecting the design of a multilevel structure. Although the structural parameters of each floor are the same, the experiment inevitably leads to failure from top to bottom, owing to the large difference in the number of units on each floor. However, designing the structural parameters can enable us to design structures with similar bearing effects between the three floors. Therefore, the original design sample had a uniform pyramid truss structure. The truss inclination of each floor was 45°, the truss length was 14 mm, and the unit structural parameters of each floor were consistent. Through calculations, we designed the truss inclinations of the bottom floors to be 7°, the truss inclination of the second floor to be 28°, and the truss inclination of the top floor to be 60°. The truss length changed with a change in inclination. Under this design, the load capacity of each floor was relatively close, and an overall slow failure could be realized.

The compression of a multilayer pyramidal truss structure was analyzed using the compression test data and photographs, as shown in Figure 6. For pyramidal truss structures arranged directly in the vertical direction, regardless of whether the structure had three layers of the same height or three layers with varying height gradients, it maintained a certain load capacity until the overall structure failed and deformed completely. Evidently, for pyramid structures designed at the same height, when the strain was 0.78, all the three layers underwent deformation. Additionally, the second layer of the structure was damaged first; then, the top layer structure was compressed; and finally, the third layer of the structure was fully compressed owing to the printing accuracy. However, for pyramid structures designed at different heights, it can be clearly seen that the top layer was compressed first. From the influence of the structural parameters on mechanical performance, it can be seen that the top layer's parameters lead to the failure of its structure more easily thus indicating a top-down failure form.

From the test results, it is evident that, for the second type of multilayer pyramid truss structure, layer-by-layer failure occurs in compression, with collapse deformation occurring from top to bottom. The force–displacement curves both show three peaks corresponding to the peak forces at the three layers when the damage occurs. In the case of the uniform pyramidal truss structure, the form of failure of the individual cells was the same as that of the single-layer cell structure. The load capacity of the layer rapidly decreased when the structure failed and the layer compressed to its limit. The load capacity then began to increase to the peak load capacity of the next layer, which resulted in the failure of the next layer structure, and the load capacity rapidly decreased. The ratio of the three peak forces to the number of truss units in the three-layer structure was 1:4:9. For the non-uniform structure, the failure form and overall damage process were similar to those of the uniform structure. Because of the adjustment in the inclination angle of each layer, the peak forces of the three-layer structure failure were relatively close to each other, indicating that special overall mechanical properties can be achieved through the adjustment of the unit structure. In general, this proves that it is feasible to design the structural parameters of a unit structure to affect its mechanical properties.

There may be some deviation between the predicted situation and experimental results owing to the printing error and instability of the truss connection at each floor. However, through the design of the structural parameters, it was found that the mechanical behavior of the pyramid truss realized under different conditions changed significantly and was close to the designed situation. The mechanical properties of each layer can be controlled through various aspects such as the number of units in each layer and the angle of the truss. Furthermore, a pyramidal truss structure with a specific mechanical behavior was obtained, demonstrating controllable mechanical behavior.

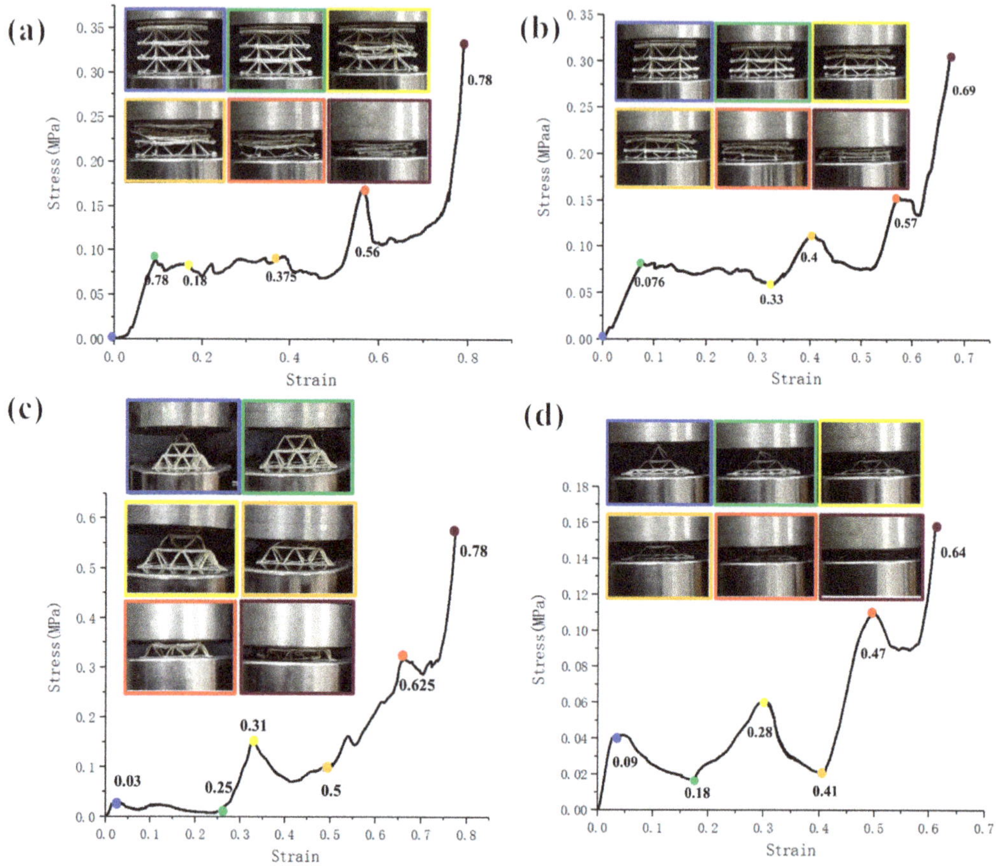

Figure 6. Compression performance curves of different multilayer pyramid structures. (**a**,**b**) compression performance curves of vertically stacked three-layer truss sample (uniform structure design and non-uniform structure design per layer), and (**c**,**d**) ompression performance curves of 45° connected stacked three-layer truss sample (uniform structure design and non-uniform structure design per layer).

3.3. *Discussion*

3.3.1. Theoretical Analysis

To obtain further quantitative effects of the structural parameters on the mechanical properties, the relationship between the structural parameters and the load bearing should be obtained. We conducted theoretical modeling for a pyramid truss structure with four inclined trusses. The mechanical properties of the continuous fiber composite space truss structure were systematically studied and analyzed, as shown in Figure 7. First, a mechanical analysis for a single truss unit in the truss structure was carried out: the length of the unit was set to L, the inclination angle was set to θ, the cross-sectional area was A, the variation in the height of the truss was Δh, and the modulus of elasticity of the material was E (the elastic modulus of the material here refers to the mechanical properties of continuous fiber-reinforced thermoplastic composites [8]).

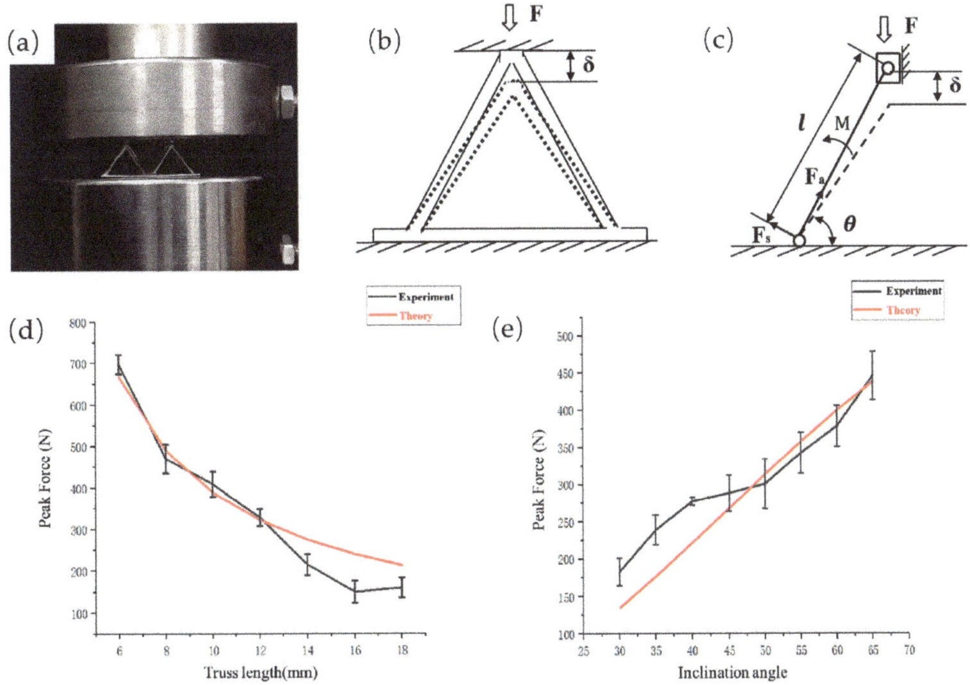

Figure 7. (**a**) Actual photo of 2*2 pyramid truss unit compression, (**b**) force analysis diagram of pyramid truss unit, (**c**) force analysis of simplified pyramid single truss, and comparison of experimental data and theoretical peak force: (**d**) different truss length and (**e**) different inclination.

Ideally, the stress of a single truss is uniform; therefore, we only analyzed the stress of one truss. The distance of movement δ downwards was decomposed into the distance along the truss direction δ_a and that perpendicular to the truss direction δ_s. Similarly, the force along the truss direction and that perpendicular to the truss direction were divided into F_a and F_s, respectively. The relationship between the stress and the strain in the elastic deformation range by Hooke's law can be obtained as follows:

The axial force F_a of a single truss:

$$F_a = \frac{EA}{L} \times \delta_a = \frac{EA}{L} \times \Delta h \times \sin\theta \quad (5)$$

The normal force F_s of a single truss:

$$F_s = \frac{12EI}{L^3} \times \delta_s = \frac{12EI}{L^3} \times \Delta h \times cos\theta \quad (6)$$

The load on a single truss:

$$F = F_a \times cos\theta + F_s \times sin\theta = \frac{EA \times L^2 \times sin^2\theta + 12EI \times cos^2\theta}{L^3} \times \Delta h \quad (7)$$

For instance, we selected a pyramid structure with a truss length of 12 mm and a tilt angle of 50° for theoretical calculation. According to the experimental results, the displacement when the maximum peak force was reached was 0.75 mm, so we calculated that the overall bearing capacity of the truss was 337 N when the displacement was 0.75 mm through Formula (7), and we considered this theoretical calculation value to be the theoretical maximum bearing capacity of the truss under the structural parameters.

From the theoretical calculation, it can be concluded that the most important physical parameters affecting its compression performance are the tilt angle θ of the truss structure and the single rod length L of the truss. These two physical parameters directly affect its mechanical performance; therefore, we studied different unit length trusses and different tilt angle trusses with the unit length L at 6 mm~18 mm and the tilt angle at 30°~65°. For different length units, although the reduction in truss length brings about greater load-carrying performance, the relative density of the truss structure further increases and the relative density of the truss structure needs to be considered simultaneously with the load-carrying performance. The theoretical calculation results and the results of the experimental data trends remained consistent, and the error was within a certain range of control. The main reason for the error could be the instability of the printing process. From the experimental data and theoretical structure, we can explain the angle and length on the mechanical properties of the structure with the specific influence. Therefore, we can design and estimate the pyramid structure with different performance outputs.

3.3.2. Failure Deformation Analysis

The response of the structure to load deformation determines the ability of the pyramidal truss structure to exhibit excellent equivalent load bearing, equivalent strength, equivalent modulus, and low relative density. Evidently, when the material properties are the same, the key structures act as stresses. In the pyramid-shaped truss structural unit proposed in this paper, the bottom truss serves to fix the position of each inclined truss while the peripheral frame is created to guarantee the continuity of the path throughout the continuous fiber printing process. The peak load of the truss without an external frame was reduced by 15% compared with the truss with an external frame, as can be seen from the mechanical characteristics of the original and frameless connection samples. Furthermore, the bottom truss undoubtedly plays an important part in the overall truss's stability during compression; the peak load of the truss without the bottom truss is exceedingly low, as illustrated in Figure 8.

Understanding the truss failure modes is crucial for developing high-performance trusses. As shown in Figure 8, when the pyramid truss structure is subjected to compression deformation, there are various deformation modes of truss bending deformation, such as fiber and resin matrix separation, truss structure fracture, and truss bending. Our failure criteria here include structural load failure (when the peak force dropped to 80%, the structural failure mode could be easily found), the various failure forms, the structural deformation buckling, the material fracture at the macro level, as well as the separation of the fiber resin interface at the micro level. The different failure modes are caused by the uneven stress of the four inclined trusses during structural compression, so the trusses in the same direction receive greater impact, resulting in more serious failure. When the four trusses are subjected to more uniform forces, all four trusses will bend in the same direction of rotation and the structure will not undergo a complete destructive failure, such as fiber and resin matrix separation, truss structure fracture, and truss bending. Among the various failure forms, the overall structural load-bearing performance is better when the four trusses constituting the pyramid structure fail uniformly at the same time.

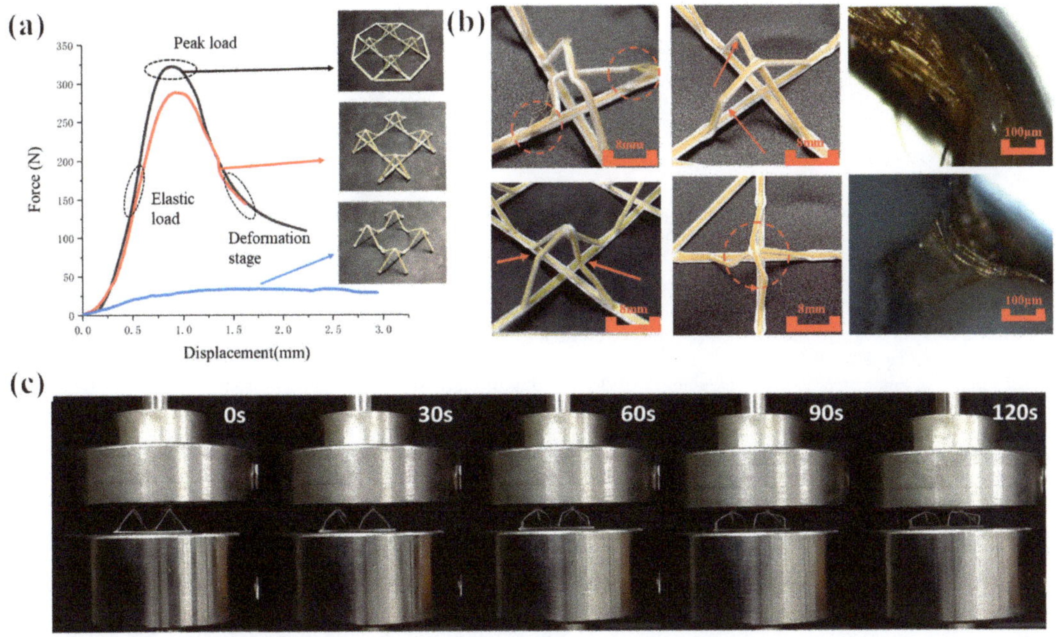

Figure 8. (**a**) Compression failure performance of different structures, (**b**) compression failure damage form of inclined trusses, and (**c**) deformation in different stages.

4. Conclusions

In this paper, a multilayer pyramid truss structure with different structural parameters was manufactured by using the method of space continuous fiber thermoplastic composite additive manufacturing. Firstly, through the optimization of the printing path and the design of the process parameters, the manufacture of the pyramid truss structure with relatively high precision was realized. On this basis, experimental and theoretical studies were carried out on the pyramid truss with different structural parameters, the relationship between the structural parameters and the bearing performance was obtained, and the manufacture of a pyramid truss structure with controllable mechanical properties was finally realized. The preliminary results indicated that a 3D-printed continuous fiber composite truss structure is a feasible and promising load-bearing structure. Truss structures with different mechanical properties can play different roles in dealing with different bearing environments. For example, in the case of heavy loads, a truss with a higher bearing performance can realize the load of the final structure, while in the case of low loads, the truss can achieve a certain deformation and energy absorption effect. It may also be possible to achieve a shock absorption effect in the field of construction. This possibility provides a potential design method for multilayer structures in the future.

Author Contributions: Conceptualization, D.Z.; methodology, D.Z.; validation, Y.Z. and W.Y.; formal analysis, D.Z. and. Y.Z.; investigation, Q.W. and A.A.Z.; resources, X.T., L.W. and D.L.; data curation, Q.W.; writing—original draft preparation, D.Z.; writing—review and editing, D.Z. and X.T.; supervision, X.T., L.W. and D.L.; project administration, X.T. and L.W.; funding acquisition, X.T. All authors have read and agreed to the published version of the manuscript.

Funding: This work was supported by the National Natural Science Foundation of China (52075422), K. C. Wong Education Foundation, The Youth Innovation Team of Shaanxi Universities, and the Fundamental Research Funds for the Central Universities.

Institutional Review Board Statement: Not applicable.

Data Availability Statement: The data is included in this article.

Conflicts of Interest: We declare that we have no financial and personal relationships with other people or organizations that can inappropriately influence our work, and there is no professional or other personal interest of any nature or kind in any product, service, and/or company that could be construed as influencing the position presented in, or the review of, the manuscript entitled.

References

1. Tian, X.; Todoroki, A.; Liu, T.; Wu, L.; Hou, Z.; Ueda, M.; Hirano, Y.; Matsuzaki, R.; Mizukami, K.; Iizuka, K.; et al. 3D Printing of Continuous Fiber Reinforced Polymer Composites: Development, Application, and Prospective. *Chin. J. Mech. Eng. Addit. Manuf. Front.* **2022**, *1*, 100016. [CrossRef]
2. Goh, G.D.; Yap, Y.L.; Agarwala, S.; Yeong, W.Y. Recent Progress in Additive Manufacturing of Fiber Reinforced Polymer Composite. *Adv. Mater. Technol.* **2018**, *4*, 1800271. [CrossRef]
3. Penumakala, P.K.; Santo, J.; Thomas, A. A critical review on the fused deposition modeling of thermoplastic polymer composites. *Compos. Part B Eng.* **2020**, *201*, 108336. [CrossRef]
4. Dickson, A.N.; Abourayana, H.M.; Dowling, D.P. 3D Printing of Fibre-Reinforced Thermoplastic Composites Using Fused Filament Fabrication—A Review. *Polymers* **2020**, *12*, 2188. [CrossRef]
5. Tian, X.; Wu, L.; Gu, D.; Yuan, S.; Zhao, Y.; Li, X.; Ouyang, L.; Song, B.; Gao, T.; He, J.; et al. Roadmap for Additive Manufacturing: Toward Intellectualization and Industrialization. *Chin. J. Mech. Eng. Addit. Manuf. Front.* **2022**, *1*, 100014. [CrossRef]
6. Boston, D.M.; Phillips, F.R.; Henry, T.C.; Arrieta, A.F. Spanwise wing morphing using multistable cellular metastructures. *Extrem. Mech. Lett.* **2022**, *53*, 101706. [CrossRef]
7. Rajak, D.K.; Pagar, D.D.; Menezes, P.L.; Linul, E. Fiber-Reinforced Polymer Composites: Manufacturing, Properties, and Applications. *Polymers* **2019**, *11*, 1667. [CrossRef]
8. Tian, X.; Liu, T.; Yang, C.; Wang, Q.; Li, D. Interface and performance of 3D printed continuous carbon fiber reinforced PLA composites. *Compos. Part A Appl. Sci. Manuf.* **2016**, *88*, 198–205. [CrossRef]
9. Liu, G.; Xiong, Y.; Zhou, L. Additive manufacturing of continuous fiber reinforced polymer composites: Design opportunities and novel applications. *Compos. Commun.* **2021**, *27*, 100907. [CrossRef]
10. Liu, G.; Zhang, X.; Chen, X.; He, Y.; Cheng, L.; Huo, M.; Yin, J.; Hao, F.; Chen, S.; Wang, P.; et al. Additive manufacturing of structural materials. *Mater. Sci. Eng. R Rep.* **2021**, *145*, 100596. [CrossRef]
11. Fernandes, R.R.; van de Werken, N.; Koirala, P.; Yap, T.; Tamijani, A.Y.; Tehrani, M. Experimental investigation of additively manufactured continuous fiber reinforced composite parts with optimized topology and fiber paths. *Addit. Manuf.* **2021**, *44*, 102056. [CrossRef]
12. Hunt, C.J.; Morabito, F.; Grace, C.; Zhao, Y.; Woods, B.K. A review of composite lattice structures. *Compos. Struct.* **2021**, *284*, 115120. [CrossRef]
13. Kang, S.; Liu, W.; Wang, J.; Song, H.; Yuan, W.; Huang, C. Self-adaptive 3D lattice for curved sandwich structures. *Addit. Manuf.* **2022**, *54*, 102761. [CrossRef]
14. Li, N.; Link, G.; Wang, T.; Ramopoulos, V.; Neumaier, D.; Hofele, J.; Walter, M.; Jelonnek, J. Path-designed 3D printing for topological optimized continuous carbon fibre reinforced composite structures. *Compos. Part B Eng.* **2020**, *182*, 107612. [CrossRef]
15. Wang, T.; Li, N.; Link, G.; Jelonnek, J.; Fleischer, J.; Dittus, J.; Kupzik, D. Load-dependent path planning method for 3D printing of continuous fiber reinforced plastics. *Compos. Part A Appl. Sci. Manuf.* **2021**, *140*, 106181. [CrossRef]
16. Huang, Y.; Tian, X.; Zheng, Z.; Li, D.; Malakhov, A.V.; Polilov, A.N. Multiscale concurrent design and 3D printing of continuous fiber reinforced thermoplastic composites with optimized fiber trajectory and topological structure. *Compos. Struct.* **2022**, *285*, 115241. [CrossRef]
17. Sugiyama, K.; Matsuzaki, R.; Ueda, M.; Todoroki, A.; Hirano, Y. 3D printing of composite sandwich structures using continuous carbon fiber and fiber tension. *Compos. Part A Appl. Sci. Manuf.* **2018**, *113*, 114–121. [CrossRef]
18. Chen, Y.; Ye, L. Designing and tailoring effective elastic modulus and negative Poisson's ratio with continuous carbon fibres using 3D printing. *Compos. Part A Appl. Sci. Manuf.* **2021**, *150*, 106625. [CrossRef]
19. Zeng, C.; Liu, L.; Bian, W.; Leng, J.; Liu, Y. Temperature-dependent mechanical response of 4D printed composite lattice structures reinforced by continuous fiber. *Compos. Struct.* **2022**, *280*, 114952. [CrossRef]
20. Hao, W.; Liu, Y.; Zhou, H.; Chen, H.; Fang, D. Preparation and characterization of 3D printed continuous carbon fiber reinforced thermosetting composites. *Polym. Test.* **2017**, *65*, 29–34. [CrossRef]
21. He, X.; Ding, Y.; Lei, Z.; Welch, S.; Zhang, W.; Dunn, M.; Yu, K. 3D printing of continuous fiber-reinforced thermoset composites. *Addit. Manuf.* **2021**, *40*, 101921. [CrossRef]
22. Eichenhofer, M.; Wong, J.C.; Ermanni, P. Continuous lattice fabrication of ultra-lightweight composite structures. *Addit. Manuf.* **2017**, *18*, 48–57. [CrossRef]
23. Liu, S.; Li, Y.; Li, N. A novel free-hanging 3D printing method for continuous carbon fiber reinforced thermoplastic lattice truss core structures. *Mater. Des.* **2018**, *137*, 235–244. [CrossRef]
24. Wang, Z.; Luan, C.; Liao, G.; Yao, X.; Fu, J. Mechanical and self-monitoring behaviors of 3d printing smart continuous carbon fiber-thermoplastic lattice truss sandwich structure. *Compos. Part B Eng.* **2019**, *176*, 107215. [CrossRef]

25. van de Werken, N.; Tekinalp, H.; Khanbolouki, P.; Ozcan, S.; Williams, A.; Tehrani, M. Additively Manufactured Carbon Fiber-Reinforced Composites: State of the Art and Perspective. *Addit. Manuf.* **2019**, *31*, 100962. [CrossRef]
26. Wickramasinghe, S.; Do, T.; Tran, P. FDM-Based 3D Printing of Polymer and Associated Composite: A Review on Mechanical Properties, Defects and Treatments. *Polymers* **2020**, *12*, 1529. [CrossRef]
27. Sun, Y.; Guo, L.-C.; Wang, T.-S.; Yao, L.-J.; Sun, X.-Y. Bending strength and failure of single-layer and double-layer sandwich structure with graded truss core. *Compos. Struct.* **2019**, *226*, 111204. [CrossRef]
28. Ye, G.; Bi, H.; Hu, Y. Compression behaviors of 3D printed pyramidal lattice truss composite structures. *Compos. Struct.* **2020**, *233*, 111706. [CrossRef]
29. Xiong, J.; Ma, L.; Wu, L.; Wang, B.; Vaziri, A. Fabrication and crushing behavior of low density carbon fiber composite pyramidal truss structures. *Compos. Struct.* **2010**, *92*, 2695–2702. [CrossRef]
30. Zhang, G.; Ma, L.; Wang, B.; Wu, L. Mechanical behaviour of CFRP sandwich structures with tetrahedral lattice truss cores. *Compos. Part B Eng.* **2012**, *43*, 471–476. [CrossRef]

polymers

MDPI

Article

Path Planning and Bending Behaviors of 3D Printed Continuous Carbon Fiber Reinforced Polymer Honeycomb Structures

Kui Wang , Depeng Wang, Yisen Liu *, Huijing Gao, Chengxing Yang * and Yong Peng

Key Laboratory of Traffic Safety on Track of Ministry of Education, School of Traffic & Transportation Engineering, Central South University, Changsha 410075, China; kui.wang@csu.edu.cn (K.W.); wangdepeng@csu.edu.cn (D.W.); ghjghj@csu.edu.cn (H.G.); yong_peng@csu.edu.cn (Y.P.)

* Correspondence: liuyisen@csu.edu.cn (Y.L.); chengxing_yang_hn@163.com (C.Y.)

Abstract: Continuous fiber reinforced polymer composites are widely used in load-bearing components and energy absorbers owing to their high specific strength and high specific modulus. The path planning of continuous fiber is closely related to its structural defects and mechanical properties. In this work, continuous fiber reinforced polymer honeycomb structures (CFRPHSs) with different printing paths were designed and fabricated via the fused deposition modeling (FDM) technique. The investigation of fiber dislocation at path corners was utilized to analyze the structural defects of nodes caused by printing paths. The lower stiffness nodes filled with pure polymer due to fiber dislocation result in uneven stiffness distribution. The bending performance and deformation modes of CFRPHSs with different printing paths and corresponding pure polymer honeycomb structures were investigated by three-point bending tests. The results showed that the enhancement effect of continuous fibers on the bending performance of honeycomb structures was significantly affected by the printing paths. The CFRPHSs with a staggered trapezoidal path exhibited the highest specific load capacity (68.33 ± 2.25 N/g) and flexural stiffness (627.70 ± 38.78 N/mm). In addition, the fiber distributions and structural defects caused by the printing paths determine the stiffness distribution of the loading region, thereby affecting the stress distribution and failure modes of CFRPHSs.

Keywords: 3D printing; continuous fiber reinforced polymer; honeycomb structures; path planning; structural defects; bending behaviors

1. Introduction

Due to their lightweight, high strength, and superior energy absorption capacity, honeycomb structures are widely used as load-bearing components and energy absorbers in the forms of sandwich structures and thin-walled structures, owing great application potential in many industrial fields such as aerospace, automobile, and construction [1–5]. The macroscopic performance of honeycomb structures depends on the structural design of their cells and material selection [6–8]. With the advancement of design, numerous cells with various forms have been applied in the evolution of honeycomb structures, such as Kagome [9], re-entrant [10], chiral honeycombs [11], etc. The materials of honeycomb structures have also expanded from traditional metals and wood to various newly developed composites, such as polymer composites with excellent mechanical properties and lightweight performance [12–14].

Adding continuous fibers as reinforcement in polymers could significantly improve the specific strength and modulus while maintaining their lightweight performance [15–18]. Traditional preparation methods of continuous fiber reinforced polymers (CFRPs), such as hot press molding [19], vacuum-assisted molding [20], and filament winding [21], require complicated multi-step processes and expensive manufacturing equipment, while the manufacturing of honeycomb structures commonly needs custom molds and subsequent bonding processes [22,23]. These factors became obstacles to the development of continuous fiber reinforced polymer honeycomb structures (CFRPHSs). However, 3D printing techniques with high design freedoms and multi-scale molding capabilities allow a non-mold-based approach to manufacturing structures with customized materials and complex geometric shapes, providing the possibility for the integrated manufacturing of CFRPHSs [24–26]. According to the ISO/ASTM 52900 standard, 3D printing technologies are divided into seven types, among which material extrusion (MEX) technology is very suitable for 3D printing of polymers [27,28]. Fused deposition modeling (FDM), as one of the most widely used MEX technologies, with rapid prototyping and low-cost characteristics, combines continuous fibers with polymer matrix [29–31]. Many researchers have introduced CFRPs with lightweight and high specific strength features into the design of honeycomb structures via the FDM technique [32–34]. For example, Kentaro et al. [35] printed a series of CFRPHSs with hexagon, rhombus, rectangle, and circle core shapes through the FDM technique. The experimental results showed that bending properties increased with the increase of core density, and the CFRPHSs with a rhombus core shape exhibited the strongest bending properties. Dou et al. [36] fabricated hexagon-filled CFRPHSs and studied their in-plane compression properties. The results showed the specific energy absorption of composite honeycomb was 186.58% and 596.84% higher than that of pure polymer matrix and aluminum alloy, respectively.

In addition, the printing path also had a crucial influence on the arrangement and orientation of continuous fibers and further affects the mechanical properties of the structure [37–39]. Some researchers have tried to improve the manufacturing process and mechanical properties of CFRPHSs by printing path planning [40,41]. For most FDM-based 3D printers without fiber cutting and jumping functions, the one-stroke printing path planning was an inevitable choice to achieve cyclic printing in the height direction [42,43]. Based on one-stroke printing path planning, Quan et al. [44] designed and fabricated a group of auxetic honeycomb-filled CFRPHSs. The compressive stiffness and energy absorption of the CFRPHSs were increased by 86.3% and 100%, respectively, compared to the structures made of pure polymer matrix. Dong et al. [45] investigated the influences of printing paths on the tensile properties of diamond-filled CFRPHSs. The improved trapezoid-like path allowed an even fiber distribution without weak nodes in the CFRPHSs, exhibiting excellent tensile strength.

Depending on the path planning, there were various situations of the fiber crossing at the nodes of the honeycomb structures [46]. The nodes without fiber interlacement might result in structurally weak regions [45], and printing path corners inevitably appeared at the nodes owing to the structural features of the honeycomb. Some printing defects, including fiber dislocation and the absence of fiber, can occur at the corner paths, which might further exacerbate the structural defects of the CFRPHSs [47–50]. The details of the fiber distributions and printing defects were effectively displayed on the honeycomb structures, especially at the nodes. These fiber distributions and structural defects caused by printing paths affect the mechanical behavior of CFRPHSs during loading. However, to the best of the authors' knowledge, only a small amount of literature has investigated how the printing path planning affects 3D printed continuous fiber reinforced structural defects and further mechanical properties.

In this study, diamond-filled CFRPHSs with four different printing paths were designed and fabricated. Three corners (45°, 90°, and 135°) that appeared in the printing paths were selected to investigate the fiber dislocation phenomena at the path corners. Three-point bending tests and morphological analysis were conducted to investigate the influence of the printing path and printing defects at path corners on the bending behaviors of CFRPHSs. This study analyzed the mapping relationship between the printing path planning, structural defects, and final structural and mechanical performance, aiming to provide guidance for the printing path planning of 3D printed CFRPHSs with high performance.

2. Materials and Methods

2.1. Design and Fabrication

In this study, a honeycomb structure with a diamond configuration was designed to investigate bending behaviors and failure mechanisms, as shown in Figure 1a. The honeycomb core consisted of a repeating arrangement of unit diamond cells. The included angle of the diamond cell was 90°; the length (*L*) and width (*W*) of the CFRPHSs were determined by the diamond cell length (*P*); the height (*H*) of the CFRPHSs was 10 mm.

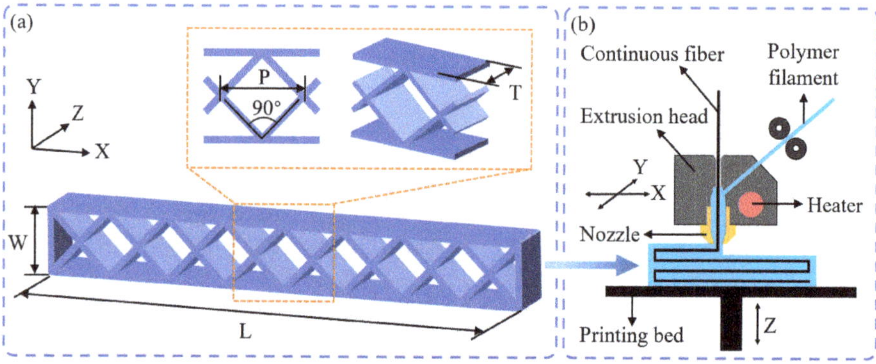

Figure 1. Schematic presentations of (**a**) the diamond-filled CFRPHSs with geometrical dimensions and (**b**) 3D printer using the in situ impregnation FDM technique.

Polyamide (PA) filament (Polymaker Inc., Changshu, China) with a diameter of 1.75 mm was adopted as the polymer matrix of the CFRPHSs. The continuous carbon fiber (CCF) (HTA40-E15-1K, Toho Tenax Co., Ltd., Tokyo, Japan) was used as the reinforced fiber, which was a flat fiber bundle with a width of 0.80 mm. Before printing, the filaments were sealed and stored at 20 °C. According to the instructions from the materials supplier, the mechanical properties of the PA filament and CCF are listed in Table 1.

Table 1. Properties of the 3D printing PA filament and CCF.

Filament Type	Density (g/cm^3)	Tensile Strength (MPa)	Tensile Modulus (GPa)	Elongation (%)
PA	1.12	31.40	1.05	216.50
CCF	1.77	4100.00	240.00	1.70

A commercial 3D printer (COMBOT-200, Fibertech Technology Development Co., Ltd., Shaanxi China), based on the FDM technique, was used to fabricate the CFRPHSs. The schematic representation of the 3D printing process is shown in Figure 1b. The PA filament (blue) and CCF (black) were simultaneously fed into the extrusion head through these two feed ports, respectively. The in situ impregnation of CFRPs was realized in the heated mixing chamber. Then, CFRPs were extruded onto the printing bed and solidified rapidly. With the relative motion of the printing nozzle and the printing bed, the subsequent CFRPs were dragged out from the printing nozzle, ensuring the continuity of the printing process. The models of the CFRPHSs were sliced before manufacturing and the slice data was recorded in the G-codes developed by our team, including the printing paths, geometrical parameters, and printing parameters. For all samples in this study, the printing nozzle temperature was 260 °C, and the printing bed temperature was 50 °C. The printing speed was 180 mm/min, and the layer height was 0.4 mm.

One-stroke printing path planning was required for the CFRPHSs to ensure the integrity of CCF due to the printer's lack of jumping and fiber-cutting features [51]. The manufacturing of 3D printed CFRPHSs was based on the layer-by-layer stacking mechanism of FDM, which was only needed to design a single-layer printing path and print the same single-layer printing path repeatedly in Z-direction. Based on the flexibility of the 3D printing path planning, the printing paths at the nodes lapped between the core and face sheets of the CFRPHSs could be crossing or non-crossing, resulting in fiber interleaving or adjacency [45,46]. Figure 2a shows three printing path strategies at the nodes. According to these different path strategies, four printing paths for the diamond-filled CFRPHSs were designed in this study, as shown in Figure 2b–e. These complete single-layer paths were decomposed into multiple branches distinguished by different colors according to the printing order; these branches were printed successively from the outside to the inside. The combined graphs of the path branches were shown at each bottom. The printing nozzle began from the start point of a single-layer printing path and moved along the arrow direction during printing. After completing the one-stroke path of each layer, the printing nozzle returned to the start point. Then, the printing bed moved downward by a layer height along the Z-direction to repeat the print for the next layer. In the whole printing process, the continuous fibers remained intact. On the right side of these printing paths, the corresponding fiber distribution modes at the nodes were displayed, including antagonistic distribution, symmetrical distribution, and cross-distribution. In addition, pure polymer honeycomb structures (PPHSs) were also prepared with the same geometrical parameters and printing parameters.

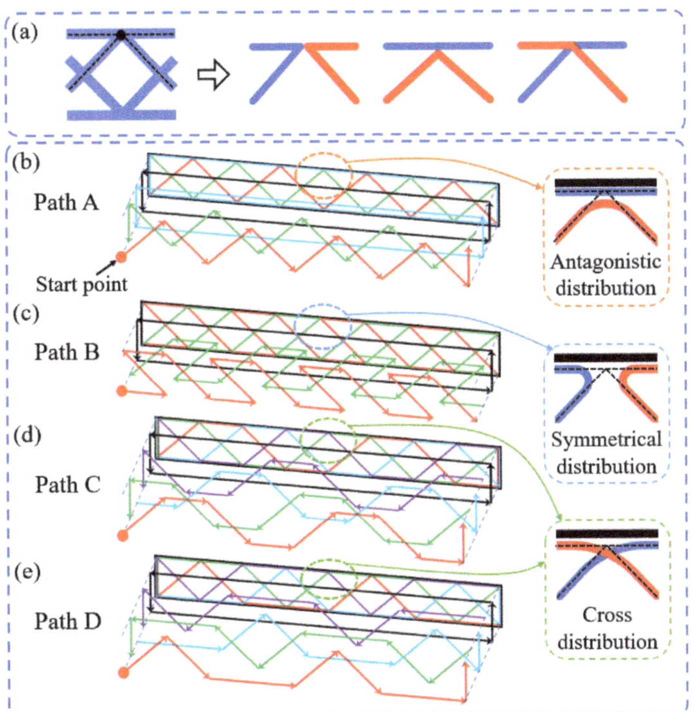

Figure 2. Printing path planning: (**a**) printing path strategies at the nodes, and (**b–e**) four printing paths of the CFRPHSs (the arrows indicate the printing directions, and the colors indicate different printing orders of continuous fibers).

2.2. Experimental Methods

The fiber dislocation inevitably appeared at the path corners in the CFRP's printing process. The morphology of the fiber dislocation at different path corners was recorded by a digital camera (5D mark IV, Canon, Tokyo, Japan). ImageJ version 1.53t software was used to measure the actual angle and position of the fibers at these corners, and a quantitative evaluation method of fiber dislocation was used to analyze the obtained data.

Honeycomb structures are subjected to a complex stress state during the bending process, and these are helpful in studying the influence of the printing path on mechanical behaviors [52,53]. Therefore, the three-point bending tests were performed on a universal material testing machine (E44, MTS Co., Eden Prairie, MN, USA) to investigate the bending properties and failure modes of 3D printed CFRPHSs with different printing paths and PPHSs. The schematic diagram of the three-point bending test is shown in Figure 3. The span between the supports was 80 mm, and the loading roller radius was 5 mm [35,54]. The displacement loading was applied to the samples with a speed of 2 mm/min, and the maximum displacement was 20 mm. The load and displacement data of all tests were automatically collected in the data acquisition system, and the whole test process was recorded by the digital camera. The failure areas of the samples were characterized through an optical microscope (AO-3M150GS, AOSVI, Shenzhen, China). According to the measured load-displacement curves, several indicators were used to quantify the bending properties, including specific load capability and flexural stiffness. The specific load capability (SLC) was calculated by dividing a maximum peak load by the structure mass:

$$\mathrm{SLC} = \frac{F_{peak}}{m} \tag{1}$$

where m is the mass of the sample, and F_{peak} is the maximum peak load. The flexural stiffness was calculated as the slope of the load-displacement curve in a low displacement regime where the curve was almost linear. Each group of tests was repeated five times to ensure reliability, and the average results are reported in this study.

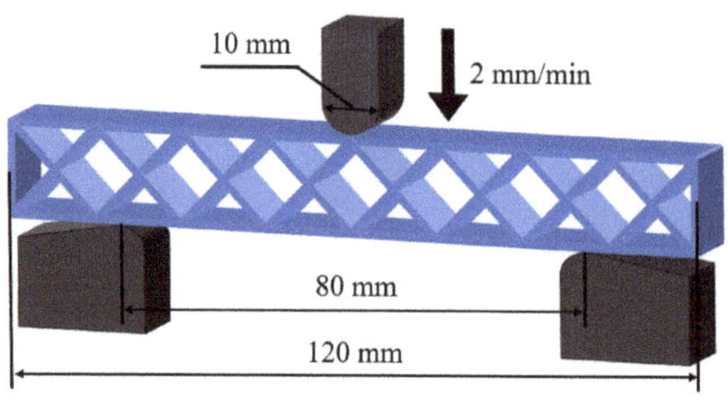

Figure 3. Schematic diagram of the three-point bending test (the arrow indicates the loading direction).

3. Results and Discussions

3.1. Investigation of Fiber Dislocation at Path Corners

The corner paths appear at the nodes (i.e., the corners of the honeycomb walls) due to the structural features of honeycomb structures. At the nodes lapped between the core and face sheets in the CFRPHSs, the printing paths are composed of different corner paths distributed in different forms (e.g., crossing or non-crossing), which are the distinctions between different printing paths of the CFRPHSs [45,46]. The printing defects of fiber dislocations are inevitably generated at the corner paths [48,49]. The corner paths with fiber dislocation may cause structural defects at the nodes of the CFRPHSs, which can further affect the mechanical properties and deformation modes.

The degree of fiber dislocation was mainly affected by the angle of path corner and the shortest length of straight paths forming the corner [47]. For the diamond-filled CFRPHSs, there were only three kinds of path corners (45°, 90°, and 135°) in all printing paths, as shown in Figure 4a. The shortest straight path (L_s) of the CFRPHSs was limited by the structural geometric parameters. Here, L_s was equal to the diamond cell length (P), as well as the width of the CFRPHSs (W). Thus, a corner path consisting of two equal straight paths (L_s) is designed in Figure 4b, which could be 45°, 90°, and 135°. Under the default printing process parameters, these corner paths were printed as L_s = 10 mm, 15 mm, and 20 mm, corresponding to Figure 4d–f, respectively. Noticeable carbon fiber dislocations at these path corners could be observed. To further quantify the degree of fiber dislocation at these corners and determine the appropriate geometric parameters for the CFRPHSs, the angle distortion ratio (φ) and the height distortion ratio (δ) were used as the quantitative evidence, and are expressed as follows:

$$\varphi = \frac{\alpha_1 - \alpha_0}{\alpha_0} \tag{2}$$

$$\delta = \frac{\varepsilon}{h} \tag{3}$$

where α_0 denotes the programmed angle of the path corner, α_1 refers to the actual angle of the CCF at the path corner, h is the theoretical height of the CCF raised at the corner, and ε is the retraction distance of the CCF at the corner, as shown in Figure 4c.

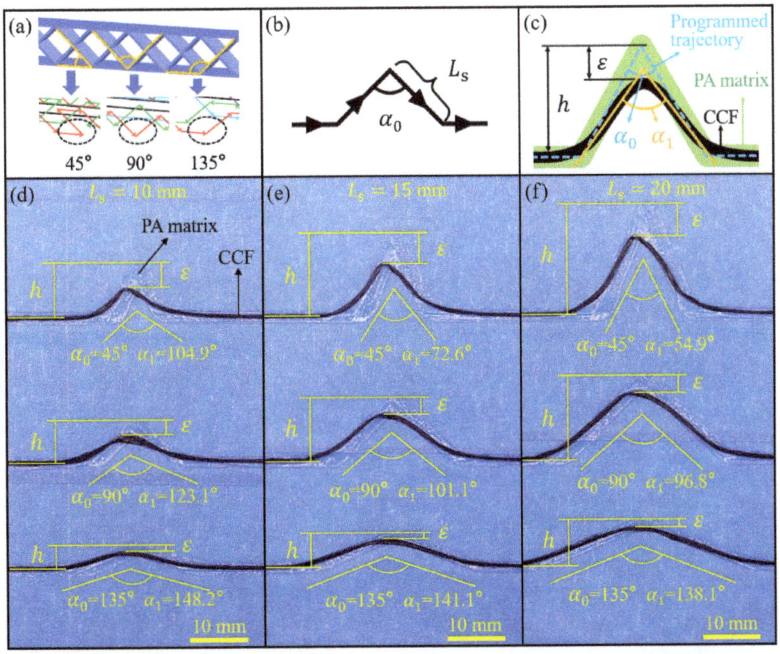

Figure 4. Fiber dislocation of the path corners: (**a**) three path corners in the printing paths: 45°, 90°, and 135°; (**b**) the printing paths for three path corners; (**c**) schematic diagram of the evaluation methods; geometric images of three path corners in different L_s, (**d**) L_s = 10 mm, (**e**) L_s = 15 mm, and (**f**) L_s = 20 mm.

According to Formulas (2) and (3), the calculated values of φ and δ of these path corners are exhibited in Figure 5a,b, respectively. It can be observed that the values of φ and δ increased with the decrease of α_0, as well as the decrease of L_s. According to the schematic diagram in Figure 6a, the incompletely solidified matrix could not constrain the fiber effectively [47], which was the main factor causing the fiber dislocation at the corner paths. A pulling force F, generated by the movement of the nozzle during printing, pulled the subsequent CCF out of the nozzle to ensure the continuity of the printing. After the nozzle turned, the direction of the pulling force F changed, which was parallel to the second straight path. However, there was some incompletely solidified matrix close to the printing nozzle due to the short cooling time and inadequate solidification, which could not provide enough constraints for the CCF at the corner. As a result, the printed CCF at the end of the first straight path was pulled out by the pulling force F. The direction of the first straight path and its vertical direction were defined as the X-direction and the Y-direction, respectively. F_x and F_y are the two vertical components of the pulling force F in the X-direction and Y-direction. When $\alpha_0 \geq 90°$, the fiber dislocation is mainly caused by F_y, while F_x along the X-axis positive direction has no negative effect. The vertical component F_y increased as α_0 decreased when F was constant, exacerbating the fiber dislocation. When $\alpha_0 < 90°$, the CCF, pulled by F_x along the X-axis positive direction, and F_y, exhibited significantly higher angle distortion and height distortion. Therefore, the values of φ and δ were increased with the decrease of α_0, and they were significantly higher when $\alpha_0 = 45°$. The length of the incompletely solidified matrix close to the nozzle was almost fixed under the same process conditions. The proportion of the fiber dislocation area in the corner increased with the decrease of L_s when the angle was constant. Therefore, both the values of φ and δ increased with the decrease of L_s, as shown in Figure 5.

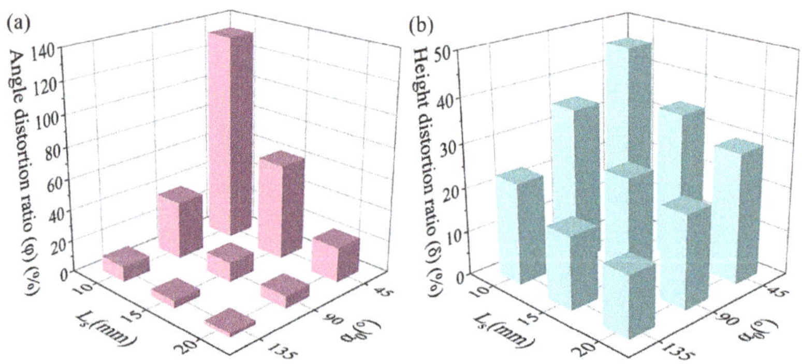

Figure 5. The degree of fiber dislocation at the path corners in different L_s and α_0: (**a**) angle distortion ratio (φ) and (**b**) height distortion ratio (δ).

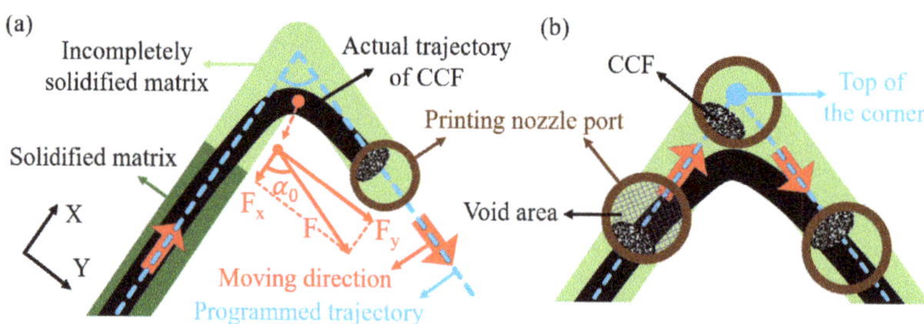

Figure 6. Schematic diagrams of the fiber dislocation at the path corners: (**a**) the incompletely solidified matrix could not constrain the CCF effectively; (**b**) the CCF could not maintain a fixed position relative to the printing nozzle port.

In addition, the fiber could not maintain a fixed position relative to the nozzle port, which was another factor causing fiber dislocation [48], as shown in Figure 6b. To reduce fiber damage during printing, there was a difference between the cross-sectional area of fiber and the nozzle port, resulting in the CCF with a high degree of freedom in the nozzle port. The relative position of CCF could not remain fixed during nozzle turning owing to the high fluidity of the molten polymer matrix. Therefore, the CCF could not reach the top of the corners.

According to the above investigation, the angle distortion ratio of the 45° corner reached 133.1%, and its height distortion ratio reached 45.7% when L_s = 10 mm, showing serious printing defects. To maintain the fiber dislocation at relatively low levels, L_s could not be less than 15 mm. Moreover, the CFRPHSs show better mechanical properties with increasing filling density, which tended to be with a smaller L_s [55]. Therefore, L_s was set as 15 mm of the CFRPHSs with different printing paths.

3.2. Effect of Printing Path on Bending Properties

The distribution and orientation of continuous fibers within the CFRPHSs were determined by their printing paths, which affects the mechanical behaviors [56]. To investigate the influence of the printing path on the bending performance of the 3D printed CFRPHSs, three-point bending tests for the CFRPHSs with four printing paths and the PPHSs were carried out. According to the discussions in Section 3.1, L and W of the CFRPHSs were set as 120 mm and 15 mm, respectively. The maximum bending displacement was 20 mm. Figure 7a,b shows the photographs of the bending test samples at displacements of 0 mm and 20 mm, respectively.

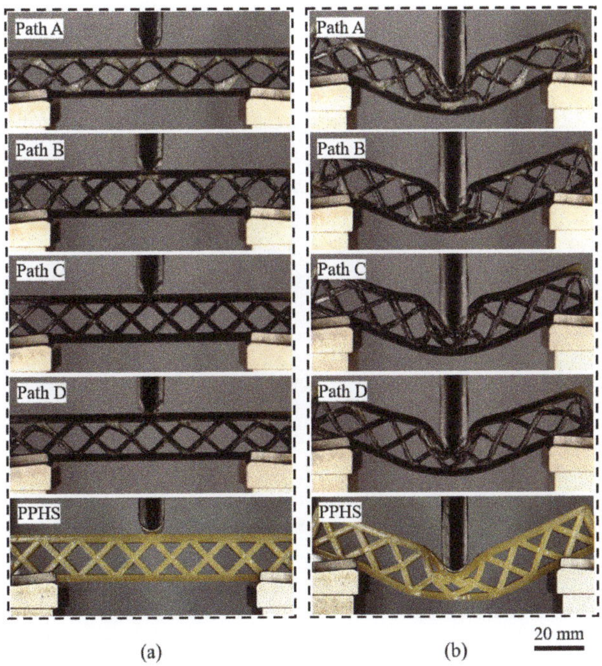

Figure 7. Photographs of the three-point bending test samples at displacements of (**a**) 0 mm and (**b**) 20 mm.

Figure 8a exhibits the load-displacement curves of the CFRPHSs with four printing paths and the PPHSs under three-point bending tests. It can be observed that the load-displacement curves of these CFRPHSs can be divided into three clear stages: the elastic stage, yield stage, and plateau stage. The CFRPHSs initially exhibited short-term elastic deformation during the bending process, and the load increased almost linearly up to the initial peak load, which was the elastic stage. After elastic deformation, the CFRPHSs quickly entered the yield stage. In this stage, the load dropped rapidly to a relatively low value with the progressive failure of structures. Subsequently, the load-displacement curves presented a plateau stage accompanied by small fluctuations in the load. Meanwhile, the deformation stages of the PPHSs were similar to those of the CFRPHSs.

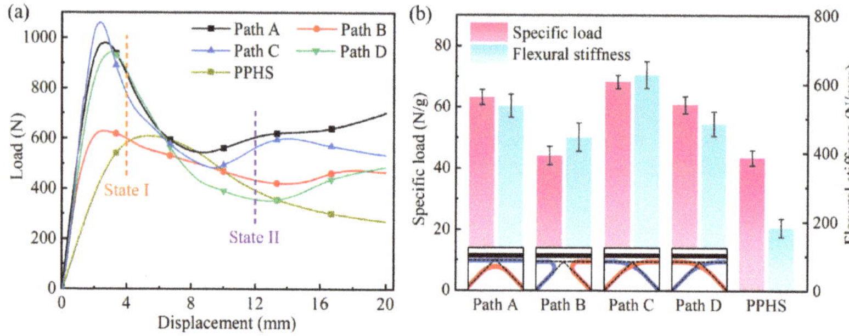

Figure 8. The bending properties of the CFRPHSs with different printing paths and the PPHS; (**a**) three-point bending test load-displacement curves; (**b**) specific load capability and flexural stiffness.

The specific load capability and flexural stiffness of these CFRPHSs and the PPHSs were calculated to further evaluate the bending properties, as shown by the results in Figure 8b and Table 2. It was observed that the addition of continuous fiber greatly improves the flexural stiffness of honeycomb structures. In particular, the CFRPHSs with path C exhibited the highest flexural stiffness (627.70 ± 38.78 N/mm), which was 241.8% higher than the PPHSs. However, the enhancement effect of continuous fiber on specific load capability was more significantly influenced by the printing paths. The CFRPHSs with path C showed the highest specific load capability (68.33 ± 2.25 N/g), while the specific load capability of the CFRPHSs with path B (44.12 ± 3.04 N/g) was close to that of the PPHSs (43.48 ± 2.46 N/g). Therefore, the printing paths would affect the improvement effect of continuous fibers on the bending performance of honeycomb structures.

Table 2. The bending properties of all samples.

Properties	Path A	Path B	Path C	Path D	PPHSs
Specific load capability (N/g)	63.23 ± 2.53	44.12 ± 3.04	68.33 ± 2.25	60.74 ± 2.85	43.48 ± 2.46
Flexural stiffness (N/mm)	627.70 ± 33.59	696.44 ± 41.28	627.70 ± 38.78	484.56 ± 35.64	183.62 ± 26.57

During the bending process, the upper face sheet was directly subjected to the load, whereas the honeycomb core showed supporting and strengthening effects for the face sheet to slow the deformation [57]. Figure 9 further exhibits the fiber distributions of the CFRPHSs in the loading region, distinguished by the color lines. There were three fiber distribution modes at the red node beneath the loading roller, including antagonistic distribution, symmetrical distribution, and cross-distribution, as shown in Figure 9a. The symmetrical distribution consisted of two opposite and symmetrical 45° path corners corresponding to path B. Based on the discussions in Section 3.1, the fiber dislocation of the 45° path corner was the most severe. Thus, a large area of polymer matrix without fiber filling formed at the red node of the CFRPHSs with path B, resulting in a lack of direct support for the compression area of the upper panel because of the low stiffness of pure polymer. It caused the lowest flexural stiffness and specific load capacity of path B. The antagonistic distribution was a 90° path corner with the tip facing the upper face sheet, corresponding to path A. And the cross-distribution was composed of two intersecting 135° path corners, corresponding to paths C and D. The red nodes with these two fiber distribution modes had higher stiffness than the nodes filled with pure polymer and could provide better support to the loading region of the upper face sheets, causing higher flexural stiffness and specific load capacity of the CFRPHSs compared with path B.

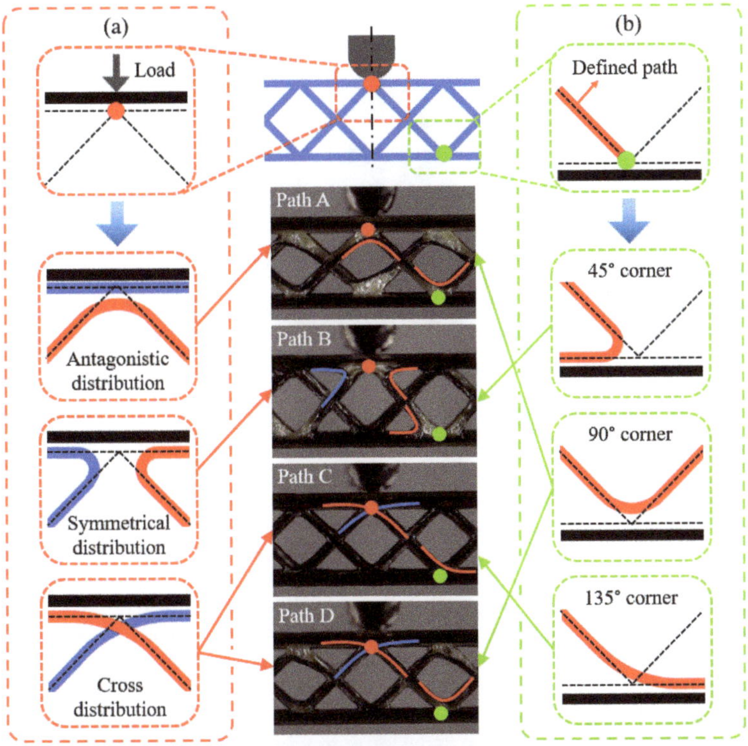

Figure 9. Fiber distribution modes at the nodes in the loading region of the CFRPHSs with different printing paths: (**a**) the red node beneath the loading roller; (**b**) the green node close to the lower face sheet (the red circles indicate the region beneath the loading roller, and the green circles indicate the region close to the lower face sheet).

In addition, the load acting on the upper face sheet was transmitted to the lower face sheet through continuous fibers distributed in the core. The fiber distribution modes (45°, 90°, and 135° corners) close to the lower face sheet would affect the load transmission from the core to the lower face sheet, as shown by the green nodes in Figure 9b. Among them, the CCF distributed at 135° corner had the largest contact area with the lower face sheet in path C, which could better transfer the load to the lower face sheet and obtain the support from the lower face sheet. Moreover, the staggered trapezoidal path C allowed CCF intersections at all nodes of the CFRPHSs, ensuring a strong connection between core and face sheets [58]. These enabled the structure to have good load transfer performance and fully utilize the supporting function of the honeycomb core. Thus, the CFRPHSs with path C showed the strongest specific load capacity and flexural stiffness compared to other CFRPHSs and PPHSs.

The core and face sheets of honeycomb structures were subjected to a complex stress state during the bending process, accompanied by potential failure modes, including face sheet indentation and core shear. Figure 10 shows the bending deformation states of the PPHSs at displacements of 6 mm and 18 mm, corresponding to the yield stage and the plateau stage, respectively. In Figure 10a, plastic yielding occurred in the lower face sheet near the node, causing the load to begin to decrease. When the displacement reached 12 mm, the PPHSs exhibited indentation failure and core shear failure induced by the core upper member buckling, as shown in Figure 10b. The plastic yielding of the lower face sheet was deepening, accompanied by fracture and delamination. Due to the contact between adjacent honeycomb core members and face sheets, the rate of load decline slowed down, and the PPHSs entered the plateau stage.

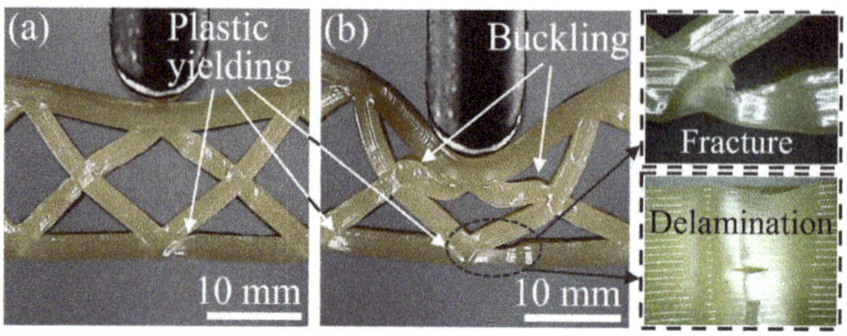

Figure 10. Deformation states of the PPHSs at displacements of (**a**) 6 mm and (**b**) 18 mm.

Figures 11 and 12 show the photographs of the bending deformation states of these CFRPHSs at two different moments during the yield and plateau stages, which corresponded to the colored lines in Figure 8a, to further analyze their failure modes. The CFRPHSs at the state I began to enter the yield stage. It can be observed in Figure 11 that the upper face sheet bending beneath the loading roller occurred in all CFRPHSs. Meanwhile, the core shear failure began to occur in the CFRPHSs, resulting in a decrease in their structural bearing capacity. Two possible core shear failure modes are caused by core member buckling or core member yielding, respectively [54]. For paths A, C, and D, the initial failure modes were core shear failure caused by core member buckling, as shown in Figure 11a,c,d. The degrees of core member buckling in these CFRPHSs were different. Path C, with the most severe core member buckling, resulting in the most rapid drop in state I in the load-displacement curve. However, for path B, two 45° corners with severe fiber dislocation were distributed close to the lower face sheet (see Figure 9b), where the nodes filled with pure polymer had low stiffness, resulting in yield failure easily occurring at the end of core members, as shown in Figure 11b. The core shear failure caused by yielding had a smaller effect on the support of the honeycomb core to the upper face sheet than that caused by buckling, maintaining the load of path B declining at a slower rate. Therefore, the core shear failure modes of the CFRPHSs were affected by fiber distribution modes at the nodes. The rate of load decline in the yield stage was determined by the mode and degree of core shear failure.

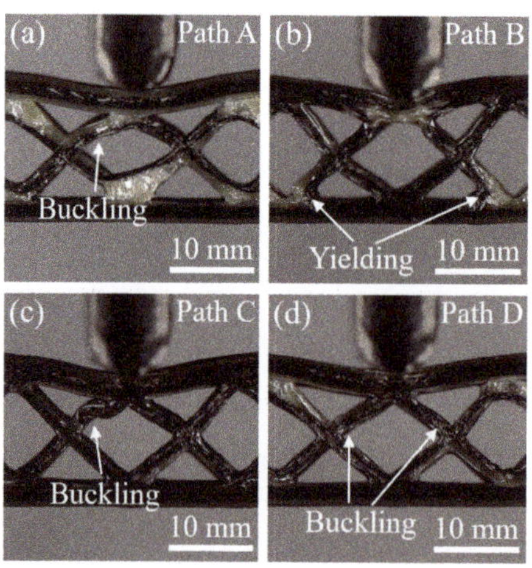

Figure 11. Deformation state I of the CFRPHSs with (**a**) path A, (**b**) path B, (**c**) path C, and (**d**) path D.

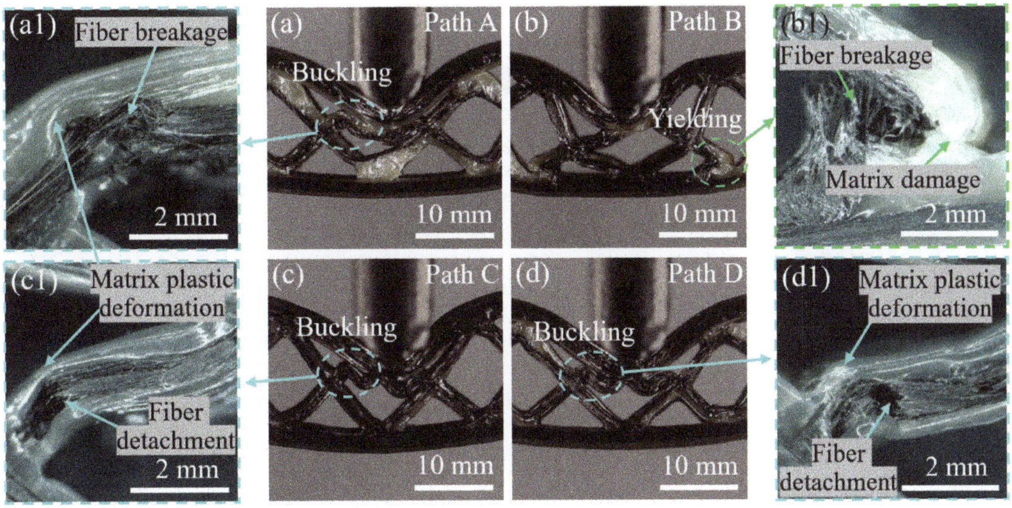

Figure 12. Deformation state II of the CFRPHSs with (**a**) path A, (**b**) path B, (**c**) path C, and (**d**) path D.

At state II, the CFRPHSs began to enter the plateau stage. Indentation failure occurred on the upper face sheets beneath the loading rollers, and core shear failure further deepened in all CFRPHSs, as shown in Figure 12. It could be observed that the CFRPHSs with paths A, C, and D exhibited indentation failure induced by the core upper member buckling (see Figure 12a,c,d), while the indentation failure of path B was induced by the core lower member yielding (see Figure 12b). The results indicated that the antagonistic distribution and cross-distribution of paths A, C, and D endowed their nodes with good stiffness, providing better support to the loading region. Thus, the stress concentration was close to the upper face sheet in the early stage in the core members, which finally resulted in core shear failure on the upper side caused by core member buckling in the plateau stage. Their failure mode was similar to that of PPHSs. However, for path B, the symmetrical distribution and the 45° corners close to the lower face sheet had severe fiber dislocation, causing the low-stiffness nodes to be filled with pure polymer and the uneven stiffness distribution in the loading region. The uneven stiffness distribution led to a stress concentration in the core members close to the lower face sheet, and the end of core members yielded easily, ultimately resulting in core shear failure on the lower side. Thus, depending on the printing paths, the fiber distribution modes, and structural defects determined the stiffness distribution in the loading region, further affecting the stress concentration regions as well as the locations and types of core shear failure in the CFRPHSs.

In addition, some core members and face sheets touched each other with the deepening indentation failure and core shear failure, causing structural densification. Therefore, the bending loads of these CFRPHSs rose in a small amplitude after state II, as shown in Figure 8a. Moreover, combined with the microscopic morphologies after unloading in Figure 12, severe plastic deformation and damage of the polymer matrix, as well as obvious fiber breakage and fiber detachment from the matrix, could be observed in the areas of core shear failure, which caused irreversible structural failure and a decrease in the bearing capacity.

4. Conclusions

In this study, diamond-filled continuous fiber reinforced polymer honeycomb structures (CFRPHSs) with different printing paths were designed and fabricated via the FDM technique. The fiber dislocation at the path corners was investigated to characterize the structural defects of the CFRPHSs caused by the printing path. The bending behaviors and failure modes of the CFRPHSs with the different printing paths and pure polymer honeycomb structures (PPHSs) were studied by three-point bending tests. The results showed that the fiber distribution modes and structural defects determined the stress concentration regions and core shear failure modes of the CFRPHSs. This study revealed the mapping relationship between the printing path planning, structural defects, and structural and mechanical performance, providing references for the printing path optimization of the 3D printed CFRPHSs with high performance. The main conclusions can be drawn as follows:

(1) There were three fiber distribution modes at the nodes lapped between the core and face sheets of the CFRPHSs, including symmetrical distribution, antagonistic distribution, and cross distribution, which were determined by the printing paths.
(2) The structural defects at the nodes of the CFRPHSs were caused by the fiber dislocations at path corners. Low stiffness nodes filled with pure polymer caused by severe fiber dislocation led to uneven stiffness distribution in the loading region of the CFRPHSs. As the angle of the corner and the length of the straight path increased, the degree of fiber dislocation at path corners would decrease.
(3) The enhancement effect of continuous fibers on the bending performance of honeycomb structures was affected by the printing paths. The path C with staggered trapezoidal distribution ensured a strong connection and good load transfer performance between the core and face sheets of the CFRPHSs, fully utilizing the supporting function of the honeycomb core, exhibiting the highest specific load capability (68.33 ± 2.25 N/g) and flexural stiffness (627.70 ± 38.78 N/mm).
(4) The nodes of the CFRPHSs with fiber antagonistic distribution and cross distribution had good stiffness, providing better support to the loading region, leading to a stress concentration in the upper core and final core shear failure on the upper side caused by core member buckling. The failure mode of these CFRPHSs was similar to that of the PPHSs.
(5) When the fibers were symmetrically distributed, the low-stiffness nodes filled with pure polymer caused uneven stiffness distribution in the loading region of the CFRPHSs, resulting in concentrated stress in the core members close to the lower face sheet, which ultimately led to yielding at the 45° corners.

This study has guiding significance for the printing path planning of other continuous fiber reinforced polymer two-dimensional cellular structures. It is necessary to carry out finite element simulation studies incorporating fiber path planning for further optimization of structural parameters in future work.

Author Contributions: Conceptualization, K.W.; methodology, Y.P.; validation, H.G.; formal analysis, Y.P.; investigation, D.W. and Y.L.; resources, C.Y.; data curation, D.W. and H.G.; writing—original draft preparation, K.W. and D.W.; writing—review and editing, Y.L.; visualization, C.Y.; supervision, K.W.; project administration, C.Y. All authors have read and agreed to the published version of the manuscript.

Funding: This research was funded by the Hu-Xiang Youth Talent Program (No. 2020RC3009).

Institutional Review Board Statement: Not applicable.

Data Availability Statement: The data presented in this study are available on request from the corresponding author.

Conflicts of Interest: The authors declare no conflict of interest.

References

1. Zhang, Q.; Yang, X.; Li, P.; Huang, G.; Feng, S.; Shen, C.; Han, B.; Zhang, X.; Jin, F.; Xu, F.; et al. Bioinspired Engineering of Honeycomb Structure—Using Nature to Inspire Human Innovation. *Prog. Mater. Sci.* **2015**, *74*, 332–400. [CrossRef]
2. Thomas, T.; Tiwari, G. Crushing Behavior of Honeycomb Structure: A Review. *Int. J. Crashworthiness* **2019**, *24*, 555–579. [CrossRef]
3. Carneiro, V.H.; Rawson, S.D.; Puga, H.; Meireles, J.; Withers, P.J. Additive Manufacturing Assisted Investment Casting: A Low-Cost Method to Fabricate Periodic Metallic Cellular Lattices. *Addit. Manuf.* **2020**, *33*, 101085. [CrossRef]

4. An, M.-R.; Wang, L.; Liu, H.-T.; Ren, F.-G. In-Plane Crushing Response of a Novel Bidirectional Re-Entrant Honeycomb with Two Plateau Stress Regions. *Thin-Walled Struct.* **2022**, *170*, 108530. [CrossRef]
5. Liu, Y.; Wang, J.; Cai, R.; Xiang, J.; Wang, K.; Yao, S.; Peng, Y. Effects of Loading Rate and Temperature on Crushing Behaviors of 3D Printed Multi-Cell Composite Tubes. *Thin-Walled Struct.* **2023**, *182*, 110311. [CrossRef]
6. Habib, F.N.; Iovenitti, P.; Masood, S.H.; Nikzad, M. Cell Geometry Effect on In-Plane Energy Absorption of Periodic Honeycomb Structures. *Int. J. Adv. Manuf. Technol.* **2018**, *94*, 2369–2380. [CrossRef]
7. Wang, Z. Recent Advances in Novel Metallic Honeycomb Structure. *Compos. Part B Eng.* **2019**, *166*, 731–741. [CrossRef]
8. Qi, C.; Jiang, F.; Yang, S. Advanced Honeycomb Designs for Improving Mechanical Properties: A Review. *Compos. Part B Eng.* **2021**, *227*, 109393. [CrossRef]
9. Wang, Z.; Lei, Z.; Li, Z.; Yuan, K.; Wang, X. Mechanical Reinforcement Mechanism of a Hierarchical Kagome Honeycomb. *Thin-Walled Struct.* **2021**, *167*, 108235. [CrossRef]
10. Zhang, X.; Yang, D. Mechanical Properties of Auxetic Cellular Material Consisting of Re-Entrant Hexagonal Honeycombs. *Materials* **2016**, *9*, 900. [CrossRef]
11. Qi, C.; Jiang, F.; Yu, C.; Yang, S. In-Plane Crushing Response of Tetra-Chiral Honeycombs. *Int. J. Impact Eng.* **2019**, *130*, 247–265. [CrossRef]
12. Rajak, D.K.; Pagar, D.D.; Menezes, P.L.; Linul, E. Fiber-Reinforced Polymer Composites: Manufacturing, Properties, and Applications. *Polymers* **2019**, *11*, 1667. [CrossRef] [PubMed]
13. Ozkan, D.; Gok, M.S.; Karaoglanli, A.C. Carbon Fiber Reinforced Polymer (CFRP) Composite Materials, Their Characteristic Properties, Industrial Application Areas and Their Machinability. In *Engineering Design Applications III: Structures, Materials and Processes*; Öchsner, A., Altenbach, H., Eds.; Advanced Structured Materials; Springer International Publishing: Cham, Switzerland, 2020; pp. 235–253. ISBN 978-3-030-39062-4.
14. Du, J.; Zhang, H.; Geng, Y.; Ming, W.; He, W.; Ma, J.; Cao, Y.; Li, X.; Liu, K. A Review on Machining of Carbon Fiber Reinforced Ceramic Matrix Composites. *Ceram. Int.* **2019**, *45*, 18155–18166. [CrossRef]
15. Li, J.; Durandet, Y.; Huang, X.; Sun, G.; Ruan, D. Additively Manufactured Fiber-Reinforced Composites: A Review of Mechanical Behavior and Opportunities. *J. Mater. Sci. Technol.* **2022**, *119*, 219–244. [CrossRef]
16. Kabir, S.M.F.; Mathur, K.; Seyam, A.-F.M. A Critical Review on 3D Printed Continuous Fiber-Reinforced Composites: History, Mechanism, Materials and Properties. *Compos. Struct.* **2020**, *232*, 111476. [CrossRef]
17. Dickson, A.N.; Barry, J.N.; McDonnell, K.A.; Dowling, D.P. Fabrication of Continuous Carbon, Glass and Kevlar Fibre Reinforced Polymer Composites Using Additive Manufacturing. *Addit. Manuf.* **2017**, *16*, 146–152. [CrossRef]
18. Xiang, J.; Liu, Y.; Wang, J.; Wang, K.; Peng, Y.; Rao, Y.; Matadi Boumbimba, R. Effect of Heat-Treatment on Compressive Response of 3D Printed Continuous Carbon Fiber Reinforced Composites under Different Loading Directions. *J. Appl. Polym. Sci.* **2023**, *140*, e53330. [CrossRef]
19. Zhang, Y.; Sun, L.; Li, L.; Wei, J. Effects of Strain Rate and High Temperature Environment on the Mechanical Performance of Carbon Fiber Reinforced Thermoplastic Composites Fabricated by Hot Press Molding. *Compos. Part A Appl. Sci. Manuf.* **2020**, *134*, 105905. [CrossRef]
20. Arulappan, C.; Duraisamy, A.; Adhikari, D.; Gururaja, S. Investigations on Pressure and Thickness Profiles in Carbon Fiber-Reinforced Polymers during Vacuum Assisted Resin Transfer Molding. *J. Reinf. Plast. Compos.* **2015**, *34*, 3–18. [CrossRef]
21. Boon, Y.D.; Joshi, S.C.; Bhudolia, S.K. Review: Filament Winding and Automated Fiber Placement with In Situ Consolidation for Fiber Reinforced Thermoplastic Polymer Composites. *Polymers* **2021**, *13*, 1951. [CrossRef]
22. Ning, H.; Janowski, G.M.; Vaidya, U.K.; Husman, G. Thermoplastic Sandwich Structure Design and Manufacturing for the Body Panel of Mass Transit Vehicle. *Compos. Struct.* **2007**, *80*, 82–91. [CrossRef]
23. Chen, X.; Yu, G.; Wang, Z.; Feng, L.; Wu, L. Enhancing Out-of-Plane Compressive Performance of Carbon Fiber Composite Honeycombs. *Compos. Struct.* **2021**, *255*, 112984. [CrossRef]
24. Wang, J.; Liu, Y.; Wang, K.; Yao, S.; Peng, Y.; Rao, Y.; Ahzi, S. Progressive Collapse Behaviors and Mechanisms of 3D Printed Thin-Walled Composite Structures under Multi-Conditional Loading. *Thin-Walled Struct.* **2022**, *171*, 108810. [CrossRef]
25. Liu, Y.; Tan, Q.; Lin, H.; Wang, J.; Wang, K.; Peng, Y.; Yao, S. Integrated Design and Additive Manufacturing of Lattice-Filled Multi-Cell Tubes. *Compos. Sci. Technol.* **2023**, *243*, 110252. [CrossRef]
26. Xie, G.; Wang, K.; Wu, X.; Wang, J.; Li, T.; Peng, Y.; Zhang, H. A Hybrid Multi-Stage Decision-Making Method with Probabilistic Interval-Valued Hesitant Fuzzy Set for 3D Printed Composite Material Selection. *Eng. Appl. Artif. Intell.* **2023**, *123*, 106483. [CrossRef]
27. Alexander, A.E.; Wake, N.; Chepelev, L.; Brantner, P.; Ryan, J.; Wang, K.C. A Guideline for 3D Printing Terminology in Biomedical Research Utilizing ISO/ASTM Standards. *3D Print. Med.* **2021**, *7*, 8. [CrossRef] [PubMed]
28. Ekinci, A.; Johnson, A.A.; Gleadall, A.; Engstrøm, D.S.; Han, X. Layer-Dependent Properties of Material Extruded Biodegradable Polylactic Acid. *J. Mech. Behav. Biomed. Mater.* **2020**, *104*, 103654. [CrossRef] [PubMed]
29. Zhang, H.; Huang, T.; Jiang, Q.; He, L.; Bismarck, A.; Hu, Q. Recent Progress of 3D Printed Continuous Fiber Reinforced Polymer Composites Based on Fused Deposition Modeling: A Review. *J. Mater. Sci.* **2021**, *56*, 12999–13022. [CrossRef]
30. Li, H.; Lou, R.; Liu, B.; Chen, Y.; Wang, Y. Research on the Fusion of Continuous Fiber Reinforced Thermoplastic Filaments for Fused Filament Fabrication. *Int. J. Solids Struct.* **2023**, *276*, 112328. [CrossRef]

31. Cheng, P.; Peng, Y.; Wang, K.; Le Duigou, A.; Yao, S.; Chen, C. Quasi-Static Penetration Property of 3D Printed Woven-like Ramie Fiber Reinforced Biocomposites. *Compos. Struct.* **2023**, *303*, 116313. [CrossRef]
32. Cheng, P.; Peng, Y.; Li, S.; Rao, Y.; Le Duigou, A.; Wang, K.; Ahzi, S. 3D Printed Continuous Fiber Reinforced Composite Lightweight Structures: A Review and Outlook. *Compos. Part B Eng.* **2023**, *250*, 110450. [CrossRef]
33. Cheng, Y.; Li, J.; Qian, X.; Rudykh, S. 3D Printed Recoverable Honeycomb Composites Reinforced by Continuous Carbon Fibers. *Compos. Struct.* **2021**, *268*, 113974. [CrossRef]
34. Dou, H.; Ye, W.; Zhang, D.; Cheng, Y.; Huang, K.; Yang, F.; Rudykh, S. Research on Drop-Weight Impact of Continuous Carbon Fiber Reinforced 3D Printed Honeycomb Structure. *Mater. Today Commun.* **2021**, *29*, 102869. [CrossRef]
35. Sugiyama, K.; Matsuzaki, R.; Ueda, M.; Todoroki, A.; Hirano, Y. 3D Printing of Composite Sandwich Structures Using Continuous Carbon Fiber and Fiber Tension. *Compos. Part A Appl. Sci. Manuf.* **2018**, *113*, 114–121. [CrossRef]
36. Dou, H.; Ye, W.; Zhang, D.; Cheng, Y.; Wu, C. Comparative Study on In-Plane Compression Properties of 3D Printed Continuous Carbon Fiber Reinforced Composite Honeycomb and Aluminum Alloy Honeycomb. *Thin-Walled Struct.* **2022**, *176*, 109335. [CrossRef]
37. Huang, Y.; Tian, X.; Wu, L.; Zia, A.A.; Liu, T.; Li, D. Progressive Concurrent Topological Optimization with Variable Fiber Orientation and Content for 3D Printed Continuous Fiber Reinforced Polymer Composites. *Compos. Part B Eng.* **2023**, *255*, 110602. [CrossRef]
38. Hou, Z.; Tian, X.; Zhang, J.; Zhe, L.; Zheng, Z.; Li, D.; Malakhov, A.V.; Polilov, A.N. Design and 3D Printing of Continuous Fiber Reinforced Heterogeneous Composites. *Compos. Struct.* **2020**, *237*, 111945. [CrossRef]
39. Zhu, W.; Li, S.; Peng, Y.; Wang, K.; Ahzi, S. Effect of Continuous Fiber Orientations on Quasi-Static Indentation Properties in 3D Printed Hybrid Continuous Carbon/Kevlar Fiber Reinforced Composites. *Polym. Adv. Technol.* **2023**, *34*, 1565–1574. [CrossRef]
40. Zhang, G.; Wang, Y.; He, J.; Xiong, Y. A Graph-Based Path Planning Method for Additive Manufacturing of Continuous Fiber-Reinforced Planar Thin-Walled Cellular Structures. *Rapid Prototyp. J.* **2023**, *29*, 344–353. [CrossRef]
41. Huang, Y.; Fang, G.; Zhang, T.; Wang, C.C.L. Turning-Angle Optimized Printing Path of Continuous Carbon Fiber for Cellular Structures. *Addit. Manuf.* **2023**, *68*, 103501. [CrossRef]
42. Zeng, C.; Liu, L.; Bian, W.; Leng, J.; Liu, Y. Compression Behavior and Energy Absorption of 3D Printed Continuous Fiber Reinforced Composite Honeycomb Structures with Shape Memory Effects. *Addit. Manuf.* **2021**, *38*, 101842. [CrossRef]
43. Feng, J.; Yao, L.; Lyu, Z.; Wu, Z.; Zhang, G.; Zhao, H. Mechanical Properties and Damage Failure of 3D-Printed Continuous Carbon Fiber-Reinforced Composite Honeycomb Sandwich Structures with Fiber-Interleaved Core. *Polym. Compos.* **2023**, *44*, 1980–1992. [CrossRef]
44. Quan, C.; Han, B.; Hou, Z.; Zhang, Q.; Tian, X.; Lu, T.J. 3D Printed Continuous Fiber Reinforced Composite Auxetic Honeycomb Structures. *Compos. Part B Eng.* **2020**, *187*, 107858. [CrossRef]
45. Dong, K.; Liu, L.; Huang, X.; Xiao, X. 3D Printing of Continuous Fiber Reinforced Diamond Cellular Structural Composites and Tensile Properties. *Compos. Struct.* **2020**, *250*, 112610. [CrossRef]
46. Yamamoto, K.; Luces, J.V.S.; Shirasu, K.; Hoshikawa, Y.; Okabe, T.; Hirata, Y. A Novel Single-Stroke Path Planning Algorithm for 3D Printers Using Continuous Carbon Fiber Reinforced Thermoplastics. *Addit. Manuf.* **2022**, *55*, 102816. [CrossRef]
47. Zhang, G.; Wang, Y.; Qiu, W.; Dong, K.; Xiong, Y. Geometric Characteristics of Single Bead Fabricated by Continuous Fiber Reinforced Polymer Composite Additive Manufacturing. *Mater. Today Proc.* **2022**, *70*, 431–437. [CrossRef]
48. Liu, J.; Kang, Y.; Ma, C.; Wang, Y. Research on a Fiber Corner Compensation Algorithm in a 3D Printing Layer of Continuous Fiber-Reinforced Composite Materials. *Appl. Sci.* **2022**, *12*, 6687. [CrossRef]
49. Zhang, H.; Chen, J.; Yang, D. Fibre Misalignment and Breakage in 3D Printing of Continuous Carbon Fibre Reinforced Thermoplastic Composites. *Addit. Manuf.* **2021**, *38*, 101775. [CrossRef]
50. Matsuzaki, R.; Nakamura, T.; Sugiyama, K.; Ueda, M.; Todoroki, A.; Hirano, Y.; Yamagata, Y. Effects of Set Curvature and Fiber Bundle Size on the Printed Radius of Curvature by a Continuous Carbon Fiber Composite 3D Printer. *Addit. Manuf.* **2018**, *24*, 93–102. [CrossRef]
51. Cui, Z.; Huang, X.; Jia, M.; Panahi-Sarmad, M.; Hossen, M.I.; Dong, K.; Xiao, X. 3D Printing of Continuous Fiber Reinforced Cellular Structural Composites for the Study of Bending Performance. *J. Reinf. Plast. Compos.* **2023**, *42*, 673–684. [CrossRef]
52. Li, T.; Wang, L. Bending Behavior of Sandwich Composite Structures with Tunable 3D-Printed Core Materials. *Compos. Struct.* **2017**, *175*, 46–57. [CrossRef]
53. Liu, Z.; Chen, H.; Xing, S. Mechanical Performances of Metal-Polymer Sandwich Structures with 3D-Printed Lattice Cores Subjected to Bending Load. *Arch. Civ. Mech. Eng.* **2020**, *20*, 89. [CrossRef]
54. Zeng, C.; Liu, L.; Bian, W.; Leng, J.; Liu, Y. Bending Performance and Failure Behavior of 3D Printed Continuous Fiber Reinforced Composite Corrugated Sandwich Structures with Shape Memory Capability. *Compos. Struct.* **2021**, *262*, 113626. [CrossRef]
55. Zárybnická, L.; Machotová, J.; Pagáč, M.; Rychlý, J.; Vykydalová, A. The Effect of Filling Density on Flammability and Mechanical Properties of 3D-Printed Carbon Fiber-Reinforced Nylon. *Polym. Test.* **2023**, *120*, 107944. [CrossRef]
56. Wang, T.; Li, N.; Link, G.; Jelonnek, J.; Fleischer, J.; Dittus, J.; Kupzik, D. Load-Dependent Path Planning Method for 3D Printing of Continuous Fiber Reinforced Plastics. *Compos. Part A Appl. Sci. Manuf.* **2021**, *140*, 106181. [CrossRef]

57. Cai, L.; Zhang, D.; Zhou, S.; Xu, W. Investigation on Mechanical Properties and Equivalent Model of Aluminum Honeycomb Sandwich Panels. *J. Mater. Eng. Perform.* **2018**, *27*, 6585–6596. [CrossRef]
58. Wang, B.; Ming, Y.; Zhou, J.; Xiao, H.; Wang, F.; Duan, Y.; Kazancı, Z. Fabrication of Triangular Corrugated Structure Using 3D Printed Continuous Carbon Fiber-Reinforced Thermosetting Epoxy Composites. *Polym. Test.* **2022**, *106*, 107469. [CrossRef]

polymers

MDPI

Article

Study on the Properties of Multi-Walled Carbon Nanotubes (MWCNTs)/Polypropylene Fiber (PP Fiber) Cement-Based Materials

Xiangjie Niu [1], Yuanzhao Chen [1,2,3,*], Zhenxia Li [1,2,3], Tengteng Guo [1,2,3], Meng Ren [1] and Yanyan Chen [1]

1 School of Civil Engineering and Communication, North China University of Water Resources and Electric Power, Zhengzhou 450045, China; xiangjieniu@163.com (X.N.)
2 Technology Innovation Center of Henan Transport Industry of Utilization of Solid Waste Resources in Traffic Engineering, North China University of Water Resources and Electric Power, Zhengzhou 450045, China
3 Henan Province Engineering Technology Research Center of Environment Friendly and High-Performance Pavement Materials, Zhengzhou 450045, China
* Correspondence: cyz740513@ncwu.edu.cn

Abstract: In order to improve the mechanical properties and durability of cement-based materials, a certain amount of multi-walled carbon nanotubes (MWCNTs) and polypropylene fiber (PP fiber) were incorporated into cement-based materials. The mechanical properties of the multi-walled carbon nanotubes/polypropylene fiber cement-based materials were evaluated using flexural strength tests, compressive strength tests, and splitting tensile tests. The effects of multi-walled carbon nanotubes and polypropylene fiber on the durability of cement-based materials were studied using drying shrinkage tests and freeze–thaw cycle tests. The effects of the multi-walled carbon nanotubes and polypropylene fibers on the microstructure and pore structure of the cement-based materials were compared and analyzed using scanning electron microscopy and mercury intrusion tests. The results showed that the mechanical properties and durability of cement-based materials can be significantly improved when the content of multi-walled carbon nanotubes is 0.1–0.15%. The compressive strength can be increased by 9.5% and the mass loss rate is reduced by 27.9%. Polypropylene fiber has little effect on the compressive strength of the cement-based materials, but it significantly enhances the toughness of the cement-based materials. When its content is 0.2–0.3%, it has the best effect on improving the mechanical properties and durability of the cement-based materials. The flexural strength is increased by 19.1%, and the dry shrinkage rate and water loss rate are reduced by 14.3% and 16.1%, respectively. The three-dimensional network structure formed by the polypropylene fiber in the composite material plays a role in toughening and cracking resistance, but it has a certain negative impact on the pore structure of the composite material. The incorporation of multi-walled carbon nanotubes can improve the bonding performance of the polypropylene fiber and cement matrix, make up for the internal defects caused by the polypropylene fiber, and reduce the number of harmful holes and multiple harmful holes so that the cement-based composite material not only has a significant increase in toughness but also has a denser internal structure.

Keywords: multi-walled carbon nanotubes; polypropylene fiber; cement-based materials; mechanical properties; durability performance; microstructure; pore structure

1. Introduction

As a traditional building material, cement-based materials play a vital role in various fields. However, their own brittleness, poor crack resistance, and other shortcomings cannot be ignored [1,2]. These shortcomings limit the application of cement-based materials to a certain extent. In order to improve the shortcomings of cement-based materials and improve the application range, after a large number of experiments and studies by many scholars, it has been found that the mechanical properties and durability of composite materials have been significantly improved by adding fibers or nanomaterials to cement-based materials [3–6].

As a hollow carbon crystal curled by multilayer graphite sheets, carbon nanotubes can be divided into single-walled carbon nanotubes and multi-walled carbon nanotubes [7]. Multi-walled carbon nanotubes (MWCNTs) are the most commonly used in the field of cement-based materials. Generally, the diameter of single-walled carbon nanotubes is between 0.4–2 nm, and the diameter of multi-walled carbon nanotubes is 1.4–100 nm, and they have extremely high strength, toughness, and elastic modulus [8]. However, there are strong van der Waals forces between the tubes; therefore, carbon nanotubes are generally agglomerated. If they are directly added to the cement matrix, it is easy to cause stress concentration, which will not only fail to enhance the corresponding effect, but will also reduce the performance of the cement matrix [9,10]. Based on this consideration, researchers usually use mechanical stirring dispersion, ultrasonic dispersion, electric field-induced dispersion, surfactant modification, and other methods to achieve uniform dispersion of carbon nanotubes in the cement matrix [11–13].

Compared with auxiliary cementitious materials such as silica fume and fly ash, a very low content of carbon nanotube materials can achieve higher mechanical properties and durability [14–17]. For example, Cerro-Prada E et al. [18] found that only a small amount of multi-walled carbon nanotubes can increase the compressive strength and flexural strength of cement-based materials by 25% and 20%, respectively, within 90 days using conventional mechanical tests. Qin et al. [19] found that multi-walled carbon nanotubes changed the fracture process in the cement matrix and significantly increased the fracture energy when micro-cracks were initiated using laboratory tests and molecular dynamics simulations, thereby improving the overall mechanical properties. In addition, multi-walled carbon nanotubes with different sizes and morphologies have significant differences in their effects on the properties of cement-based materials. For example, Gao et al. [20] studied the effects of multi-walled carbon nanotubes with different diameters (10–20 nm, 20–40 nm and 40–60 nm) on the mechanical properties of cement-based materials. It was found that multi-walled carbon nanotubes with a diameter of 40–60 nm had the best effect on improving the flexural strength of cement, and 10–20 nm had the best effect on improving the compressive strength of cement-based materials. Ramezani et al. [21] found that the average length and average diameter of carbon nanotubes were 10–20 μm and 20–32.5 nm, respectively, which were most favorable for improving the mechanical properties of cement-based materials. Li et al. [22] used machine learning methods to predict the single-system and multi-system compressive strength of cement-based materials doped with carbon nanotubes. Some optimal parameters of carbon nanotubes were determined by the multi-system SHAP results: the optimal length and diameter of carbon nanotubes are 20 μm and 25 nm, respectively.

The improvement effect of carbon nanotubes on the mechanical properties and durability of cement-based materials is mainly attributed to its size effect and filling effect [23]. Chen et al. [24] studied the effect of carbon nanotubes on the structure of hydration products using nanoindentation tests. It was found that carbon nanotubes can bridge the pores in cement paste, promote the formation of calcium hydroxide, and promote the conversion of low-density calcium silicate hydrate to high-density calcium silicate hydrate. Wang et al. [25] used a variety of microscopic analysis methods to study multi-walled carbon nanotube cement-based materials. It was found that multi-walled carbon nanotubes can improve the hydration process, increase the number of hydration products, and reduce the porosity of cement-based materials. It has the effect of inhibiting crack propagation and improving compactness. Naqi A et al. [26] found that properly dispersed MWCNTs provide additional nucleation sites for the formation of hydrated calcium silicate (C-S-H), filling the fine pores in the cement matrix, resulting in a denser microstructure, thereby increasing strength and limiting autogenous shrinkage.

In the study of fiber-reinforced cementitious composites, steel fiber [27–29], basalt fiber [30,31], glass fiber [32,33], and polypropylene fiber [34,35] are often mentioned. However, steel fiber is easily agglomerated during the mixing process, and it is heavy and easy to corrode, which will incur significant cost and affect all aspects of the performance of the matrix material [36]. Polypropylene fiber has the advantages of a high tensile strength, low density, good thermal stability, and low cost [37,38]. It can significantly improve the crack resistance, toughness, and control ability of cracks caused by early shrinkage of the cement matrix [39–41]. For polypropylene fibers of different sizes, different contents enable them to be effectively distributed in the matrix, thereby giving the matrix a higher flexural strength [42–44]. The crack resistance of fiber-reinforced cement-based materials is closely related to the interface bonding between the fiber and matrix [45]. However, the application of polypropylene fiber is greatly affected by the smooth surface of polypropylene fibers, poor hydrophilicity, poor dispersion, and insufficient matrix bonding in cement mortar [46]. Researchers often use nano-coating, chemical modification, surface oxidation, etching, grafting, etc. to achieve adhesion between polypropylene fibers and matrix materials [47,48]. This method can further improve the mechanical properties and freeze–thaw resistance of composites. For example, Jia et al. [49] prepared a series of polypropylene composite fibers by melt-spinning using micro-silicon as a hydrophilic modifier and studied their structure and properties. It was found that, compared with unmodified polypropylene fibers, the composite fibers showed a rougher and more hydrophilic surface, and the interfacial bonding strength between the polypropylene fibers and cement matrix was significantly improved. Feng et al. [50] used nano-calcium carbonate to modify the surface of polypropylene fibers and incorporated them into cement-based materials. They found that the modified polypropylene fiber cement-based materials had better bending resistance. However, the above surface treatment methods for polypropylene fibers also have problems, such as a complicated operation, high equipment requirements, and poor stability.

According to the survey results, it was found that the size effect and filling effect of carbon nanotubes can effectively promote the hydration process of cement, fill holes, and make the matrix more dense, thus effectively improving the mechanical properties and durability of the cement matrix. Although polypropylene fiber can improve the flexural and tensile properties of cement-based materials, due to the lack of hydrophilic groups in its molecular chain, its bonding with the cement matrix is poor, and cracks are prone to occur at the fiber–matrix interface, thus reducing the reinforcement effects of the polypropylene fibers. Therefore, carbon nanotubes and polypropylene fibers were selected as modifiers. Multi-walled carbon nanotubes/polypropylene fiber cement-based composites were prepared via double mixing in order to demonstrate the excellent properties of multi-walled carbon nanotubes, improve the bonding performance between polypropylene fibers and the cement matrix, and achieve the purpose of improving the mechanical properties and durability of cement-based composites. The effects of multi-walled carbon nanotubes and polypropylene fibers on the mechanical properties of the cement-based materials were studied using flexural tests, compressive tests, and splitting tensile tests. The durability of the multi-walled carbon nanotubes/polypropylene fiber cement-based materials was evaluated using drying shrinkage tests and freeze–thaw cycle tests. The microstructure and pore structure of the multi-walled carbon nanotubes/polypropylene fiber cement-based materials were characterized using scanning electron microscopy and mercury intrusion tests, and the synergistic mechanisms of the multi-walled carbon nanotubes and polypropylene fibers were revealed.

2. Materials and Methods

2.1. Raw Materials

The multi-walled carbon nanotubes (MWCNTs) were produced by Suzhou Carbonfeng Technology Co., Ltd. (Suzhou, China), and their related technical indexes are shown in Table 1. The technical indexes of the polypropylene fibers (PP fibers) are shown in Table 2. The polypropylene fiber has a smooth surface, stable chemical properties, and a high aspect ratio. Its microstructure is shown in Figure 1. Polyvinylpyrrolidone (PVP) was selected as the dispersant, which is an amphiphilic polymer with a special structure. The pyrrolidone group is a hydrophilic group, and the main chain is a hydrophobic segment of the C–C bond. Introducing it into the surface of multi-walled carbon nanotubes can significantly improve their dispersion. The PVP and other raw materials and their technical indexes are shown in Table 3, which meet the requirements of the specifications.

Table 1. Technical indexes of multi-walled carbon nanotubes.

Item	Diameter (nm)	Length (μm)	Purity (wt%)	Specific Surface Area (%)
MWCNTs	10–20	5–15	>97%	90–120 m^2/g

Table 2. Technical indexes of polypropylene fibers.

Item	Indexes
Fiber type	Bunchy monofilament
Tensile strength (MPa)	>486
Elastic modulus (GPa)	>4.8
Melting point (°C)	169
Density (g/cm^3)	0.91
Length (mm)	9
Diameter (μm)	18–48

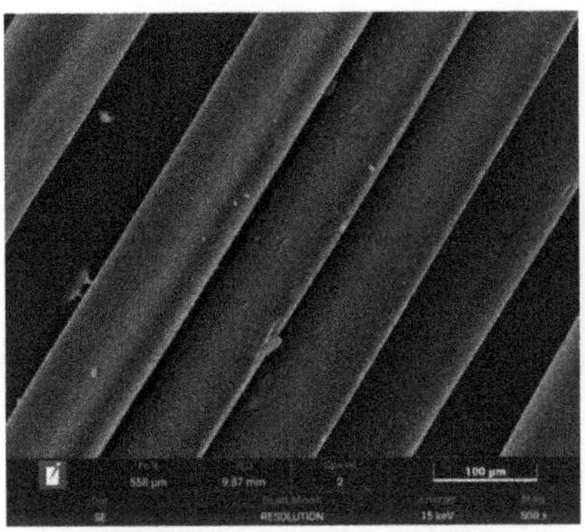

(**a**) Appearance morphology of PP fibers

(**b**) PP fiber cross-section

Figure 1. SEM of PP fibers.

Table 3. Related raw materials and indexes.

No.	Types	Materials	Indexes
1	Binding material	P.O 42.5 Cement	Fineness, 5.4%; standard consistency water consumption, 26.1%; initial setting time, 255 min; final setting time, 365 min; soundness, 1.2 mm
2	Binding material	Fly ash (I grade)	Mean diameter, 20.13 μm; fineness, 9.8%; water demand ratio, 93.1%; moisture content, 0.2%; loss on ignition, 1.35%
3	Dispersant	Polyvinyl pyrrolidone	White powder; K-value, 27.0–32.4; pH value, 3.0–5.0; total nitrogen content, 11.5–12.8%; ignition residue, ≤0.1%; aldehyde content, ≤0.05%; formic acid content, ≤0.5%; vinylpyrrolidone content, ≤0.001%; plumbum content, ≤0.001%; water content ≤ 5.0%
4	Auxiliary material	Naphthalene water reducer	Yellow–brown powder; water reduction, 8–14%; bleeding rate, 55%; gas content, 3.0%; 28 d shrinkage ratio, 110%
5	Sand	ISO standard sand	Grain diameter, 0.08–2 mm

2.2. Preparation of Cement-Based Composite Materials

Through many experiments in the laboratory, the following preparation process was determined. First, a certain amount of polyvinylpyrrolidone (PVP) was weighed, and the mass ratio of MWCNTs to PVP was 1:2. The amounts were added to a beaker and completely dissolved in water via magnetic stirring. Then, the MWCNTs were weighed (weighed according to the mass of the cementitious material) and added to the mixed solution via magnetic stirring. Finally, the solution was placed in the ultrasonic dispersion device, and the oscillation time was 40 min to obtain the MWCNT dispersion suspension. By consulting the relevant literature and through many experimental studies, the water-to-cementitious ratio (W/C) was determined to be 0.43. Then, polypropylene fiber (volume content) was mixed with standard sand dryly, in which the cement–sand ratio was 1:3, and cement and fly ash (fly ash replaces 20% cement quality) were added. Then, water, water reducer, and suspension were added and stirred to a uniform fluid state, and the amount of water reducer was obtained by referring to the relevant literature and debugging according to the actual mix ratio. Finally, the mixture was slowly filled into a mold that had been painted with oil, and the surface was smoothed. After being placed in an environment at 20 °C for 24 h, the mold was removed and the specimens were cured under standard curing conditions. After the curing was completed, the corresponding tests were performed, as shown in the Figure 2. The mix proportions of the cement-based composite materials are shown in Table 4.

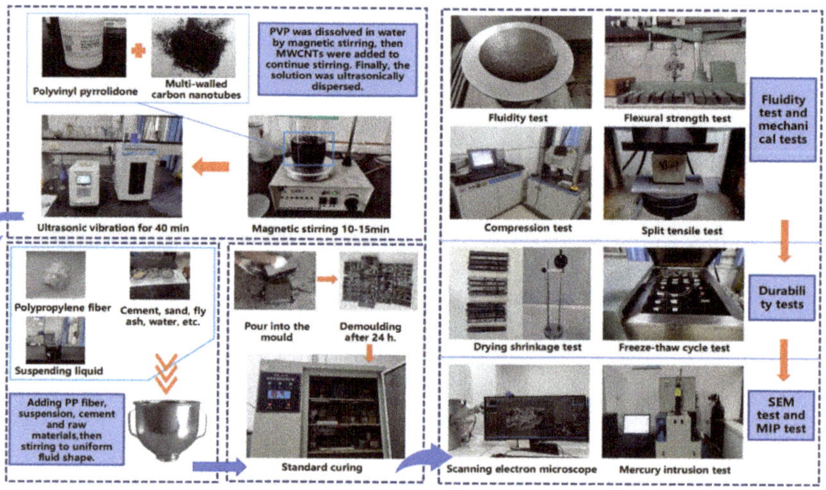

Figure 2. Preparation process and test flow chart.

Table 4. Mix proportions of cement-based composite materials.

Scheme	W/C	MWCNTs (wt%)	PP Fiber (%)
T0	0.43	0	0
T1	0.43	0.05	0
T2	0.43	0.1	0
T3	0.43	0.15	0
T4	0.43	0.2	0
T5	0.43	0	0.1
T6	0.43	0	0.2
T7	0.43	0	0.3
T8	0.43	0	0.4
T9	0.43	0.1	0.1

Table 4. Cont.

Scheme	W/C	MWCNTs (wt%)	PP Fiber (%)
T10	0.43	0.1	0.2
T11	0.43	0.1	0.3
T12	0.43	0.1	0.4

2.3. Test Methods

2.3.1. Fluidity Test and Mechanical Tests

The fluidity tests were carried out according to GB/T 2419-2005, and the fluidity was measured using the jump table test. According to GB/T 17671-1999, the mechanical properties of the cement-based composites were tested, including the use of flexural strength tests and compressive strength tests. The curing ages of the specimens were 3 d, 7 d, and 28 d. The flexural strength test adopted the central loading method. One side of the test body was placed on the support cylinder of the test machine. The long axis of the specimen was perpendicular to the support cylinder. A load was uniformly applied vertically to the relative side of the prism at 50 N/s through the loading cylinder until it broke. Then, the compressive strength test was carried out on the samples after the flexural test. The loading rate of the compressive strength test was 2400 N/s, and the average speed was loaded until the specimen was destroyed. The splitting tensile test was used to study the tensile properties of the cement-based materials. The specimen size was 70.7 mm × 70.7 mm × 70.7 mm. After curing for 28 days, the splitting tensile test was carried out. The qualified specimens were placed in the center of the pressure plate, the arc pads were placed above and below the pressure plate, and the loading rate was set to 0.08 MPa/s.

2.3.2. Durability Tests

(1) Drying shrinkage test

The drying shrinkage test was carried out according to JC/T 603-2004. The size of the specimen was 25 mm × 25 mm × 280 mm. After curing in the standard curing box for 24 h, the mold was removed, and the specimens were placed in water at 20 °C for 48 h. After 2 days of curing, the initial length and initial weight of the specimens were tested using a specific length meter and balance. After the measurements, the specimens were placed in the curing box for drying and curing and measured again at each age (3 d, 7 d, 14 d, 21 d, 28 d, 56 d, 90 d). Finally, the dry shrinkage rate and water loss rate were calculated.

(2) Freeze–thaw cycle test

The freeze–thaw cycle test was carried out according to JGJ/T 70-2009. The size of the specimen was 40 mm × 40 mm × 160 mm. After removing the formwork, the specimens were cured in a standard curing box for 28 days, and then the specimens were immersed in water at a temperature of 20 °C for 4 days. The initial mass of the specimens was weighed, then the freeze–thaw cycle test was started. The freeze–thaw cycle test was completed in an automatic low-temperature freeze–thaw testing machine. The single freeze–thaw cycle test piece lasted 2 h, the melting lasted 0.5 h, the freezing temperature and melting temperature were −18 °C and 5 °C, respectively, and the number of freeze–thaw cycles was 100. The results of the freeze–thaw cycle test are expressed as the mass loss rate and strength loss rate. The strength test was carried out according to Section 2.3.1.

2.3.3. Scanning Electron Microscopy

A MAGNA emission scanning electron microscope produced by Taisiken Co., Ltd. (Shanghai, China) was used for the test. The sample size was approximately 1 cm × 1 cm × 0.3 cm. Before the test, the selected samples were soaked in anhydrous ethanol to terminate the hydration of the cement samples. The soaking time was 24 h. After soaking, the samples were dried. Due to the poor conductivity of cement-based materials, the samples were polished and sprayed with gold. Finally, the samples were placed in the scanning chamber for observation.

2.3.4. Mercury Intrusion Test

The mercury injection test was carried out using AutoPore9600/9510 automatic pore size analyzer. The maximum working pressure of the mercury injection instrument was 400 MPa, and the measurable pore diameter range was 0.003~950 μm. Firstly, after 28 days of standard curing, a long strip of 0.5–1.5 g of each of the samples was taken and immersed in anhydrous ethanol for 24 h. Then, the hydration was terminated, and the samples were dried, loaded into an expansion meter, and tested under low pressure and high pressure conditions.

3. Results and Analysis

3.1. Workability and Mechanical Properties Analysis

3.1.1. Workability Analysis

The fluidity was used to evaluate the workability of the cement-based composites. The test results are shown in Table 5, and the test process is shown in Figure 3.

Table 5. Fluidity of the cement-based composites.

Scheme	T0	T1	T2	T3	T4	T5	T6	T7	T8	T9	T10	T11	T12
Fluidity (mm)	225	210	195	183	180	207	193	181	170	204	184	174	165

Figure 3. Fluidity test.

It can be seen from Table 5 that the addition of MWCNTs and PP fibers reduced the fluidity of the cement-based composite mixture. The contents of the MWCNTs were 0.05%, 0.1%, 0.15%, and 0.2%, and they could be decreased by 6.7%, 13.3%, 18.7%, and 20%, respectively, compared with T0. The PP fiber contents were 0.1%, 0.2%, 0.3%, and 0.4%, and the fluidity decreased by 8.0%, 14.2%, 19.6%, and 24.4%, respectively, compared with T0. When the content of PP fiber was 0.4%, the fluidity did not meet the requirements of the specifications. Compared with T0, the fluidity of T9–T12 decreased by 9.3%, 18.2%, 22.7%, and 26.7%, respectively. The fluidity of the composite materials was negatively correlated with the content of MWCNTs, and the fluidity gradually decreased with increasing contents of MWCNTs. This is due to the large specific surface area of MWCNTs. After adding the MWCNTs, the total specific surface area of the cement-based materials increased, and the free water adsorbed on the surface of the MWCNTs increased significantly. This made the cement paste become sticky, and with the increase in the MWCNT content, the fluidity of the composite material decreased more obviously. The addition of PP fiber has an adverse effect on the fluidity of the composites. When PP fiber is added to the cement-based material, a three-dimensional network structure is formed inside the mixture, which increases the friction between the aggregates. At the same time, there is a certain viscous effect between the PP fiber and the slurry, resulting in a decrease in the fluidity of the cement mixture. With the increase in fiber content, it is difficult to disperse in cement-based composites, and the distribution spacing of PP fibers becomes smaller. Therefore, the fibers were easily agglomerated during the mixing process, which hindered the fluidity of the cement paste.

3.1.2. Flexural Strength and Compressive Strength

The flexural strength and compressive strength at 3 d, 7 d, and 28 d were selected to characterize the mechanical properties of the cement-based materials. The macroscopic morphology of the specimens after the flexural strength tests and compressive strength tests are shown in Figure 4. The test results are shown in Figures 5 and 6.

(**a**) (**b**)

Figure 4. The macroscopic morphology of the specimens. (**a**) The macroscopic morphology of the specimens after the flexural strength test. (**b**) The macroscopic morphology of the specimens after the compressive strength test.

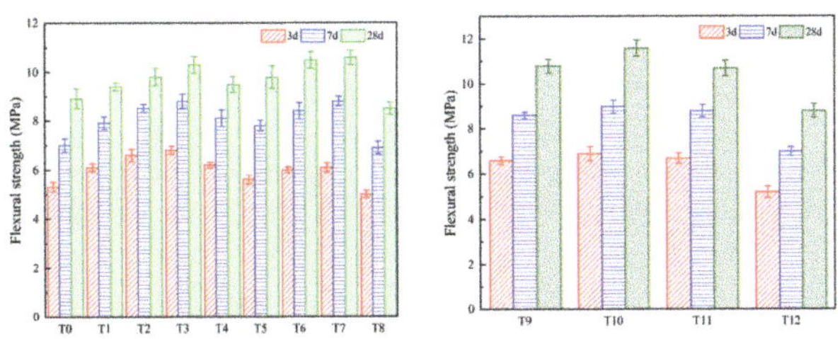

(**a**) T0–T8 (**b**) T9–T12

Figure 5. Flexural strength test results.

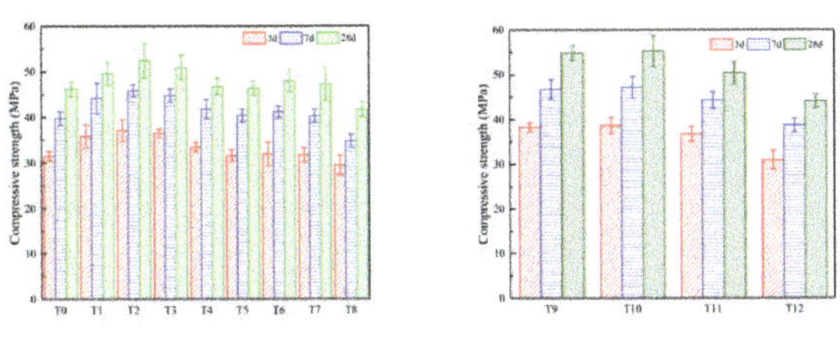

(**a**) T0–T8 (**b**) T9–T12

Figure 6. Compressive strength test results.

It can be seen from Figure 5a that the incorporation of MWCNTs has a certain effect on the flexural strength of the composites. The enhancement effect increases first and then decreases with increasing contents. At the same time, the increases at 3 d and 7 d were significantly higher than that at 28 d. Compared with T0, the incorporation of MWCNTs can promote the hydration reaction of cement and have obvious improvement effects on the flexural strength of the composite materials in the early stages. When the age of the specimen was 28 d, compared with T0, the flexural strength of T1, T2, T3, and T4 increased by 5.6%, 10.1%, 15.7%, and 6.7%, respectively, and the flexural strength of T3 increased the most. It can be seen from Figure 5a that when the age of the specimen was 28 d, the flexural strength of the material increased by 10.1%, 18.0%, and 19.1%, respectively, when the PP fiber content was 0.1%, 0.2%, and 0.3% compared with T0. When the PP fiber content was 0.3%, the improvement effect was the best, and the improvement effect of the PP fiber on the flexural strength of the composite was better than that of carbon nanotubes. When the PP fiber content was 0.4%, it was reduced by 4.5% compared with the reference group. This is because the appropriate amount of PP fiber can enhance the toughness of the composite material and play a role in reinforcement and crack resistance. When the amount of PP fiber is greater, the dispersion of the fiber in the cement matrix is poor, and the stress concentration phenomenon occurs, which seriously weakens the flexural performance of the cement-based material. These results are consistent with previous research conclusions [38]. From Figure 5b, it can be seen that when the specimen age is 28 d, compared with T2 (0.1% MWCNTs), the flexural strength of the carbon nanotube/polypropylene fiber composites of T9, T10, and T11 increase by 10.2%, 18.4%, and 9.2%, respectively, while the flexural strength of the T12 scheme decreases by 10.2%. Combined with scanning electron microscopy and mercury injection tests, the incorporation of MWCNTs can improve the internal structure of PP fiber composites and reduce the porosity of the PP fiber composites. At the same time, due to the nucleation of MWCNTs [26], the hydration of cement is promoted, the bonding performance between fiber and matrix is enhanced, and the toughening effect of PP fiber is further improved. However, the excessive amount of fiber results in a decrease in the mechanical properties.

From Figure 6a, it can be seen that MWCNTs have a certain effect on the compressive strength of cement-based materials. When their contents are increased, the compressive strength shows a trend of first increasing and then decreasing. When the age of the specimen is 28 d and the MWCNT content is 0.05%, 0.1%, 0.15%, 0.2%, the compressive strength of the composite material is increased by 7.4%, 13.4%, 10.2%, and 1.3%, respectively, compared with the reference group. This is because MWCNTs have a size effect and filling effect, optimizing the pore structure of the composite material, enhancing the compactness of the structure, and giving the composite material superior compressive properties [23,24].

When the age of the specimen was 28 days, compared with the T0 group, the compressive strength of the composite was increased by 0.4%, 3.9%, and 2.4% when the PP fiber content was 0.1%, 0.2% and 0.3%, respectively. When the PP fiber content was 0.4%, the compressive strength decreased by 9.7% compared with the T0 group. It was found that when the content of PP fiber was too low, the fibers could not form a strong support system, having little effect on the compressive strength. When the fiber content was too large, the fibers agglomerated in the composite material and could not inhibit the development of small cracks. At the same time, the fiber agglomeration increased the number of internal pores and reduced the density of the composite material, thus causing the compressive strength of the composite material to decrease. These findings are consistent with the results of Li [39] and Al-Katib [41]. It can be seen from Figure 6b that when the age of the specimen is 28 days, compared with the T2 group, the compressive strength of the carbon nanotube/polypropylene fiber composites of the T9 and T10 schemes is increased by 8.5% and 9.3%, respectively. When the content of PP fiber in the composite group is 0.3% and 0.4%, it is reduced by 0.4% and 12.8%, respectively. This shows that, although the combined addition of MWCNTs and PP fiber can further improve the compressive strength of the composite, the enhancement effect cannot be completely superimposed. When the content is too high, the compressive strength of the composite is weakened.

3.1.3. Split Tensile Strength

The splitting tensile strength at 28 d was used to characterize the tensile stress resistance of the cement-based materials. The macroscopic morphology of the specimens after the split tensile strength test is shown in Figure 7. The splitting tensile test results are shown in Figure 8.

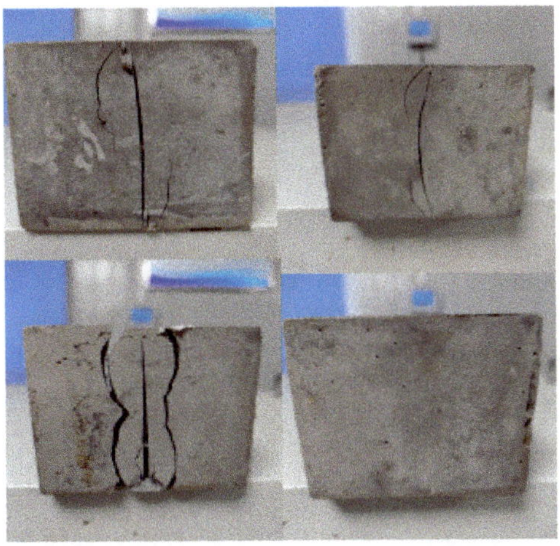

Figure 7. The macroscopic morphology of the specimens after the split tensile strength test.

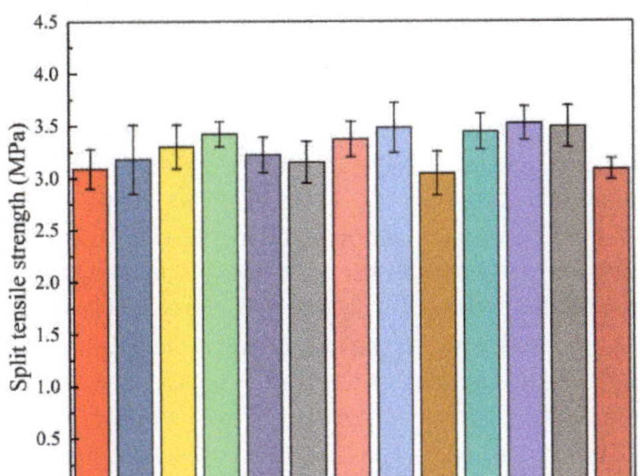

Figure 8. T0–T12 results of the splitting tensile tests.

It can be seen from Figure 8 that when the content of MWCNTs is 0.05%, 0.1%, 0.15%, and 0.2%, the splitting tensile strength of the cement-based composites increases by 2.9%, 6.8%, 10.7%, and 4.2%, respectively, compared with the reference group. MWCNTs have a nanoscale size and excellent tensile strength, which have good filling effects on the internal micro-pores of the composite material. They also improve the compactness of the matrix and play a bridging role to a certain extent, which has a certain inhibitory effect on the development of cracks [19,24]. When the content of PP fiber was 0.1%, 0.2%, and 0.3%, the splitting tensile strength of the cement-based composites increased by 1.9%, 9.1%, and 12.6%, respectively, compared with the reference group, while it was decreased by 1.6% when the fiber content was 0.4%. The toughening and crack resistance effects of PP fiber have a significant effect on the splitting tensile strength of the composite material. These results are consistent with previous studies [40,44]. When the matrix reaches the critical failure load, due to the existence of the fiber, the concentrated stress at the crack is transferred to the other interface of the matrix through the fiber, which reduces the stress concentration at the crack end and inhibits the continuous development of the crack. Compared with 0.1% MWCNTs, the splitting tensile strength of the 0.1%, 0.2%, and 0.3% composites can be increased by 4.2%, 6.7%, and 5.8%, respectively. When the PP fiber content is 0.4%, it can be decreased by 6.7%. Compared with the single-doped method, the composite-doped method shows more excellent tensile strength. Combined with the results of the scanning electron microscopy and mercury intrusion tests, it can be assumed that the addition of MWCNTs improves the bonding degree of the PP fiber–cement interface and ensures the toughening and crack resistance effects of the PP fiber on composite materials.

3.2. Durability Analysis

3.2.1. Drying Shrinkage Test

The durability of the cement-based composites was characterized by the dry shrinkage and water loss rate. The test results are shown in Figure 9.

From Figure 9a, it can be seen that when the MWCNT content was 0.1%, the reduction in the drying shrinkage rate of the composite material was the best. With increasing curing ages, it showed a trend of first increasing and then decreasing. Compared with the reference group, the drying shrinkage rate of the composite with 0.1% MWCNT was decreased by 15.7%, 16.1%, 13.5%, 13.1%, 12.5%, and 12.1% at 3 d, 7 d, 14 d, 28 d, 60 d, and 90 d, respectively. The incorporation of MWCNTs significantly improves the drying shrinkage of the composite. This is because, on the one hand, the size effect and filling effect of MWCNTs significantly improve the pore structure of the composite material. At the same time, when 0.1%MWCNTs is added to the cement-based material, the water loss rate of the composite material improves the most. The water loss rate showed a trend of decreasing first and then increasing with increasing MWCNT contents, and the improvements in the water loss rate gradually decreased with increasing age. When the content of the MWCNTs is too much, it agglomerates locally in the matrix, destroys the capillary pore structure of the composite, and accelerates the water loss. From Figure 9b, it can be seen that when the content of PP fiber is 0.3%, the drying shrinkage rate of the composite material reduces the most. The drying shrinkage rate of the composite material with 0.3% PP fiber at the ages of 3 d, 7 d, 14 d, 28 d, 60 d, and 90 d was reduced by 13.3%, 13.8%, 14.5%, 16.4%, 14.4%, and 14.3%, respectively, compared with the T0 group. In addition, the PP fiber reduced the water loss rate of the composites. At the early ages, with the increase in curing age, the reduction effect is more significant. When the curing age exceeded 28 d, the water loss rate of the composite material decreased, and the water inside the composite material was lost. When the content of PP fiber was 0.3%, the water loss rate of the composite was the lowest. When the fiber content is low, the fiber is distributed inside the cement paste, and the fiber is almost not observed on the surface of the matrix. This phenomenon results in an increase in the water loss channels generated by the fiber and the slurry, resulting in an increase in the water loss rate. With the increasing fiber contents, the water loss rate gradually decreases. This is because the fiber is not only dispersed inside, but it is also distributed on the surface of the matrix. Because the water loss channel is blocked by the fiber, the water loss path inside the composite is reduced, and the water evaporation rate is reduced. It can be seen in Figure 9c that the dry shrinkage rates of the cement-based composites with carbon nanotubes/PP fibers at 3 d, 7 d, 14 d, 28 d, 60 d, and 90 d were 4.3%, 8.7%, 2.1%, 7.0%, 5.0%, and 4.6% lower than those of the 0.1% MWCNT group, respectively. When the MWCNTs and PP fibers act simultaneously in the composite, the drying shrinkage of the composite is further improved, and the two materials jointly bear the anti-drying shrinkage ability of the composite. Using the compound blending combination method, when the PP fiber content was 0.2%, the water loss rate of the composite material was further reduced, and the drying shrinkage improvement effect was also optimal.

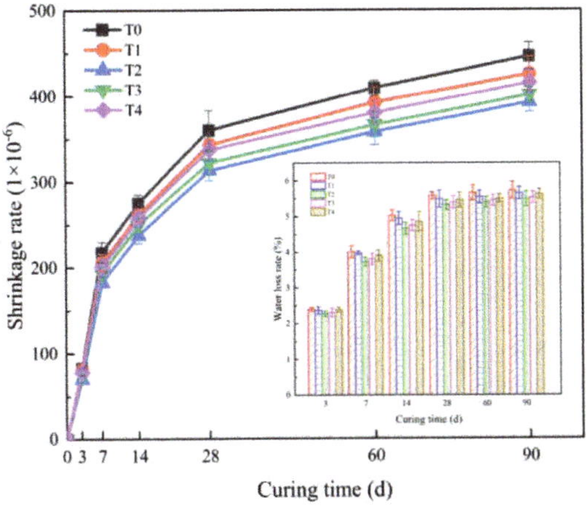

(**a**) T0–T4

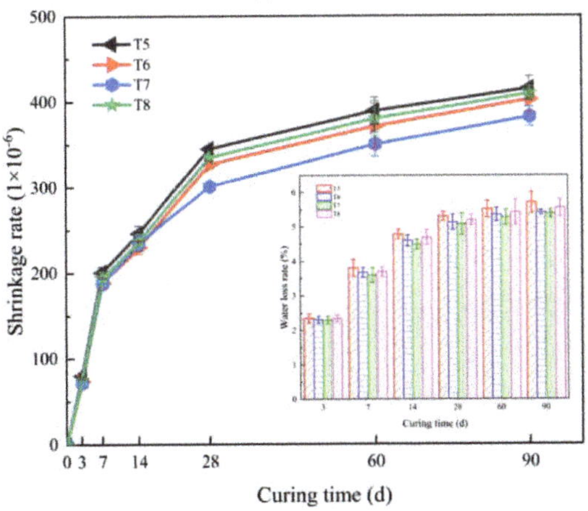

(**b**) T5–T8

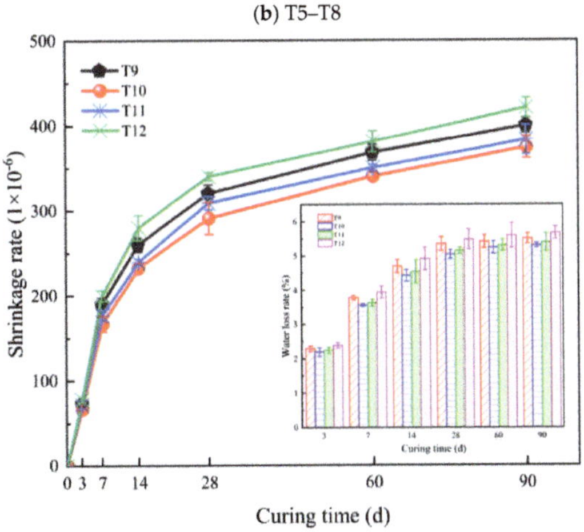

(**c**) T9–T12

Figure 9. Results of the drying shrinkage tests.

3.2.2. Freeze–Thaw Cycle Test

The mass loss rate and strength loss rate were used to characterize the durability of the cement-based materials. The macroscopic morphology of the specimen after the test is shown in Figure 10, and the test results are shown in Table 6 and Figure 11.

Figure 10. The macroscopic morphology of the specimens after the freeze–thaw cycle test.

Table 6. The mass loss rate test results.

Scheme	Freeze–Thaw Cycle (Cycles)			
	25	50	75	100
T0	0.11	0.85	2.45	4.41
T1	−0.08	0.30	2.10	3.65
T2	−0.23	0.19	1.90	3.18
T3	−0.18	0.26	2.01	3.39
T4	−0.14	0.46	2.20	3.74
T5	−0.02	0.37	2.11	3.78
T6	−0.05	0.21	1.95	3.30
T7	−0.03	0.30	2.02	3.65
T8	−0.01	0.55	2.23	3.95
T9	−0.10	0.41	2.10	3.42
T10	−0.25	0.14	1.78	2.87
T11	−0.13	0.36	2.01	3.25
T12	−0.05	0.54	2.28	3.76

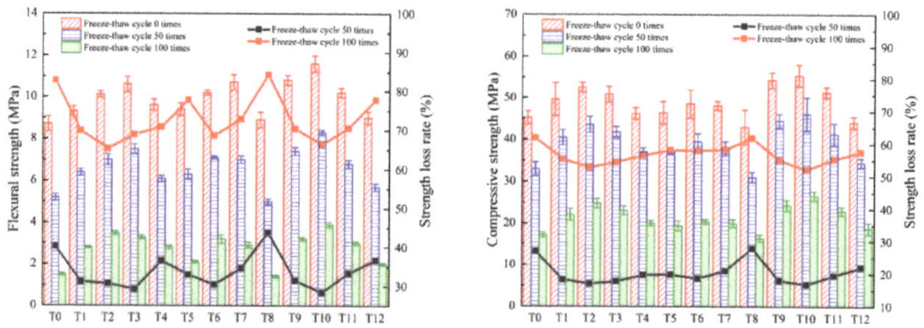

(**a**) Loss rate of flexural strength (**b**) Compressive strength loss rate

Figure 11. Strength loss rate.

It can be seen from Table 6 that, except for the T0 group, the mass loss rates of the other 12 groups of specimens were negative after 25 freeze–thaw cycles. The reason for this phenomenon may be due to the water absorption of the carbon nanotubes and PP fibers. When the freeze–thaw cycle was completed 50 times, the cement was fully hydrated, the adsorbed water inside the material was saturated, and the mass loss rate began to be greater than the mass growth rate due to freeze–thaw damage. As the freeze–thaw test continued, the degree of damage to the specimen gradually increased, and the mass loss rate increased sharply. The MWCNTs were added to the cement-based materials and subjected to 100 freeze–thaw cycles. The mass loss rate of the composite material can be reduced to a certain extent compared with the reference group, and the reduction is different for different dosages. When the MWCNT contents were 0.05%, 0.1%, 0.15%, and 0.2%, the mass loss rate was reduced by 17.2%, 27.9%, 23.1%, and 15.2%, respectively, compared with the reference group. It can be seen from the mass loss rate of the composite material that the optimal content of the MWCNTs was 0.1%. According to the results of the scanning electron microscopy and mercury intrusion tests, it can be seen that MWCNTs can optimize the pore structure, enhance the density of the material, and improve the freeze–thaw resistance. In addition, when the composite material produces expansion pressure during the freeze–thaw cycle, the bridging effect and excellent mechanical properties of the MWCNTs can resist part of the pressure and inhibit the generation of cracks. When the content of the PP fiber was 0.1%, 0.2%, 0.3%, and 0.4%, the mass loss rate was 14.3%, 25.2%, 17.2%, and 10.4% lower than that of the reference group, respectively. It can be seen from the mass loss of the composite material that the optimum content of PP fiber is 0.2%. During the freeze–thaw cycle, when the cement matrix is subjected to expansion pressure, cracks begin to appear inside the matrix, and the bridging effect of the fiber inhibits the expansion of the cracks to a certain extent. At the same time, due to the binding effects of the PP fiber, the integrity of the composite material is ensured, and the anti-stripping ability is improved. In the composite group (T9–T12), when the content of PP fiber was 0.1%, 0.2%, 0.3%, and 0.4%, the mass loss rate was reduced by 22.4%, 34.9%, 26.3%, and 14.7%, respectively, after 100 freeze–thaw cycles compared with the reference group. When the content of PP fiber was 0.2%, the improvement effect on the mass loss of the composite material was the best.

It can be seen from Figure 11 that the strength loss of composite materials in all the test schemes increased with increasing numbers of freeze–thaw cycles. This is because with increasing numbers of freeze–thaw cycles, the matrix continues to be subjected to freeze–thaw damage, the internal micro-cracks continue to expand, and the compactness decreases, resulting in a sharp decline in the mechanical properties. When the number of freeze–thaw cycles is 50 and the MWCNT content is 0%, 0.05%, 0.1%, 0.15%, and 0.2%, the flexural strength loss rate of the composite is 40.2%, 31.2%, 30.7%, 29.2%, and 36.5%, respectively, and the compressive strength loss rate of the composite is 27.0%, 18.3%, 17.0%, 17.7%, and 19.7%, respectively. When the number of freeze–thaw cycles reaches 100 and the MWCNT content is 0%, 0.05%, 0.1%, 0.15%, and 0.2%, the loss rates of the flexural strength of the composites are 82.8%, 73.1%, 69.3%, 73.6%, and 77.1%, respectively, and the loss rates of the compressive strength of the composites are 61.9%, 55.4%, 52.8%, 54.5%, and 56.5%, respectively. It can be found that the addition of MWCNTs significantly improved the strength loss of the composites under the freeze–thaw cycles. When the number of freeze–thaw cycles is 50 and the content of PP fiber is 0%, 0.1%, 0.2%, 0.3%, and 0.4%, the flexural strength loss rates of the composites are 40.2%, 33.0%, 27.5%, 34.6%, and 43.8%, respectively, and the compressive strength loss rates of the composites are 27.0%, 19.8%, 18.6%, 21.0%, and 27.9%, respectively. When the number of freeze–thaw cycles reaches 100 and the PP fiber content is 0%, 0.1%, 0.2%, 0.3%, and 0.4%, the flexural strength loss rate of the composite is 82.8%, 77.7%, 68.6%, 72.9%, and 84.3%, respectively, and the compressive strength loss rate of the composite is 61.9%, 58.2%, 57.9%, 58.4%, and 61.9%, respectively.

The incorporation of PP fiber has a significant improvement in the toughness of the cement-based materials. At the same time, the three-dimensional network structure formed by the fiber inside the composite material improves the overall stability of the composite material. Therefore, it has a higher resistance to freeze–thaw cycles than the reference group, and the strength loss is also reduced. When the number of freeze–thaw cycles was 50, the flexural strength loss rates of the T9–T12 cement-based composites were 31.5%, 28.4%, 33.3%, and 36.7%, respectively, and the compressive strength loss rates of the composites were 18.0%, 16.8%, 19.5%, and 22.0%, respectively. When the number of freeze–thaw cycles reached 100, the flexural strength loss rates of the T9–T12 cement-based composites were 70.4%, 66.4%, 70.6%, and 77.8%, respectively, and the compressive strength of the composites were 55.2%, 52.2%, 55.4%, and 57.6%, respectively. Compared with the single-doped combination method, the composite-doped method enables the cement matrix to maintain higher mechanical properties under freeze–thaw cycles.

3.3. *Scanning Electron Microscopy Test Analysis*

The microstructures of T0, T2, T6, and T10 were observed using scanning electron microscopy. The test results are shown in Figures 12–15.

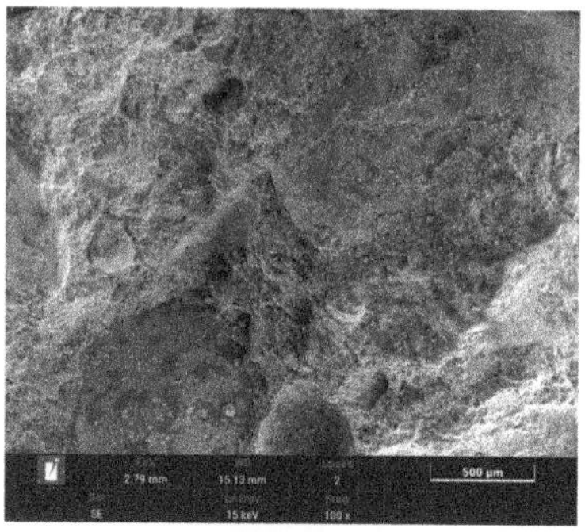

(**a**) 100 times

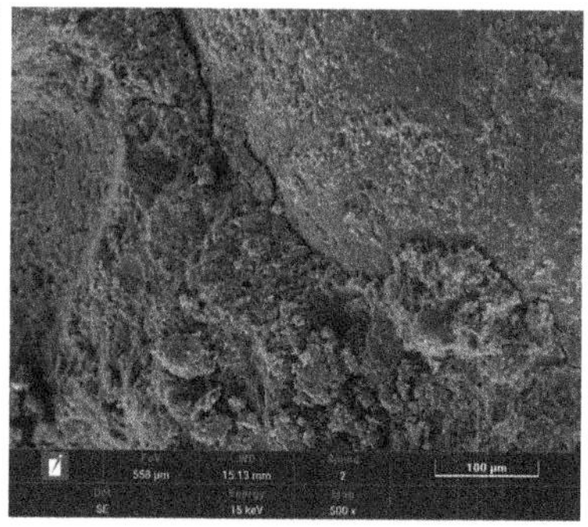

(**b**) 500 times

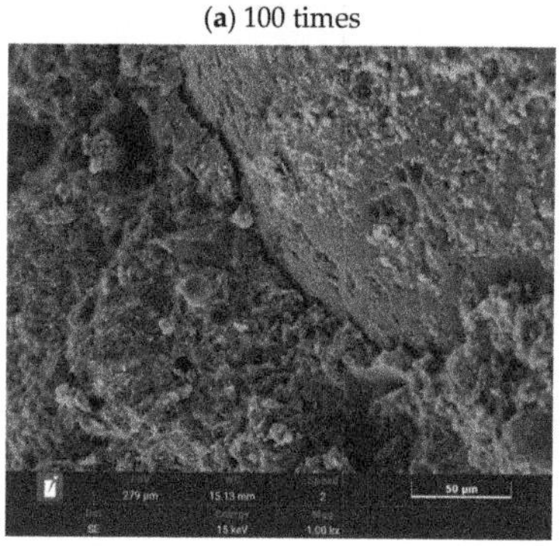

(**c**) 1000 times

(**d**) 10,000 times

Figure 12. T0 microtopography.

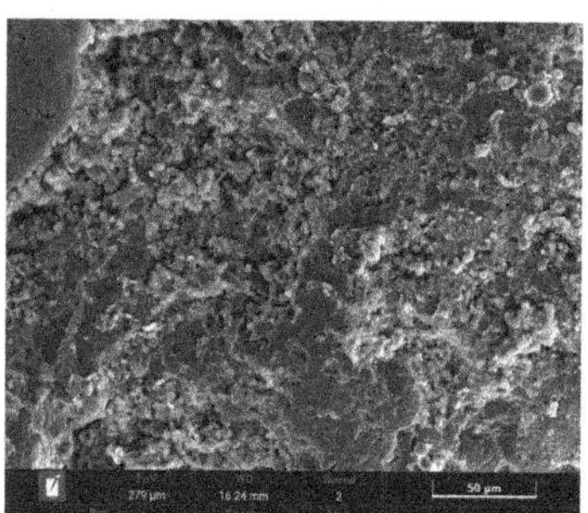

(**a**) 1000 times

(**b**) 5000 times

Figure 13. *Cont.*

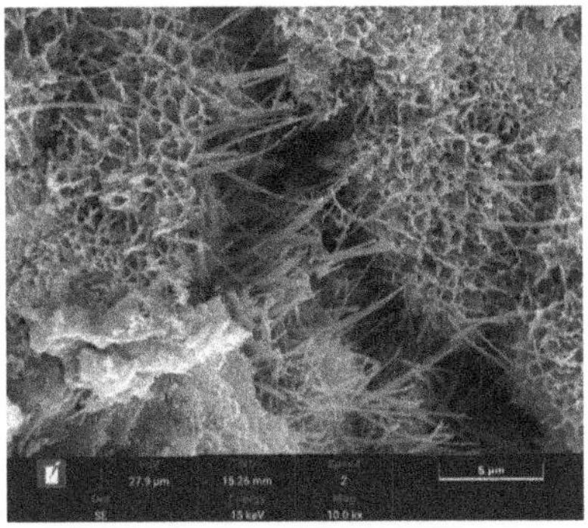

(**c**) 10,000 times

Figure 13. T2 microtopography.

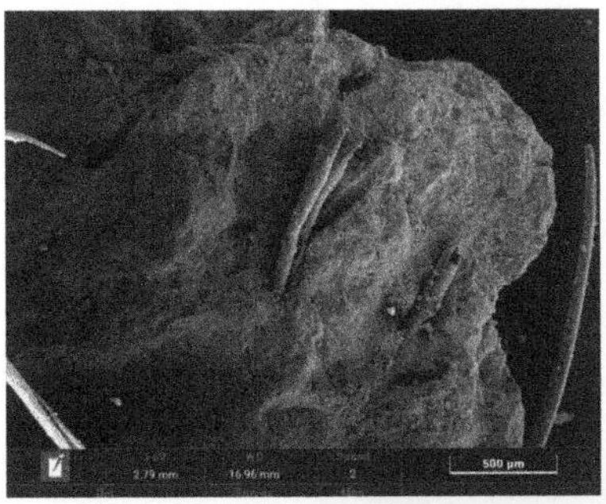

(**a**) 100 times

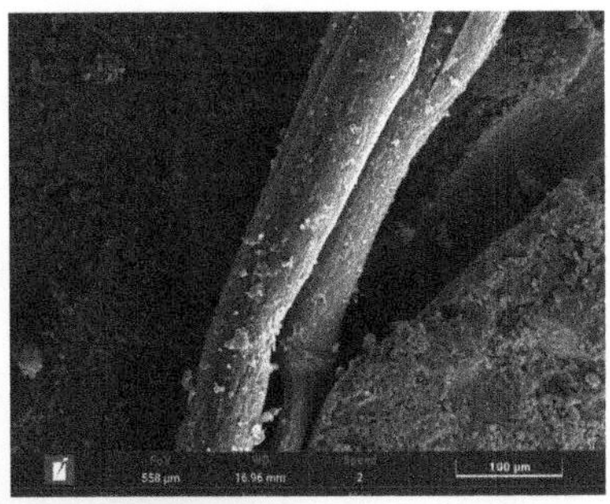

(**b**) 5000 times

(**c**) 10,000 times

Figure 14. T6 microtopography.

By comparing Figures 12 and 13, it can be seen that a large number of hydration products were attached to the surface of the composites doped with MWCNTs, and the number of tiny pores was significantly reduced. This shows that MWCNTs have a significant filling effect on the micropores of the composite material, exerting an inhibitory effect on the generation of micro cracks and making the composite material more dense. Figure 13c shows the micromorphology of the crack when it is magnified 10,000 times. It can be found that some MWCNTs are lapped at both ends of the crack, and some MWCNTs are pulled off. When the crack appears and further expands, the presence of MWCNTs delays the continued expansion of the crack and transfers the stress of the crack propagation to different spaces inside the material.

(**a**) 500 times (**b**) 10,000 times

Figure 15. T10 microtopography.

From Figure 14a, it can be seen that some fibers are in a state of tensile fracture. When the fiber is subjected to tensile stress, tensile deformation occurs. When it exceeds its own tensile strength, the fiber is broken. The PP fiber reduces the brittleness of the composite material through the bridging effect. Figure 14b shows the morphology of the joint of PP fiber and cement matrix under 5000 times magnification. It can be seen that PP fiber plays a certain role in hindering the expansion of cracks. When the crack appears and begins to develop, the crack propagation is hindered due to the presence of the fiber. When the crack changes the development direction or passes through the fiber, the internal stress field of the composite material is weakened to a certain extent, and the concentrated stress at the crack tip makes it difficult to support the crack propagation. At the same time, it can be seen that there are small cracks in the interface transition zone between the fiber and the cement matrix. However, there are some hydration products around the fiber, indicating that the combination of PP fiber and cement paste has certain deficiencies, and the toughening effect of the composite material cannot be fully reflected. From Figure 14c, it can be found that a certain amount of C-S-H is attached to the surface of the fiber. These hydration products have a strong bonding force with each other, and the fiber and the matrix are anchored together. The more hydration products, the better the anchoring effect.

By comparing Figures 14 and 15, it can be seen that there is a better bonding effect between the fiber and the matrix. Figure 15b shows a microscopic diagram of the connection between the PP fiber and cement matrix under 10,000 times magnification. It can be seen that the multi-walled carbon nanotube/polypropylene fiber cement-based material has more hydration products than the polypropylene fiber cement-based material. This is due to the fact that PP fiber has no obvious improvement effects on cement hydration, while the addition of MWCNTs promotes an increase in hydration products. The distribution of C-S-H gel is more dense, and a large number of hydration products wrap the PP fiber, which strengthens the bonding ability between the fiber and matrix. At the same time, MWCNTs can fill the tiny pores generated when PP fibers are combined with the matrix, enhancing the compactness of the internal structure, improving the binding capacity of PP fiber and cement matrix, and enhancing the toughening effect of PPA.

3.4. *Mercury Intrusion Test*

Mercury injection experiments were carried out on the T0, T2, T6, and T10 specimens, respectively, to study the influence of multi-walled carbon nanotubes and polypropylene fibers on the pore structure of the cement-based materials. The experimental results are shown in Table 7 and Figure 16. The internal pores of the composite material were divided into four categories [51]: harmless pores less than 20 nm, less harmful pores between 20 nm and 50 nm, harmful pores between 50 nm and 200 nm, and more harmful pores greater than 200 nm. The porosity of the different types of pores was counted, and the pore size distribution is shown in Table 8.

Table 7. Mercury injection test results.

Scheme	Average Pore Size (nm)	Medium Pore Diameter (nm)	Median Volume (cc g^{-1})	Median Surface Area (m^2 g^{-1})
T0	73.31	128.23	0.0287	2.253
T2	59.95	91.33	0.0175	1.452
T6	87.94	135.21	0.0301	2.295
T10	69.82	108.45	0.0195	1.784

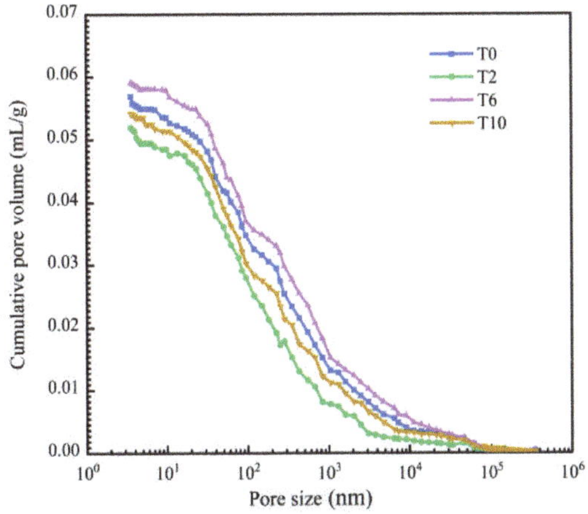

Figure 16. Cumulative pore volume.

Table 8. Pore size distribution test results.

Scheme	Pore Size Distribution (%)				Most Probable Pore Size (nm)	Porosity (%)
	<20 nm	20–50 nm	50–200 nm	>200 nm		
T0	2.3183	3.0358	1.2712	2.1205	40.2718	8.7458
T2	2.3472	3.0868	0.8142	1.2002	26.2991	7.4484
T6	2.2501	2.8872	1.4742	2.1928	45.7543	8.8043
T10	2.2748	2.9015	0.9177	1.8334	36.3927	7.9274

It can be seen from Figure 16 that the porosity curve of the T2 group was lower than that of the T0 group, and it gradually tended to 0 with increasing pore sizes. At the same time, according to the Table 7, the pore structure parameters, such as the average pore size and median pore size, of the T2 group were lower than those of the T0 group. This indicated that the multi-walled carbon nanotubes could effectively change the pore structure of the cement-based materials. By comparing T0 and T6, it can be seen that the addition of PP fibers had an adverse effect on the overall pore structure of the composite. However, compared with the T6 group, the pore structure parameters of the T10 group were improved to some extent, indicating that the incorporation of MWCNTs had a certain refinement effect on the pore size of the PP fiber composite. This is due to the filling effect of MWCNTs and the promotion of cement hydration. The improvement in the pore structure improves the compactness of the composite. It can be seen from Table 8 that the number of harmful pores and multi-harmful pores of the composites was reduced through the optimization and improvements in the pore structures of MWCNTs, and the proportion of harmless pores and less harmful pores was increased. Since PP fiber may bring in trace air when it is added to cement-based materials, bubbles are generated in the composite material, resulting in an increase in harmful pores. At the same time, due to the insufficient bonding ability of the fiber in the composite material with the cement slurry, the porosity of the composite material increases after the fiber is added. The number of harmful pores and multi-harmful pores in the T10 group was lower than that in the T6 group. Therefore, the addition of MWCNTs in the PP fiber composite improved the pore structure to a certain extent, reduced the adverse effects of PP fibers on the internal structure of the composite, and enhanced the compactness of the composite.

4. Conclusions

(1) When the content of MWCNTs is 0.15%, the 28 d flexural strength and splitting tensile strength of the composites are the best, which are 15.7% and 10.7% higher than those of the reference group, respectively. Compared with the reference group, when the content of MWCNTs is 0.1%, the compressive strength of the composite is improved by 9.5%. At the same time, the durability of the composite is also significantly improved. The 90 d dry shrinkage rate and water loss rate are reduced by 12.1% and 4.2%, respectively. After 100 freeze–thaw cycles, the mass loss rate, flexural strength loss rate, and compressive strength loss rate are reduced by 27.9%, 16.3%, and 14.7%, respectively.

(2) When the content of PP fiber is 0.3%, compared with the reference group, the flexural strength and splitting tensile strength of the composites at 28 d can be increased by 19.1% and 12.6%, respectively, and the dry shrinkage rate and water loss rate at 90 d can be decreased by 14.3% and 6.1%, respectively. When the content of PP fiber is 0.2%, the 28 d compressive strength can be increased by 3.9% compared with the reference group. After 100 freeze–thaw cycles, the mass loss rate, compressive strength loss rate, and flexural strength loss rate can be decreased by 25.2%, 17.1%, and 6.5%, respectively, compared with the reference group.

(3) The mechanical properties and durability of the 0.1% MWCNTs/0.2% PP fiber composites are further improved. At the age of 28 d, the flexural strength, compressive strength, and splitting tensile strength can be increased by 18.4%, 9.3%, and 6.7%, respectively. The dry shrinkage rate and water loss rate of the composites at 90 d are 4.6% and 3.3% lower than those of the single-doped MWCNTs, respectively. After 100 freeze–thaw cycles, the mass loss rate, flexural strength loss rate, and compressive strength loss rate are reduced by 9.7%, 4.2%, and 1.1%, respectively, compared with the single-doped MWCNTs.

(4) MWCNTs significantly improve the internal structure of the composite material, promote the production of cement hydration products, and improve the compactness of the composite material. The porosity of the single-doped MWCNTs is 14.8% lower than that of the T0 group. PP fiber has obvious toughening and crack resistance effects on the composites. However, the porosity of the single-doped PP fiber is 0.7% higher than that of the T0 group. When the MWCNTs and PP fibers are mixed, the porosity of the composite material is 10.0% lower than that of the single-doped PP fiber. This shows that adding MWCNTs to the PP fiber composite material can make up for the adverse effects of the PP fiber, enhance the bonding ability between the fiber and the cement matrix, and improve the compactness of the composite material.

Author Contributions: X.N.: Writing—original draft, writing—review and editing, visualization, validation, investigation. Y.C. (Yuanzhao Chen): Conceptualization, project administration, supervision, funding acquisition. Z.L.: Data curation, writing—review and editing. T.G.: Methodology. M.R.: Writing—original draft, investigation, validation. Y.C. (Yanyan Chen): Investigation, validation. All authors have read and agreed to the published version of the manuscript.

Funding: This work was supported by the Planned Project supported by the Science and Technology Innovation Team of universities in Henan province (24IRTSTHN011).

Institutional Review Board Statement: Not applicable.

Data Availability Statement: Some or all of the data, models, or code that support the findings of this study are available from the corresponding author upon reasonable request.

Conflicts of Interest: The authors declare no conflicts of interest.

References

1. Ahmad, W.; Khan, M.; Smarzewski, P. Effect of short fiber reinforcements on fracture performance of cement-based materials: A systematic review approach. *Materials* **2021**, *14*, 1745. [CrossRef] [PubMed]
2. Ma, C.; Fan, F.; Chen, M.; Li, S.; Chen, Y.; Pan, Z.; Liu, R. Preparation of a novel superabsorbent fiber–cement composite and evaluation of its self-healing performance. *Cem. Concr. Compos.* **2022**, *133*, 104713. [CrossRef]
3. Lei, D.; Li, M.; Zhang, P.; Yin, B.; Li, Y.; Song, L.; Zhang, J.; Zhang, X.; Lu, G.; Qian, K.; et al. Research progress on multi-scale anti-cracking and toughening mechanism of cement-based materials. *Acta Silic. Sin.* **2023**, *51*, 2876–2889.
4. Wang, X.; Feng, D. Research progress of nano-carbon/cement-based composites. *Mater. Rep.* **2023**, *37*, 120–135.
5. Balea, A.; Fuente, E.; Monte, M.C.; Blanco, A.; Negro, C. Recycled fibers for sustainable hybrid fiber cement based material: A review. *Materials* **2021**, *14*, 2408. [CrossRef] [PubMed]
6. Balea, A.; Fuente, E.; Blanco, A.; Negro, C. Nanocelluloses: Natural-based materials for fiber-reinforced cement composites. A critical review. *Polymers* **2019**, *11*, 518. [CrossRef]
7. Niu, D.; He, J.; Fu, Q.; Li, D.; Guo, B. The effect of carbon nanotubes on the microstructure and durability of cement-based materials. *Acta Silic. Sin.* **2020**, *48*, 705–717.
8. Pan, R.; Zhang, S.; Zheng, D.; Cui, H.; Li, D. Research progress of multi-dimensional nano-reinforced cement-based composites. *Mater. Introd.* **2017**, *31*, 97–103.
9. Rashad, A.M. Effect of carbon nanotubes (CNTs) on the properties of traditional cementitious materials. *Constr. Build. Mater.* **2017**, *153*, 81–101. [CrossRef]
10. Zhang, J.; Ke, Y.; Zhang, J.; Han, Q.; Dong, B. Cement paste with well-dispersed multi-walled carbon nanotubes: Mechanism and performance. *Constr. Build. Mater.* **2020**, *262*, 120746. [CrossRef]
11. Li, S.; Zhang, Y.; Lin, S.; Yan, J.; Du, S. Effects of nano-SiO_2 coated multi-walled carbon nanotubes on mechanical properties of cement-based composites. *Constr. Build. Mater.* **2021**, *281*, 122577. [CrossRef]
12. Metaxa, Z.S.; Boutsioukou, S.; Amenta, M.; Favvas, E.P.; Kourkoulis, S.K.; Alexopoulos, N.D. Dispersion of Multi-Walled Carbon Nanotubes into White Cement Mortars: The Effect of Concentration and Surfactants. *Nanomaterials* **2022**, *12*, 1031. [CrossRef] [PubMed]
13. Cui, H.; Yang, J.; Lin, J. Research progress of carbon nanotube dispersion technology and carbon nanotube-cement matrix composites. *Mater. Introd.* **2016**, *30*, 91–95.
14. Nadiv, R.; Shtein, M.; Refaeli, M.; Peled, A.; Regev, O. The critical role of nanotube shape in cement composites. *Cem. Concr. Compos.* **2016**, *71*, 166–174. [CrossRef]
15. Du, Y.; Gao, P.; Yang, J.; Shi, F.; Shabaz, M. Experimental analysis of mechanical properties and durability of cement-based composite with carbon nanotube. *Adv. Mater. Sci. Eng.* **2021**, *2021*, 8777613. [CrossRef]
16. Li, W.W.; Ji, W.M.; Wang, Y.C.; Liu, Y.; Shen, R.X.; Xing, F. Investigation on the mechanical properties of a cement-based material containing carbon nanotube under drying and freeze-thaw conditions. *Materials* **2015**, *8*, 8780–8792. [CrossRef]
17. Silvestro, L.; Gleize, P.J.P. Effect of carbon nanotubes on compressive, flexural and tensile strengths of Portland cement-based materials: A systematic literature review. *Constr. Build. Mater.* **2020**, *264*, 120237. [CrossRef]
18. Cerro-Prada, E.; Pacheco-Torres, R.; Varela, F. Effect of multi-walled carbon nanotubes on strength and electrical properties of cement mortar. *Materials* **2020**, *14*, 79. [CrossRef]
19. Qin, R.; Zhou, A.; Yu, Z.; Wang, Q.; Lau, D. Role of carbon nanotube in reinforcing cementitious materials: An experimental and coarse-grained molecular dynamics study. *Cem. Concr. Res.* **2021**, *147*, 106517. [CrossRef]
20. Gao, F.; Tian, W.; Wang, Z.; Wang, F. Effect of diameter of multi-walled carbon nanotubes on mechanical properties and microstructure of the cement-based materials. *Constr. Build. Mater.* **2020**, *260*, 120452. [CrossRef]
21. Ramezani, M.; Kim, Y.H.; Sun, Z. Mechanical properties of carbon-nanotube-reinforced cementitious materials: Database and statistical analysis. *Mag. Concr. Res.* **2020**, *72*, 1047–1071. [CrossRef]
22. Li, Y.; Li, H.; Shen, J. The study of effect of carbon nanotubes on the compressive strength of cement-based materials based on machine learning. *Constr. Build. Mater.* **2022**, *358*, 129435. [CrossRef]
23. Sarvandani, M.M.; Mahdikhani, M.; Aghabarati, H.; Fatmehsari, M.H. Effect of functionalized multi-walled carbon nanotubes on mechanical properties and durability of cement mortars. *J. Build. Eng.* **2021**, *41*, 102407. [CrossRef]
24. Chen, J.; Akono, A.T. Influence of multi-walled carbon nanotubes on the hydration products of ordinary Portland cement paste. *Cem. Concr. Res.* **2020**, *137*, 106197. [CrossRef]
25. Wang, B.; Pang, B. Properties improvement of multiwall carbon nanotubes-reinforced cement-based composites. *J. Compos. Mater.* **2020**, *54*, 2379–2387. [CrossRef]
26. Naqi, A.; Abbas, N.; Zahra, N.; Hussain, A.; Shabbir, S.Q. Effect of multi-walled carbon nanotubes (MWCNTs) on the strength development of cementitious materials. *J. Mater. Res. Technol.* **2019**, *8*, 1203–1211. [CrossRef]
27. Qing, L.; Yang, Z.; Mu, R.; Zhang, J.; Ji, X. Mesoscopic numerical simulation of fracture of aligned steel fiber reinforced cementitious composites. *J. Build. Mater.* **2023**, *26*, 111–121.
28. Yang, K.; Tang, Z.; Cheng, Z.; Zhao, H.; Feng, R.; Long, G. Mechanical properties of ultra-high strength cement-based materials (UHSC) incorporating metal powders and steel fibers. *Constr. Build. Mater.* **2022**, *318*, 125926. [CrossRef]

29. Bai, G.; Wang, L.; Wang, F.; Cheng, X. Preparation and mechanical properties testing of 3D printed UHPC. *Mater. Introd.* **2021**, *35*, 12063–12069.
30. Cao, K.; Liu, G.; Li, H.; Huang, Z. Mechanical properties and microstructures of Steel-basalt hybrid fibers reinforced Cement-based composites exposed to high temperatures. *Constr. Build. Mater.* **2022**, *341*, 127730. [CrossRef]
31. Monaldo, E.; Nerilli, F.; Vairo, G. Basalt-based fiber-reinforced materials and structural applications in civil engineering. *Compos. Struct.* **2019**, *214*, 246–263. [CrossRef]
32. Chen, H.; Wang, P.; Pan, J.; Lawi, A.S.; Zhu, Y. Effect of alkali-resistant glass fiber and silica fume on mechanical and shrinkage properties of cement-based mortars. *Constr. Build. Mater.* **2021**, *307*, 125054. [CrossRef]
33. Zhang, W.; Zhang, Y.; Wu, Z.; Liu, N.; Yuan, D. Optimization design and performance of glass fiber reinforced cementitious materials. *Mater. Rev.* **2019**, *33*, 2331–2336.
34. Yao, X.; Han, Y.; Shen, L.; Zhu, D.; Cao, M. Multi scale method for thermal conductivity of polypropylene fiber-reinforced cement-based composites after high temperature. *J. Compos. Mater.* **2021**, *38*, 3531–3542.
35. Jin, H.; Li, F.; He, X.; Wang, J.; Hu, D.; Hu, Z. Study on the frost resistance of polypropylene fiber cement-based composites. *Mater. Introd.* **2020**, *34*, 8071–8076+8082.
36. Liang, N.; Zhou, K.; Lan, F.; Liu, X.; Deng, Z. Experimental study on bearing capacity of basalt-polypropylene coarse fiber reinforced concrete pipe. *Mater. Bull.* **2023**, *37*, 104–111.
37. Li, X.; Xu, J.; Xie, W.; Liu, W.; Yu, F.; Sun, X.; Luo, Y.; Chen, Q. Cost analysis of polypropylene fiber reinforced cement-based building materials based on full lifecycle cost. *J. Fujian Norm. Univ. (Nat. Sci. Ed.)* **2023**, *39*, 25–31.
38. Małek, M.; Jackowski, M.; Łasica, W.; Kadela, M. Characteristics of recycled polypropylene fibers as an addition to concrete fabrication based on portland cement. *Materials* **2020**, *13*, 1827. [CrossRef]
39. Li, J.J.; Niu, J.G.; Wan, C.J.; Jin, B.; Yin, Y.L. Investigation on mechanical properties and microstructure of high performance polypropylene fiber reinforced lightweight aggregate concrete. *Constr. Build. Mater.* **2016**, *118*, 27–35. [CrossRef]
40. Ahmed, T.W.; Ali, A.A.M.; Zidan, R.S. Properties of high strength polypropylene fiber concrete containing recycled aggregate. *Constr. Build. Mater.* **2020**, *241*, 118010. [CrossRef]
41. Al-Katib, H.A.A.; Alkhudery, H.H.; Al-Tameemi, H.A. Behavior of polypropylene fibers reinforced concrete modified with high performance cement. *Int. J. Civ. Eng. Technol. (IJCIET)* **2018**, *9*, 1066–1074.
42. Alamshahi, V.; Taeb, A.; Ghaffarzadeh, R.; Rezaee, M.A. Effect of composition and length of PP and polyester fibres on mechanical properties of cement based composites. *Constr. Build. Mater.* **2012**, *36*, 534–537. [CrossRef]
43. Liang, N.; Mao, J.; You, X.; Liu, X.; Zhou, K. Multiscale bending fatigue life test and numerical simulation of polypropylene fiber reinforced concrete. *Mater. Rev.* 2023, pp. 1–17. Available online: http://kns.cnki.net/kcms/detail/50.1078.tb.20230414.1003.002.html (accessed on 8 December 2023).
44. Teixeira, R.S.; Santos, S.F.; Christoforo, A.L.; Savastano Jr, H.; Lahr, F.A.R. Extrudability of cement-based composites reinforced with curauá (*Ananas erectifolius*) or polypropylene fibers. *Constr. Build. Mater.* **2019**, *205*, 97–110. [CrossRef]
45. Li, C.; Chen, X.; Zhang, X.; Wang, H. Mechanical properties of interface bonding between synthetic coarse polypropylene fibers and cement mortar. *J. Compos. Mater.* **2023**, *40*, 2427–2440.
46. Wu, B.; Qiu, J. Enhancing the hydrophobic PP fiber/cement matrix interface by coating nano-AlOOH to the fiber surface in a facile method. *Cem. Concr. Compos.* **2022**, *125*, 104297. [CrossRef]
47. Ahmadi, K.; Mousavi, S.S.; Dehestani, M. Influence of nano-coated micro steel fibers on mechanical and self-healing properties of 3D printable concrete using graphene oxide and polyvinyl alcohol. *J. Adhes. Sci. Technol.* **2023**, 1–22. [CrossRef]
48. Feng, G.; Wang, X.; Zhang, D.; Cao, H.; Qian, K.; Xiao, X. A comparative study on mechanical properties of surface modified polypropylene (PP) fabric reinforced concrete composites. *Constr. Build. Mater.* **2017**, *157*, 372–381. [CrossRef]
49. Jia, E.; Mou, H.; Liu, Z.; Wang, J.; Zeng, L.; Yang, X.; Liu, P. Surface hydrophilic modification of polypropylene fibers and their application in fiber-reinforced cement-based materials. *J. Macromol. Sci. Part B* **2020**, *60*, 286–298. [CrossRef]
50. Feng, J.; Yang, F.; Qian, S. Improving the bond between polypropylene fiber and cement matrix by nano calcium carbonate modification. *Constr. Build. Mater.* **2021**, *269*, 121249. [CrossRef]
51. Wu, Z. Exploration of the Recent Development Direction of Concrete Science and Technology. *Acta Sin. Sin.* **1979**, 262–270.

polymers

MDPI

Article

PLA-Based Composite Panels Prepared via Multi-Material Fused Filament Fabrication and Associated Investigation of Process Parameters on Flexural Properties of the Fabricated Composite

Zhaogui Wang [1,*], Lihan Wang [1], Feng Tang [2] and Chengyang Shen [2]

[1] Department of Mechanical Engineering, Naval Architecture and Ocean Engineering College, Dalian Maritime University, Dalian 116026, China
[2] Houston International Institute, Dalian Maritime University, Dalian 116026, China
* Correspondence: zhaogui_wang@dlmu.edu.cn

Abstract: This study prepares composite panels with three Polylactic acid (PLA)-based materials via the multi-material fused filament fabrication method. The influences of four processing parameters on the mechanical properties of 3D-printed samples are investigated employing the Taguchi method. These parameters include the relative volume ratio, material printing order, filling pattern, and filling density. A "larger is better" signal-to-noise analysis is performed to identify the optimal combination of printing parameters that yield maximum bending strength and bending modulus of elasticity. The results reveal that the optimal combination of printing parameters that maximizes the bending strength involves a volume ratio of 1:1:2, a material sequence of PLA/foam-agent-modified eco-friendly PLA (ePLA-LW)/glass fiber-reinforced eco-friendly PLA (ePLA-GF), a Gyroid filling pattern, and a filling density of 80%, and the optimal combination of printing parameters for maximum bending modulus involves a volume ratio of 1:2:1 with a material sequence of PLA/ePLA-LW/ePLA-GF, a Grid filling pattern, and 80% filling density. The Taguchi prediction method is utilized to determine an optimal combination of processing parameters for achieving optimal flexural performances, and predicted outcomes are validated through related experiments. The experimental values of strength and modulus are 43.91 MPa and 1.23 GPa, respectively, both very close to the predicted values of 46.87 MPa and 1.2 GPa for strength and modulus. The Taguchi experiments indicate that the material sequence is the most crucial factor influencing the flexural strength of the composite panels. The experiment result demonstrates that the flexural strength and modulus of the first material sequence are 67.72 MPa and 1.53 GPa, while the flexural strength and modulus of the third material sequence are reduced to 27.09 MPa and 0.72 GPa, respectively, only 42% and 47% of the first material sequence. The above findings provide an important reference for improving the performance of multi-material 3D-printed products.

Keywords: multi-material fused filament fabrication; PLA-based composites; sandwich panels; flexural properties; Taguchi method

1. Introduction

Additive manufacturing (also known as 3D printing) creates complex geometries by depositing materials in a layer-by-layer manner [1]. Unlike traditional subtractive methods, Additive Manufacturing (AM) reduces material and time costs in parts and tooling rapid-prototyping [2], making it a widely used technique across various industries [3–5]. Among the AM approaches, material extrusion AM is one of the most popular methods due to its low cost in hardware and high ability to deal with a wide range of polymeric materials. Fused Filament Fabrication (FFF) is a common approach within material extrusion AM, where a continuous thermoplastic filament is melted via a heated nozzle and then deposited onto a pre-heating print-bed to form 3D objects. Engineering plastics, including Acrylonitrile Butadiene Styrene (ABS), Polylactic Acid (PLA), Polycarbonate (PC), and others, can be easily processed by FFF systems. Among these, PLA has garnered significant attention due to its high degree of biodegradability, which is environmentally friendly [6,7]. Nevertheless, virgin PLA polymers exhibit relatively low mechanical properties as compared to other competitive plastics products (e.g., ABS, PC) due to their linear molecular structures. To overcome this limitation, fibrous reinforcements are incorporated into the PLA matrix to produce PLA composites with superior mechanical performances [7–9]. Dong et al. [10] investigated the effects of coconut shell fiber content and alkali treatment on the mechanical, thermal, and biodegradable properties of PLA/coconut shell fiber bio-composites fabricated by the compression molding process. Their study found that the treated fiber composites had improved the tensile and flexural moduli. Although natural fiber-reinforced PLA materials exhibit higher environmental friendliness, natural fibers are less strong and more hydrophilic, which can accelerate the degradation of PLA materials, thereby hindering their practical application in engineering. In contrast, synthetic fibers such as carbon fibers and glass fibers [9,11] possess superior mechanical properties. Ruz-Cruz et al. [12] fabricated cellulose microfibers (MFCs)-reinforced PLA composites and a blend of MFCs and cellulose nanocrystals (CNCs)-reinforced PLA composites. The thermomechanical properties of these multi-scale PLA biocomposites were investigated using thermogravimetry (TGA), differential scanning calorimetry (DSC), bending mechanics, and dynamic mechanics tests (DMA), revealing that replacing MFCs with CNCs within the range of 1–5% effectively enhances the thermal stability of the materials, improves the crystallization of PLA materials, and increases the bending mechanical properties at room temperature by ~40%. The crystallinity of a polymer affects its physical properties, such as hardness, modulus, tensile strength, stiffness, and melting point, and for PLA, to some extent, its degradation rate [13]. Glass fibers are the most extensively used fibers for reinforcing polymers due to their excellent mechanical properties, low cost, and high heat resistance [11,14]. Chicos et al. [15] evaluated the impact of 3D printed packing density on the mechanical and thermal properties of short glass fiber-reinforced PLA specimens. Their findings indicated that the specimens exhibited optimal mechanical properties at 100% filling density, while those with 50% and 75% filling density demonstrated higher toughness compared to their 100% filling density counterparts. In addition to fiber fillers, several studies also reported that the polymer additives were effective in modifying the material properties of PLA [16–20]. The addition of foaming agents resulted in the production of lightweight PLA materials with high porosity and a high strength-to-weight ratio. Kanani et al. [21] prepared foamed-PLA using FFF systems tensile, where they observed that the foaming expansion was directly related to the extrusion temperature and nozzle moving speed. Damanpack et al. [22] conducted a parallel study and reported that the microscale material bonding of FFF-deposited foamed PLA materials was significantly influenced by the printing temperature.

The aforementioned observations indicate that the reinforced PLA materials exhibit enhanced mechanical performances when compared to the virgin polymer. Furthermore, in light of the recent advanced multi-material FFF systems, it is now feasible to 3D print composite panels utilizing various feedstock materials, thereby incorporating multiple functionally reinforced composites into a single entity (e.g., [23–26]). Kamaal et al. [27] aimed at the mechanical properties of 3D-printed carbon fiber-reinforced polylactic acid composite (CF-PLA) in the fused deposition modeling (FDM) method and attained the best parameter set that provides the maximum strength using the minimum material. In addition, regarding PLA-TPU (Thermoplastic polyurethanes) and PLA-ABS, Yavas et al. [28] explored the microstructural and mixed-mode fracture characteristics of the PLA-TPU interfaces and improved the energy absorption capacity. Rasheed et al. [29] utilized Taguchi orthogonal optimization to study the influence of distinct FDM process parameters on the mechanical properties, and they determined the optimal parameters for better tensile strength for a bi-layered composite of PLA-ABS. Another set of studies on PLA-ABS introduced the idea of laying down a high-toughness material on the 3D-printed carbon fiber-reinforced polymer composite sheet. Ahmed et al. [30] made a hybrid composite of laminar structures and printed the CF-PLA/ABS hybrid laminar composite. The impact toughness was optimized by adjusting the 3D printing parameters, and the fracture surface was characterized. Recently, short-carbon fiber-filled PLA and toughness-enhanced-additive-modified PLA were integrated into layered composite specimens. The resulting integrated specimens displayed a combined and adjustable tensile strength contingent on the relative fraction of the two constituents [31]. To further explore the functional complexity of the composite panels fabricated using multi-material FFF systems, composite sandwich panels composed of three types of PLA-based ingredients are prepared, including virgin PLA, rigid glass fiber reinforced PLA (PLA-GF), and foam-agent-modified PLA (also referred to as lightweight PLA, i.e., PLA-LW). The effects of various processing parameters on the composite panel preparation are evaluated, such as the constituent material stacking-order, relative fraction, printing infill pattern, and infill density. The Design of Experiments The taguchi method is employed to reduce the extensive number of experiments. The flexural properties of the specimens are utilized as representatives of material performance. By comparing the measured data, the optimal processing parameters for achieving superior material properties are recommended.

The presented study highlights that the sequence of material printing is a crucial determinant of the bending performance of sandwich specimens, emphasizing its significance for future studies of multi-material 3D printing. Regarding the scope of this study, the manufacturing approach offers a novel perspective for the design and production of protective materials. These sandwich panels can be employed in wearable protective equipment such as helmets, sports knee pads, and medical braces. Our study also fills the gap in the current research on the preparation parameters of sandwich composite panels made of PLA-based composites.

2. Materials and Methods

The multi-material FFF system applied for the specimen fabrication is also presented. Furthermore, the procedure of our Design of Experiment (DOE) with the Taguchi method is discussed.

2.1. Materials

In this study, PLA and its composite materials, ePLA-LW and ePLA-GF, are employed (note that ePLA refers to an eco-friendly PLA resin that is designed to enhance the bio-degradabilities of PLA composites). The feedstock filaments are supplied by Esun (Shenzhen Esun Industrial Co., Ltd., Shenzhen, China). As the selected materials are PLA-based, we would expect that the multi-material 3D-printed specimens exhibit a high quality of interlayer adhesion. Therefore, in the results discussion, we could ignore the effect of interlayer bonding issues. Finally, the material properties provided by the supplier are given in Table 1. Differential scanning calorimetry (DSC-500C.Jiezhun Instrument Equipment Co., Ltd., Shanghai, China) is employed to test three materials, and the DSC curves for these materials are presented in Figure 1. Compared to pure PLA, the glass transition temperature, cold crystallization temperature, and melting temperature of ePLA-GF have all decreased. The cold crystallization and melting peaks of the ePLA-LW material also diminish. This study indicates that the crystallinity of PLA material is influenced by chemical modification during the foaming process [32]. This may be attributed to the addition of the foaming agent, which hinders the development of the crystalline structure in PLA.

Notably, ePLA-GF filament is reinforced with 16 wt.% short glass fiber, resulting in enhanced stiffness and strength. ePLA-LW filament is modified by adding a thermally sensitive foaming agent, i.e., allowing for increased porosity with elevated nozzle temperatures (the recommended temperature range is from 210 to 270 °C). From pre-experiments on the printing temperature of ePLA-LW, it was found that the temperature of 230 °C with an extrusion rate of 75% exhibits the best expected performance. The detailed results from these preliminary experiments are provided in Appendix B for reference. The resulting foamed PLA exhibits lower material density and rigidity, but higher toughness. Finally, a multi-material FFF system is employed, utilizing the constituent materials in a prescribed layer-stacking formation. It is anticipated that the designed composite panel will exhibit favorable material properties.

Table 1. Material properties of constituent materials from the supplier.

Material	PLA	ePLA-LW	ePLA-GF
Density (g/cm^3)	1.2	1.2	1.31
Tensile Strength (MPa)	72	32.2	59.27
Flexural Strength (MPa)	90	41.31	85
Elongation at Break	11.8	68.9	7.99
Flexural Modulus (MPa)	1915	1701	4414.89
Heat Distortion Temp. (°C)	53	53	56
Melt Flow Index (g/10 min)	3.5 (190 °C/2.16 kg)	8.1 (190 °C/2.16 kg)	6.36 (190 °C/2.16 kg)

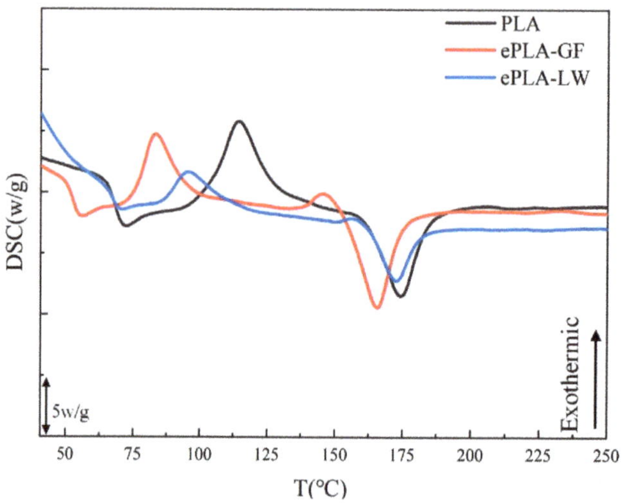

Figure 1. Three materials: DSC Curves.

2.2. Three-Point Bending Specimen Preparation with Multi-Mat FFF

The three-point bending test specimens are designed in accordance with the ASTM D790 standard [33]. SolidWorks 2020 (Dassault Systèmes Corp., Vélizy-Villacoublay, France) is used for generating the STL digital file. Cura 4.13.0 (Ultimaker Corp., Utrecht, The Netherlands) slicing software is employed to configure the printing parameters and generate the corresponding G-code file for the fabrication process. As previously mentioned, three distinct constituent materials are incorporated within the specimen. Given the layer-by-layer manufacturing nature of FFF, the specimen is ultimately fabricated as a sandwich panel. A GEEETECH A10T (Shenzhen Geeetech Technology Co., Ltd., Shenzhen, China) 3-in/1-out multi-material FFF 3D printer is utilized to print sandwich specimens. The 3D printer featured three extruders, enabling control over up to three material filaments extruded from a single nozzle. It should be noted that the original G-code file generated in Cura 4.13.0 does not guarantee successful printing of the sandwich structure composed of three materials by the GEEETECH A10T. To address this issue, the number of layers, printing temperature, and extrusion rate for each material are modified in the code according to the printing parameters of the three materials and then imported into the 3D printer for printing.

To properly establish the printing parameters of the constituent materials, pre-experiments are conducted. To ensure optimal conditions for the fabrication of the bending sandwich specimen, the filaments are uniformly dried at 50 °C for 5 h. This pre-drying process helps establish an ideal condition for the filaments prior to their use in the printing process. A series of pre-tests are performed on specimens with varying parameters, specifically noting that the heat-foaming ePLA-LW exhibits the best mechanical properties prior to foaming (i.e., nozzle temperature at ~210 °C). Conversely, the density of ePLA-LW decreases as the degree of foaming increases. In our case, the nozzle temperature of the ePLA-LW is set at 230 °C with an extrusion rate of 75%, resulting in a relatively high strength-to-weight ratio property. The general printing parameters are finalized as appearing in Table 2, aiming to achieve consistent and superior flexural strength.

Table 2. General printing parameters are applied for specimen preparation.

Material	Nozzle Temp. (°C)	Relative Feeding Rate (%)	Number of Outline Shell	Layer Thickness (mm)	Printing Speed (mm/s)	Print-Bed Temp. (°C)
PLA	210	100	2	0.2	50	60
ePLA-LW	230	75	2	0.2	50	60
ePLA-GF	230	100	2	0.2	50	60

2.3. Design of Experiments

The influence of four essential processing parameters on the flexural properties of printed composite panels is examined in this study. These parameters include constituent material volume fraction ratio, material stacking order, infill pattern, and filling density. For open-ended FFF 3D printing, previous studies often employed the Design of Experiments (DoE) Taguchi method to optimize the printing parameters when dealing with varying materials (e.g., [34–39]). Consequently, the Taguchi method was also employed to efficiently perform the experimental tests. Accordingly, a L_9 orthogonal array was generated to test the effects of the four parameters at three levels, reducing the originally large number of tests to nine trials in total [35]. As shown in Table 3, we established three-level values for each parameter factor. The volume ratio indicated the volume of each material in the sandwich specimen, which is divided into four equal parts according to the deposition direction and then partitioned into three sections in the form of Figure 2a–c. The material stacking order refers to the sequence of the three materials in the deposition direction, and in this study, three distinct orders are adopted, as demonstrated in Figure 2d–f. The printing pattern employed in this study corresponds to the internal filling structure pattern within the specimen. Each sandwich specimen is filled with a specific infill pattern throughout the entire sample. Within Cura 4.13.0 software, three distinct filling patterns are chosen: Gyroid, Grid, and 45°/135° Lines (as depicted in Figure 3). Additionally, a brim configuration is implemented in Cura 4.13.0 (highlighted in the blue section of Figure 3) to enhance material adhesion to the printing platform. Note that the 45°/135° Lines filling pattern represents an overlap of lines between different layers, while the Grid filling pattern involves an overlap within the same layer. To preclude material intrusion caused by specimen deformation, top-bottom layers between different materials are incorporated to separate them. This design increases the contact area and facilitates intermolecular diffusion, thereby bolstering interlayer bonding strength and effectively ameliorating the mechanical properties of the specimen. Furthermore, as illustrated in Figure 4, a printed sandwich specimen example is demonstrated where all three materials are employed [25].

Table 3. Design of experiments with an L9 orthogonal array.

Exp. Number	Volume Ratio	Material Sequence	Filling Pattern	Filling Density
1	2:1:1	1	Gyroid	40%
2	2:1:1	2	Lines	60%
3	2:1:1	3	Grid	80%
4	1:2:1	1	Lines	80%
5	1:2:1	2	Grid	40%
6	1:2:1	3	Gyroid	60%
7	1:1:2	1	Grid	60%
8	1:1:2	2	Gyroid	80%
9	1:1:2	3	Lines	40%

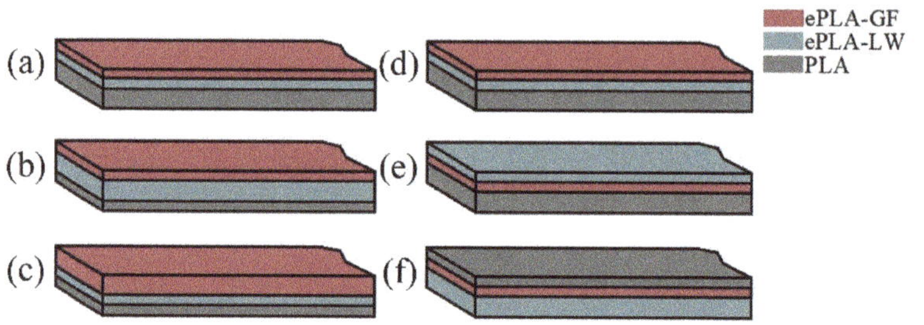

Figure 2. Diagram of Volume Ratio: (**a**) 2:1:1, (**b**) 1:2:1, (**c**) 1:1:2; The schematic diagram refers to the ordering of materials based on a sample volume ratio of 2:1:1. (**d**) PLA/ePLA-LW/ePLA-GF; (**e**) PLA/ePLA-GF/ePLA-LW; (**f**) PLA/ePLA-LW/ePLA-GF/PLA.

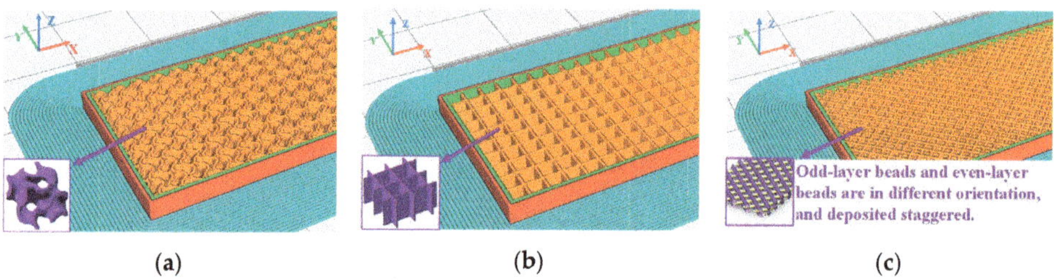

Figure 3. Filling pattern when filling density is 40%: (**a**) Gyroid, (**b**) Grid, (**c**) 45°/135° Lines (Different colors indicate the overlap of 45°/135° straight lines between layers).

Figure 4. Example of the printed sandwich specimen (volume fraction ratio 1:2:1, material stacking order: PLA/ePLA-LW/ePLA-GF).

3. Results

The Taguchi method was employed in this study to efficiently identify the optimal combination of processing parameters by reducing the number of trials. Initially, ANOVA analyses were performed on our measured flexural properties. The results revealed that all the investigated parameters (cf. Table 3) highly contribute to the experimental results in a similar way, making the differences between the parameters indistinguishable (to be concise, the ANOVA results were provided in the Appendix A). Alternatively, the Taguchi method categorizes the experimental factors into controllable and noisy components. Controllable factors are parameters or variables that can be controlled and remain constant following selection. Noisy factors refer to variables that are uncontrollable under normal conditions, i.e., random variables. The Signal-to-Noise ratio (S/N) is the resistance to noise interference. A higher S/N ratio indicates a control factor setting that minimizes the impact of the noise factor, resulting in the specimen having more consistent and stable performance. The S/N ratio quantifies the effect of multiple parameters on the experimental output values by calculating the signal-to-noise value to identify the optimal parameter settings [40]. In Taguchi experiments, there are three types of expectations: expecting small (smaller is better), expecting large (larger is better), and expecting eye type (nominal is better) [38]. Herein, the "larger is better" analysis for optimization of stiffness and strength is adopted. The equations for "larger is better" can be written as

$$S/N = -10\log\left[\frac{1}{n}\sum_{i=1}^{n}\frac{1}{y_i^2}\right] \tag{1}$$

where y_i refers to the experimental results of the i-th test, and n is the total number of the tests. Herein, five specimens are prepared for each experimental group to undergo bending testing. The mean value of the orthogonal experiment test is derived by eliminating the maximum and minimum values from the results of each experimental group, and thus $n = 1$. Additionally, taking the average of the other three specimens as the experimental values, the experimentally measured flexural properties of the FFF-produced sandwich specimens and the S/N analysis are presented.

3.1. S/N Analyses on Measured Flexural Properties

According to the orthogonal experimental sequence presented in Table 3, each group of experimental samples undergoes a three-point bending test utilizing a GTM universal material testing machine capable of exerting a maximum force of 10 KN (Xie-Qiang Instruments Manufacturer Corp., Shanghai, China). The bending span measures 50 mm, and a loading speed of 2 mm/min is applied, as illustrated in Figure 5 The bending strength is analyzed employing the "larger is better" signal-to-noise ratio analysis in Minitab 19 (Minitab, LLC, State College, PA, USA). The experimental values and signal-to-noise ratio values are provided in Table 4. The average value of the experimental results of the three samples within each group is computed. Table 5 displays the signal-to-noise ratio values at various levels of each parameter. Delta signifies the range of signal-to-noise ratios at different levels of the same factor. The width of the range indicates the extent of impact distinct levels of alteration have on the empirical outcomes. Considering the size of the range is crucial in determining the level of impact, as a broader range indicates a greater degree of influence. Consequently, evaluating the range size is crucial when determining the magnitude of change required to achieve a significant effect. Referring to the data presented in Table 5, it is evident that the material sequence yields the most pronounced influence on the bending strength, whose delta reaches 7.15, followed by the filling density with a delta of 3.29. In contrast, the filling pattern exhibits the least impact on the bending strength among the processing parameters evaluated in this study, with the smallest delta of 0.96. Drawing conclusions from the main effect plots of the signal-to-noise ratio as depicted in Figure 6, the combination of processing parameters leading to the maximum bending strength consists of a material volume ratio of 1:1:2, a material sequence of PLA/ePLA-LW/ePLA-GF, a Gyroid filling pattern, and a filling density of 80%.

Figure 5. Three-point bending test.

Table 4. Bending strength and S/N ratio values.

Number	Volume Ratio	Material Sequence	Filling Pattern	Filling Density	Bending Strength (MPa)	Standard Deviation	S/N (dB)
1	2:1:1	1	Gyroid	40%	30.44	0.45	29.67
2	2:1:1	2	Lines	60%	21.49	0.14	26.65
3	2:1:1	3	Grid	80%	17.47	0.35	24.84
4	1:2:1	1	Lines	80%	44.73	0.39	33.01
5	1:2:1	2	Grid	40%	19.31	0.29	25.71
6	1:2:1	3	Gyroid	60%	18.58	0.93	25.36
7	1:1:2	1	Grid	60%	39.65	0.15	31.96
8	1:1:2	2	Gyroid	80%	33.30	2.93	30.38
9	1:1:2	3	Lines	40%	14.09	0.16	22.98

Table 5. The S/N ratio response value to bending strength.

Level	Volume Ratio	Material Sequence	Filling Pattern	Filling Density
1	27.05	31.55	28.47	26.12
2	28.03	27.58	27.54	27.99
3	28.44	24.39	27.51	29.41
Delta	1.39	7.15	0.96	3.29
Ranking	3	1	4	2

Moreover, the signal-to-noise ratio analysis is conducted on the flexural modulus of the composite penal samples. Table 6 shows the nine experimental results and signal-to-noise ratio values listed in the orthogonal experiment. In Table 7, we further provide the response values, where the degree of influence of different levels between individual factors on the flexural modulus can be clearly seen. The largest range of signal-to-noise ratios was observed under the factor of material sequence at three levels, up to 6.59. This indicates variations in material sequence can significantly affect the bulk stiffness of the composite panels. It is noted that the material sequence is a crucial factor affecting bending strength, followed by filling density (The range is 3.44). However, unlike the trend of strength data, the volume ratio demonstrated minimal impact on the results of elastic modulus, whose delta are only 1.28. Overall, our findings indicate that the material sequence parameter exerts the most significant influence on both the strength and stiffness of samples in this experimental model. Furthermore, filling density also significantly impacts the flexural performance of samples. By analyzing the signal-to-noise ratio at each level, an optimal combination can be identified via Figure 7, which maximizes the stiffness of the samples. Ultimately, a favorable set of processing parameters is suggested as follows: The volume ratio is 1:2:1, the material order is PLA/ePLA-LW/ePLA-GF, the fill pattern is Grid, and the fill density is 80%.

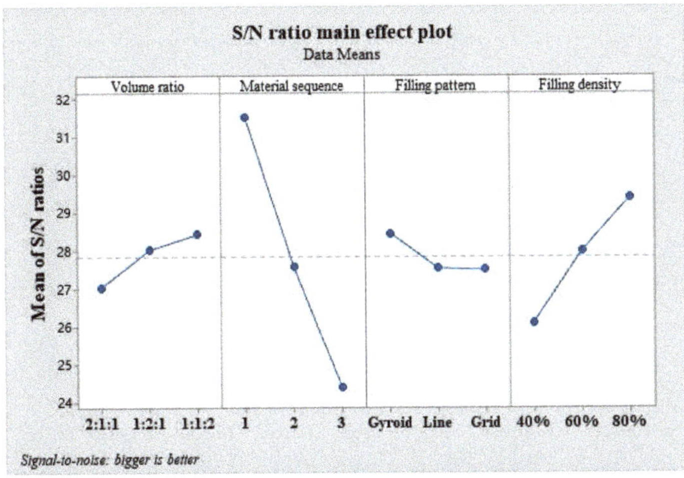

Figure 6. Main effects plot of the levels of each factor on the bending strength.

Table 6. Bending modulus and S/N ratio values.

Number	Volume Ratio	Material Sequence	Filling Pattern	Filling Density	Flexural Modulus (AVG) (GPa)	Standard Deviation ($\times 10^{-2}$)	S/N (dB)
1	2:1:1	1	Gyroid	40%	0.75	0.37	−2.48
2	2:1:1	2	Lines	60%	0.48	0.63	−6.38
3	2:1:1	3	Grid	80%	0.54	0.58	−5.40
4	1:2:1	1	Lines	80%	1.14	2.40	1.13
5	1:2:1	2	Grid	40%	0.52	1.13	−5.62
6	1:2:1	3	Gyroid	60%	0.51	0.09	−5.93
7	1:1:2	1	Grid	60%	1.10	1.02	0.81
8	1:1:2	2	Gyroid	80%	0.75	1.72	−2.49
9	1:1:2	3	Lines	40%	0.36	0.42	−8.97

Table 7. The S/N ratio response value to bending modulus.

Level	Volume Ratio	Material Sequence	Filling Pattern	Filling Density
1	−4.75	−0.18	−3.63	−5.69
2	−3.47	−4.83	−4.74	−3.83
3	−3.55	−6.77	−3.41	−2.25
Delta	1.28	6.59	1.34	3.44
Ranking	4	1	3	2

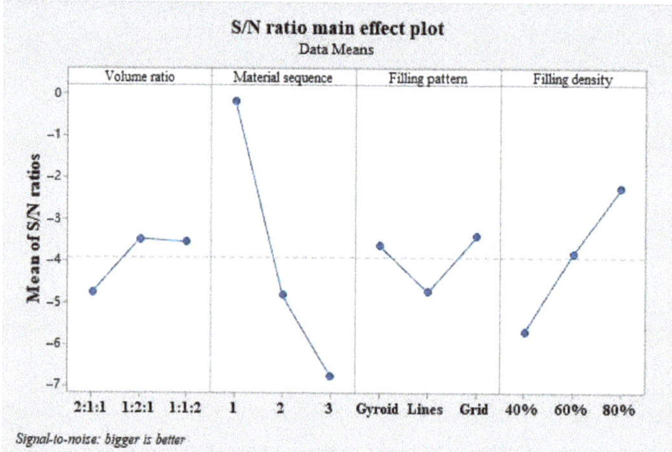

Figure 7. The main effects plot shows the levels of each factor on the bending modulus.

Furthermore, optimizing the combination of processing parameters beyond the selected levels, as previously studied, can lead to enhanced flexural strength and modulus properties, as predicted by the Taguchi method. We validate the predicted values through relevant experiments, as illustrated in Figure 8. In each experiment, three-panel samples are prepared to calculate the mean value and standard deviation of the measured properties. The small values of the standard deviation indicate the high stability of the experimental data. By comparing the validation experimental measurements with the predicted values (as in [39,41]), it is observed that this experimental model aligns with the prediction of the Taguchi method. The experimental values are 1.23 GPa and 43.91 MPa for modulus and strength, which approximate the predicted values of 1.2 GPa and 46.87 MPa, respectively. The Taguchi method can predict an additional 72 experimental data points outside of the 9 groups of experiments presented in the orthogonal table of this paper, thereby expanding the optimization area of the processing parameters.

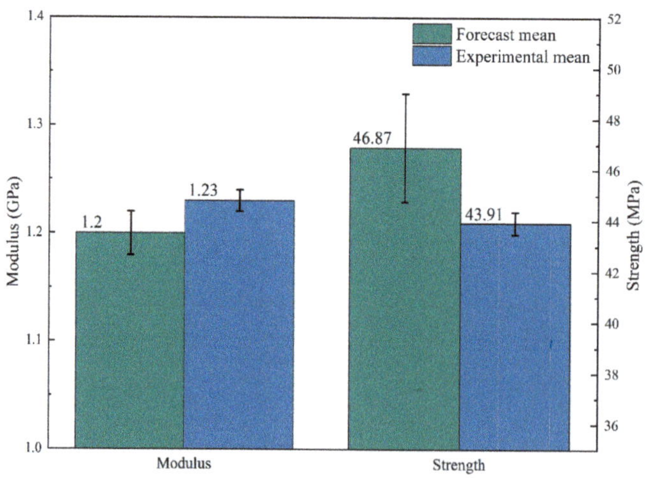

Figure 8. Taguchi comparison of predicted values and experimental values.

3.2. Meso-Structural Analysis of Failure Mechanisms

Based on the results of Section 3.1, we further analyzed the fracture mechanism of the sandwich structure under bending load. Using the Dino-lite (AnMo Electronics Corp., Taiwan, China) handheld microscope, the morphology of a single sample crack in each group of experiments in Table 3 is observed from the Side view and Bird view perspectives, as shown in Figures 9 and 10. The Keyence VHX-7000 optical microscope was employed to observe the material interfaces between PLA and ePLA-GF, PLA and ePLA-LW, as well as ePLA-GF and ePLA-LW, as shown in Figure 11.

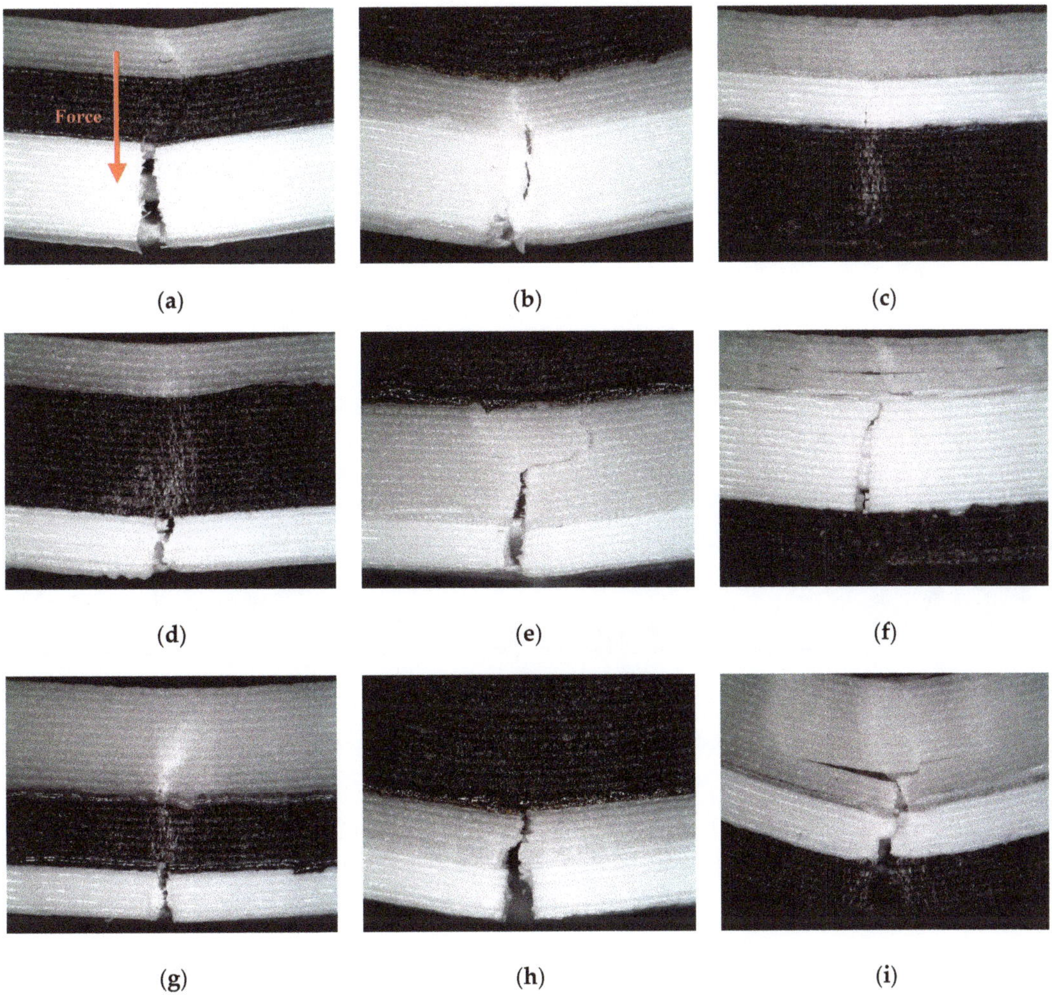

Figure 9. Crack propagation from a side view (The loading direction for all specimens in the figures is consistent with (**a**)). Note that numbers 1–9 correspond to the experimental cases appearing in Table 3. (**a**) Number 1; (**b**) Number 2; (**c**) Number 3; (**d**) Number 4; (**e**) Number 5; (**f**) Number 6; (**g**) Number 7; (**h**) Number 8; (**i**) Number 9.

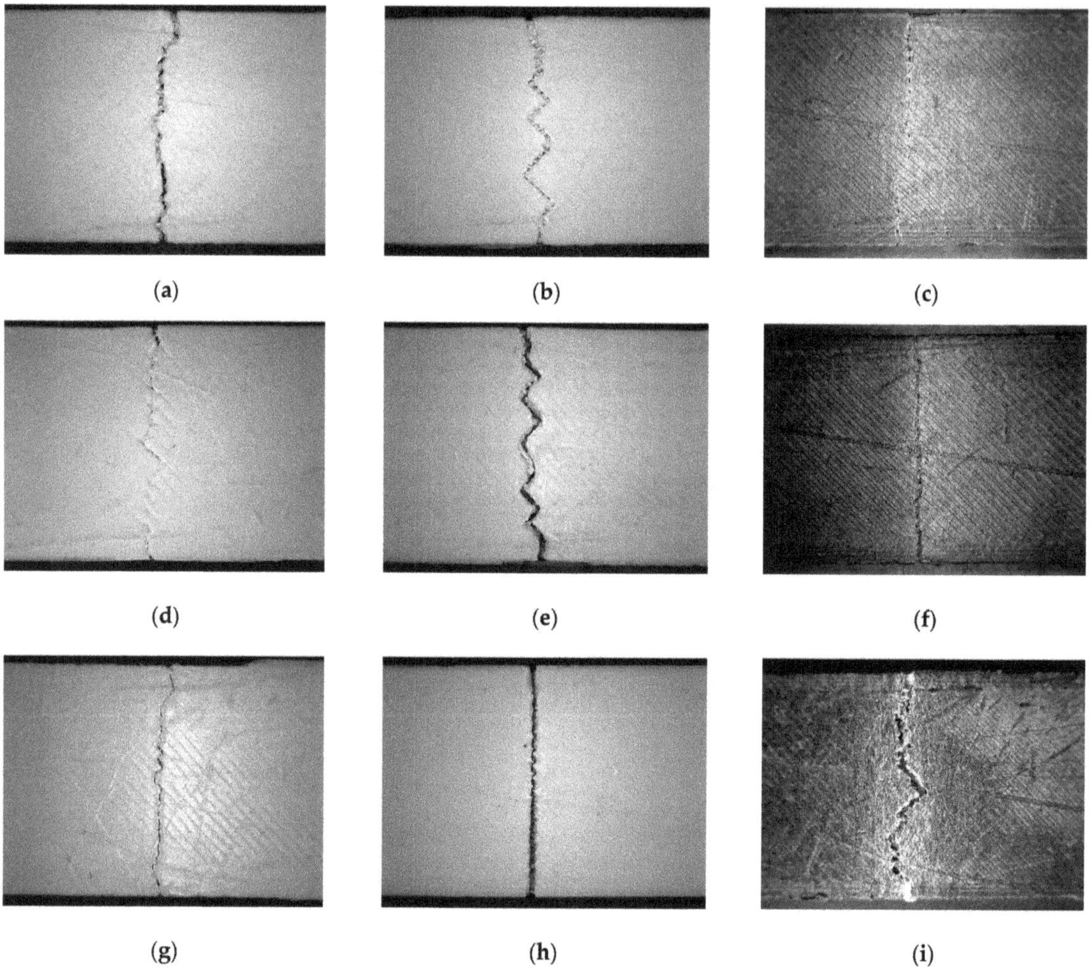

Figure 10. Crack propagation from a bird's-eye view. Note that numbers 1–9 correspond to the experimental cases appearing in Table 3. (**a**) Number 1; (**b**) Number 2; (**c**) Number 3; (**d**) Number 4; (**e**) Number 5; (**f**) Number 6; (**g**) Number 7; (**h**) Number 8; (**i**) Number 9.

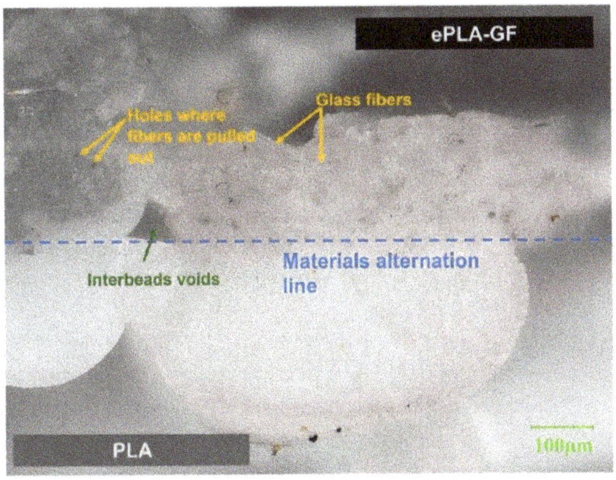

(a)

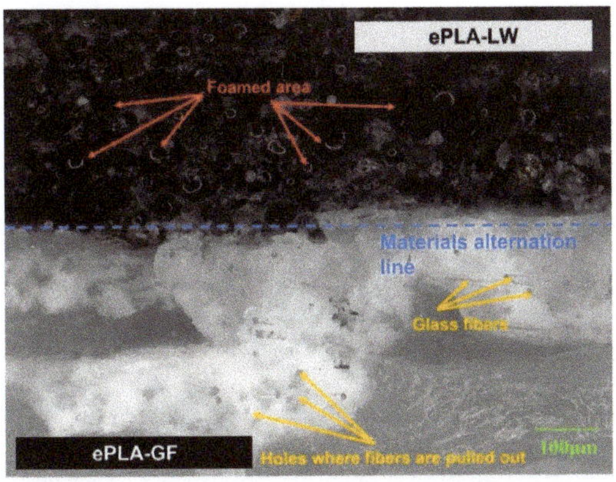

(b)

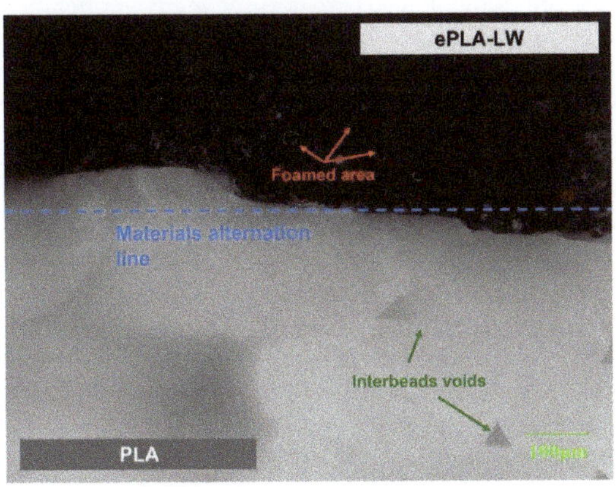

(c)

Figure 11. Microscopic imaging of interfaces between different materials: (**a**) PLA and ePLA-GF; (**b**) ePLA-LW and ePLA-GF; (**c**) PLA and ePLA-LW.

The infill density determines the amount of material present in the printed part, which directly affects its strength and stiffness. Higher infill density generally leads to higher strength, as there are more material layers and interconnections between them. Moreover, the arrangement and orientation of the printed material are determined by the infill pattern. The chosen infill pattern can influence interlayer adhesion and bonding, thereby impacting the overall mechanical performance of the printed part. Consequently, adjusting both infill density and pattern becomes crucial as they directly affect material distribution within the printed object, subsequently influencing its mechanical characteristics such as strength and stiffness [42]. We can assess the mechanical properties and further adjust the infill density and pattern based on crack propagation in the sample.

By examining the crack propagations of samples from the side view (cf. Figure 9), it is observed that fracture cracks predominantly initiate and propagate in the PLA and ePLA-GF layers. The ePLA-LW layer, however, demonstrates a lower likelihood of failure due to its high toughness. Consequently, an increased relative volume of the ePLA-LW layer can decrease the likelihood of crack propagation (e.g., Figure 9d). Alternatively, the ePLA-GF layer exhibits a high probability of fracture, often characterized by delamination within the inter-beads of the layer, causing the crack to extend laterally along the inter-bead boards (e.g., Figure 9f). Furthermore, the crack expansion between inter-beads within the PLA layer is relatively parallel to the direction of the applied bending force. However, the PLA layer demonstrates a relatively low capacity for resisting cracks once the crack initiates (e.g., Figure 9a). Figure 10 offers a bridged view of the crack propagations (i.e., of the outermost layer of the samples' bottom surface). The filling pattern and filling density have a significant impact on the crack. Under the same filling density, samples with a 45°/135° Lines filling pattern exhibit a zigzag line at the bottom of their cracks, leading to irregular serrations (taking 60% filling density as an example, such as Figure 10b,f,g), while Gyroid-filled samples display a relatively straight line of bottom cracks. Notably, although different filling patterns are employed, the shape of the crack tends to be straighter as the sample filling density increases. Furthermore, the smaller the density, the higher the degree of zigzag in the resulting bottom crack, suggesting that the filling pattern predominantly governs the crack line formation at relatively low infill density, with the infill density becoming a more dominant factor as it increases to a larger value.

Furthermore, the characterization of the microstructures depicted in Figure 11 reveals that the interfacial adhesion between each pair of the constituent materials is in excellent condition, which may be likely attributed to our pre-designed top-bottom layers for each sub-part of the sandwich material, i.e., the outermost side of each material layer is printed with fully saturated materials. This configuration also enhances the interlayer bonding quality between distinct materials. Given these configurations, it is postulated that the interlayer adhesions between different constituent materials should not detract from the bulk mechanical performance of the 3D-printed sandwiches when subject to external bending loads. However, it is noted that the impact of interlayer bonding quality is a crucial factor in determining the material properties of the sandwich samples under other loading conditions. Future in-depth studies are desired to further explore these aspects.

3.3. Controlled Variables Method on the Material Stacking Order

It should be highlighted that the interactive effects among the four investigated parameters on the printed components have been disregarded, assuming that each processing parameter produces independent effects on the resulting material properties. From the experimental results, it is evident that the arrangement of materials has a significant influence on the flexural strength and modulus of composite panels fabricated using multi-material FFF. To further delve into this factor, an additional study under the controlled variables method is conducted to assess the impact of material sequence on the strength and modulus of printed samples. The printing parameters for the samples are specified as follows: 45°/135° Line filling pattern and 100% filling density, with other general conditions identical to those outlined in Table 2. The samples in this controlled variables method utilize the same ASTM-D790 standard, except for a slight modification in thickness to 5.4 mm, resulting in equal-length pieces for the three layers of constituent materials. Note: Figure 2d–f depicts the three levels employed in this experiment. Figure 12 exhibits the stage curves of three groups of samples during three-point bending tests [38] (note that the curves are terminated at the maximum bending strength). We combined three samples at each level into one curve and depicted the discrepancy between different sample curves through error bars. The results indicate that the first-level sample exhibits both maximum stiffness and strength, while the plastic deformation area of the third-level sample is larger than its elastic deformation area, suggesting superior ductility. The stiffness of the second-level sample surpasses that of the third-level sample, yet its ductility is better than that of the first-level sample.

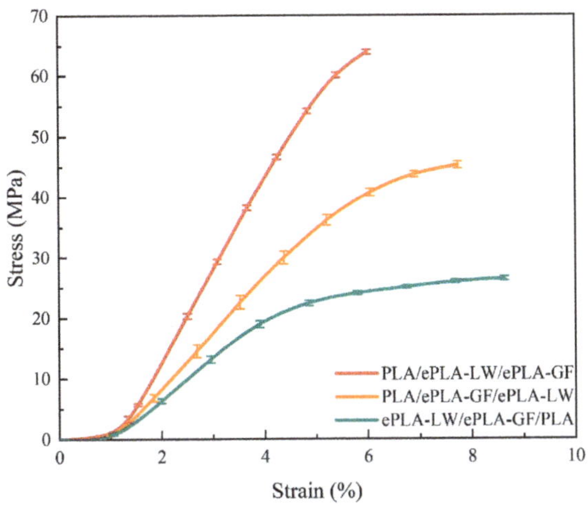

Figure 12. Three-point bending mechanical curve.

Furthermore, we compute the mean values of the maximum strength and modulus in the three experimental groups, along with the error bars of the data, as depicted in Figure 13. The results indicate that the first level of material order exhibits the highest strength and stiffness. Samples prepared using the "PLA/ePLA-LW/ePLA-GF" sequence exhibited smaller error bars, suggesting more stable mechanical properties. As illustrated in Figure 13, the first level demonstrates the highest bending strength, reaching 64.72 MPa, while the third level exhibits the lowest bending strength (only 27.09 MPa), at ~42% of the bending strength of the first level samples. The second level exhibits a higher bending strength than the third level, up to 45.63 MPa, at ~71% of the first level. However, the mechanical properties of the second-level sample are less stable than those of the other two levels, as indicated by the error bar. In addition, Figure 13 depicts the maximum flexural modulus of the three levels of samples. The first level sample exhibits the highest modulus at 1.53 GPa, followed by the second sample, which decreases to 0.96 GPa at only ~63% of the first level sample. The error bar of the second level sample is the largest, while the third level exhibits the lowest modulus of 0.72 GPa at ~47% of the first level sample.

Under the controlled variables method, our data reveal that the distinct printing sequences of the three materials have a substantial impact on the mechanical properties of the samples, with the maximum disparities in bending strength and stiffness among different stacking orders being approximately 58% and 53%, respectively. It is worth noting that the interface adhesion quality among the constituent materials could significantly contribute to determining the bulk flexural properties of the printed panels (e.g., [43]). Consequently, the material stacking order is deemed the most critical processing parameter in the 3D printing preparation of the composite panels. It is important to highlight that the interlayer bonding strength between different materials is intricately associated with several factors, such as thermodynamics [23], chemical affinity [44], and FFF process parameters [45]. It is anticipated that complexity will be delved into further through separate, in-depth studies.

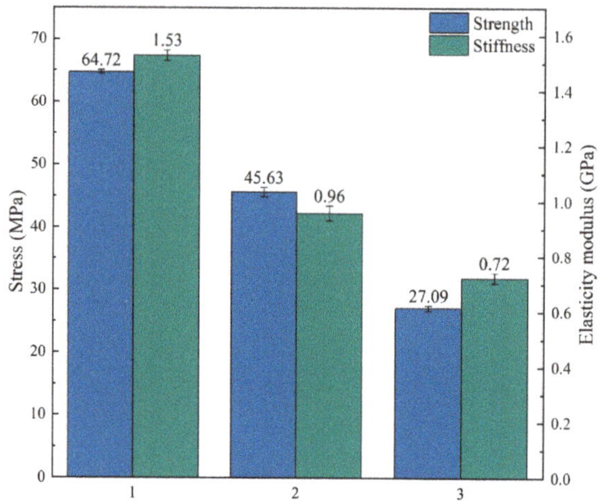

Figure 13. Histogram of average mechanical data for each horizontal bend.

4. Conclusions

Utilizing the Taguchi method, we design an orthogonal experiment designed to investigate the impact of four processing parameters on the mechanical properties of the prepared samples, namely volume ratio, material printing sequence, filling pattern, and filling density.

By employing a larger-is-better signal-to-noise ratio analysis, we identify the optimal parameter combinations for achieving maximum flexural strength and stiffness properties. The combination that maximizes flexural strength includes a volume ratio of 1:1:2, a material sequence of PLA/ePLA-LW/ePLA-GF, a Gyroid filling pattern, and an 80% filling density. Similarly, the combination that maximizes flexural modulus consists of a volume ratio of 1:2:1, a material sequence of PLA/ePLA-LW/ePLA-GF, a Grid filling pattern, and 80% filling density. They have the largest S/N ratios of 7.15 and 6.59 compared to other combinations.

The Taguchi prediction method is applied to forecast the performance of an optimized processing parameter combination obtained from signal-to-noise ratio analyses. Subsequently, the predicted results are validated through related experiments, demonstrating the accuracy of the Taguchi prediction with a minimal deviation between experimental and true values of less than 10%. The experimental values of modulus and strength are 1.23 GPa and 43.91 MPa, which are more or less the same as the predicted values of 1.2 GPa and 46.87 MPa, respectively.

Through the analysis of signal-to-noise ratios across various processing parameter levels, we discover that the material stacking order (the range reaching 6.59) is the most significant factor influencing the flexural performance of the PLA composite panels. Consequently, we conducted a controlled variables method experiment as applied to the three levels of the material stacking order parameter, revealing that the first material sequence exhibits the highest flexural strength and modulus with the smallest error bar, at 67.72 MPa for strength and 1.53 GPa for modulus. In contrast, the third material sequence demonstrates the lowest strength and stiffness while displaying higher ductility in mechanical performance. Its flexural strength and modulus, declining to 27.09 MPa and 0.72 GPa, respectively, are 42% and 47%, respectively, compared to the first material sequence.

This research concluded that the Taguchi method is a useful tool for optimizing printing parameters and highlighted the importance of material stacking sequence in determining the bending performance of multi-material 3D-printed components. By adjusting the mechanical properties, printed components can be customized to suit specific applications. The manufacturing approach employed in this study introduces a novel concept for designing and fabricating protective materials with potential applications in wearable protective equipment such as helmets, sports knee pads, and medical guards.

Author Contributions: Conceptualization, Z.W.; methodology, Z.W. and L.W.; software, L.W.; validation, Z.W., L.W., F.T. and C.S.; data curation, L.W. and F.T.; writing—original draft preparation, Z.W. and L.W.; writing—review and editing, Z.W., L.W. and C.S. All authors have read and agreed to the published version of the manuscript.

Funding: This research received external funding from the National Natural Science Foundation of China (52101381), the China Postdoctoral Science Foundation (2023M730455), the Department of Education of Liaoning Province (LJKMZ20220366), and the Dalian Human Resources and Social Security Bureau (Dalian City innovation and entrepreneurship support program for returned overseas students).

Data Availability Statement: Data are contained within the article.

Conflicts of Interest: The authors declare no conflicts of interest.

Appendix A

The ANOVA analyses on the measured flexural strength and modulus data are performed, and the results are given in Tables A1 and A2 as follows.

Table A1. Analysis of the variance of strength.

Source	DOF	SS	MS	F	P
Volume ratio	2	168.29	84.14	75.33	0.000
Material sequence	2	2137.91	1068.95	956.97	0.000
Filling pattern	2	17.92	8.96	8.02	0.003
Filling density	2	501.21	250.61	224.35	0.000
Error	18	20.11	1.12		
Total	26	2845.44			

Table A2. Analysis of the variance of modulus.

Source	DOF	SS	MS	F	P
Volume ratio	2	0.12	0.06	437.53	0.000
Material sequence	2	1.39	0.70	5184.48	0.000
Filling pattern	2	0.02	0.01	71.10	0.000
Filling density	2	0.32	0.16	1186.47	0.000
Error	18	0.00	0.00		
Total	26	1.85			

Appendix B

Considering the limitations of the existing equipment in handling high temperatures, preliminary experiments are carried out according to the extrusion parameters provided by the material supplier. In order to avoid the temperature fluctuation of the nozzle caused by high-temperature printing, the test temperatures were selected at 220 °C, 230 °C, and 240 °C. For each temperature condition, we print three groups of samples and perform three-point bending experiments after obtaining the weight of each sample. We average the bending strength, bending modulus, and weight of each group of specimens as reference values. Then, we divide the average value of flexural strength and modulus by the average weight to obtain the performance-weight ratio for each group of specimens. In the following,

Figure A1 represents the strength-to-mass ratio of the specimens prepared with the three printing parameters, and Figure A2 illustrates the modulus-to-mass ratio. When the printing temperature is 220 °C, the foaming effect of this sample is very poor, and there is almost no foaming. As a result, its sample performance is largely superior to other samples that have been foamed, which also leads to its higher performance-to-weight ratio. When the temperature is above 230 °C, the sample appears to be in an obvious foaming state. When the material extrusion parameters are 230 °C and the extrusion rate is 75%, the foaming effect is the best, and the performance-to-mass ratio is much higher than that of the sample printed at 240 °C.

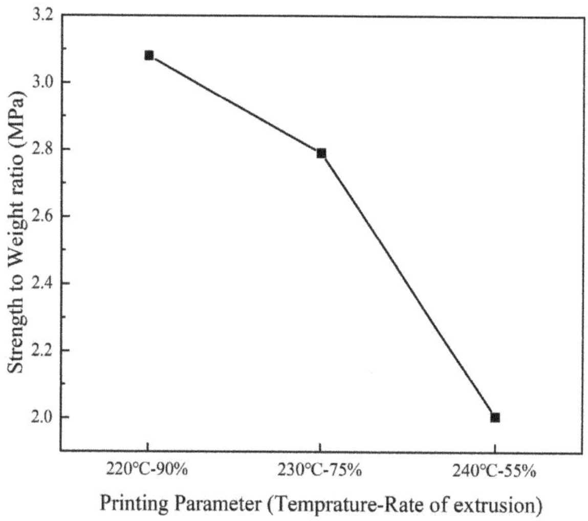

Figure A1. Strength-to-Weight ratio.

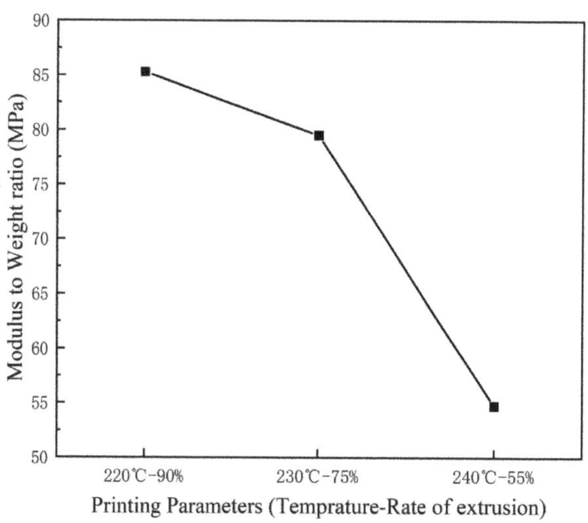

Figure A2. Modulus-to-Weight ratio.

References

1. Jandyal, A.; Chaturvedi, I.; Wazir, I. 3D printing—A review of processes, materials and applications in industry 4.0. *Sustain. Oper. Comput.* **2022**, *3*, 33–42. [CrossRef]
2. Kreiger, M.; Pearce, J.M. Environmental impacts of distributed manufacturing from 3-D printing of polymer components and products. *MRS Online Proc. Libr. (OPL)* **2013**, *1492*, 85–90. [CrossRef]
3. Aimar, A.; Palermo, A.; Innocenti, B. The role of 3D printing in medical applications: A state of the art. *J. Healthc. Eng.* **2019**, *2019*, 5340616. [CrossRef] [PubMed]
4. Tian, Y.; Chen, C.X.; Xu, X. A review of 3D printing in dentistry: Technologies, affecting factors, and applications. *Scanning* **2021**, *2021*, 1–19. [CrossRef] [PubMed]
5. Karkun, M.S.; Dharmalingam, S. 3D Printing Technology in Aerospace Industry—A Review. *Int. J. Aviat. Aeronaut. Aerosp.* **2022**, *9*, 4. [CrossRef]
6. Murariu, M.; Dubois, P. PLA composites: From production to properties. *Adv. Drug Deliv. Rev.* **2016**, *107*, 17–46. [CrossRef] [PubMed]
7. Rydz, J.; Sikorska, W.; Kyulavska, M. Polyester-based (bio) degradable polymers as environmentally friendly materials for sustainable development. *Int. J. Mol. Sci.* **2014**, *16*, 564–596. [CrossRef]
8. Tümer, E.H.; Erbil, H.Y. Extrusion-based 3d printing applications of pla composites: A review. *Coatings* **2021**, *11*, 390. [CrossRef]
9. Liu, Z.; Wang, Y.; Wu, B. A critical review of fused deposition modeling 3D printing technology in manufacturing polylactic acid parts. *Int. J. Adv. Manuf. Technol.* **2019**, *102*, 2877–2889. [CrossRef]
10. Dong, Y.; Ghataura, A.; Takagi, H. Polylactic acid (PLA) biocomposites reinforced with coir fibres: Evaluation of mechanical performance and multifunctional properties. *Compos. Part A Appl. Sci. Manuf.* **2014**, *63*, 76–84. [CrossRef]
11. Yu, B.; Geng, C.; Zhou, M. Impact toughness of polypropylene/glass fiber composites: Interplay between intrinsic toughening and extrinsic toughening. *Compos. Part B Eng.* **2016**, *92*, 413–419. [CrossRef]
12. Ruz-Cruz, M.A.; Herrera-Franco, P.J.; Flores-Johnson, E.A. Thermal and mechanical properties of PLA-based multiscale cellulosic biocomposites. *J. Mater. Res. Technol.* **2022**, *18*, 485–495. [CrossRef]
13. DeStefano, V.; Khan, S.; Tabada, A. Applications of PLA in modern medicine. *Eng. Regen.* **2020**, *1*, 76–87. [CrossRef]
14. Karger-Kocsis, J.; Mahmood, H.; Pegoretti, A. Recent advances in fiber/matrix interphase engineering for polymer composites. *Prog. Mater. Sci.* **2015**, *73*, 1–43. [CrossRef]
15. Wang, Y.; Kong, D.; Zhang, Q. Process parameters and mechanical properties of continuous glass fiber reinforced composites-polylactic acid by fused deposition modeling. *J. Reinf. Plast. Compos.* **2021**, *40*, 686–698. [CrossRef]
16. Harris, M.; Potgieter, J.; Mohsin, H. Partial polymer blend for fused filament fabrication with high thermal stability. *Polymers* **2021**, *13*, 3353. [CrossRef] [PubMed]
17. Brancewicz-Steinmetz, E.; Sawicki, J. Bonding and strengthening the PLA biopolymer in multi-material additive manufacturing. *Materials* **2022**, *15*, 5563. [CrossRef] [PubMed]
18. Notta-Cuvier, D.; Odent, J.; Delille, R. Tailoring polylactide (PLA) properties for automotive applications: Effect of addition of designed additives on main mechanical properties. *Polym. Test.* **2014**, *36*, 1–9. [CrossRef]
19. Gao, X.; Qi, S.; Zhang, D. The role of poly (ethylene glycol) on crystallization, interlayer bond and mechanical performance of polylactide parts fabricated by fused filament fabrication. *Addit. Manuf.* **2020**, *35*, 101414. [CrossRef]
20. Markarian, J. Biopolymers present new market opportunities for additives in packaging. *Plast. Addit. Compd.* **2008**, *10*, 22–25. [CrossRef]
21. Yousefi Kanani, A.; Rennie, A.E.W.; Abd Rahim, S.Z.B. Additively manufactured foamed polylactic acid for lightweight structures. *Rapid Prototyp. J.* **2023**, *29*, 50–66. [CrossRef]
22. Damanpack, A.R.; Sousa, A.; Bodaghi, M. Porous PLAs with controllable density by FDM 3D printing and chemical foaming agent. *Micromachines* **2021**, *12*, 866. [CrossRef] [PubMed]
23. Yin, J.; Lu, C.; Fu, J. Interfacial bonding during multi-material fused deposition modeling (FDM) process due to inter-molecular diffusion. *Mater. Des.* **2018**, *150*, 104–112. [CrossRef]
24. Boulaala, M.; Elmessaoudi, D.; Buj-Corral, I. Towards design of mechanical part and electronic control of multi-material/multicolor fused deposition modeling 3D printing. *Int. J. Adv. Manuf. Technol.* **2020**, *110*, 1–11. [CrossRef]
25. Baca, D.; Ahmad, R. The impact on the mechanical properties of multi-material polymers fabricated with a single mixing nozzle and multi-nozzle systems via fused deposition modeling. *Int. J. Adv. Manuf. Technol.* **2020**, *106*, 4509–4520. [CrossRef]
26. Wang, Z.; Yin, X.; Wang, L. An experimental investigation on the tensile properties of PEEK-PEI sandwich structures prepared via fused filament fabrication. *High Perform. Polym.* **2023**, *35*, 276–294. [CrossRef]
27. Kamaal, M.; Anas, M.; Rastogi, H.; Bhardwaj, N.; Rahaman, A. Effect of FDM process parameters on mechanical properties of 3D-printed carbon fibre–PLA composite. *Prog. Addit. Manuf.* **2021**, *6*, 63–69. [CrossRef]
28. Yavas, D.; Liu, Q.; Zhang, Z.; Wu, D. Design and fabrication of architected multi-material lattices with tunable stiffness, strength, and energy absorption. *Mater. Des.* **2022**, *217*, 110613. [CrossRef]
29. Rasheed, A.; Hussain, M.; Ullah, S.; Ahmad, Z.; Kakakhail, H.; Riaz, A.A.; Khan, I.; Ahmad, S.; Akram, W.; Eldin, S.M.; et al. Experimental investigation and Taguchi optimization of FDM process parameters for the enhancement of tensile properties of Bi-layered printed PLA-ABS. *Mater. Res. Express* **2023**, *10*, 095307. [CrossRef]

30. Ahmed, H.; Hussain, G.; Gohar, S.; Ali, A.; Alkahtani, M. Impact toughness of hybrid carbon fiber-PLA/ABS laminar composite produced through fused filament fabrication. *Polymers* **2021**, *13*, 3057. [CrossRef]
31. Wang, Z.; Yin, X.; Wang, L. Tensile properties of sandwich-designed carbon fiber filled PLA prepared via multi-material additive layered manufacturing and post-annealing treatment. *Int. Polym. Process.* **2023**, *38*, 343–361. [CrossRef]
32. Standau, T.; Zhao, C.; Murillo Castellón, S.; Bonten, C.; Altstädt, V. Chemical Modification and Foam Processing of Polylactide (PLA). *Polymers* **2019**, *11*, 306. [CrossRef] [PubMed]
33. *ASTM D790*; Standard Test Methods for Flexural Properties of Unreinforced and Reinforced Plastics and Electrical Insulating Materials. ASTM International: West Conshohocken, PA, USA, 2017.
34. Kam, M.; Ipekci, A.; Şengül, Ö. Investigation of the effect of FDM process parameters on mechanical properties of 3D printed PA12 samples using Taguchi method. *J. Thermoplast. Compos. Mater.* **2023**, *36*, 307–325. [CrossRef]
35. Torres, J.; Cole, M.; Owji, A. An approach for mechanical property optimization of fused deposition modeling with polylactic acid via design of experiments. *Rapid Prototyp. J.* **2016**, *22*, 387–404. [CrossRef]
36. Dong, G.; Wijaya, G.; Tang, Y. Optimizing process parameters of fused deposition modeling by Taguchi method for the fabrication of lattice structures. *Addit. Manuf.* **2018**, *19*, 62–72. [CrossRef]
37. Camposeco-Negrete, C. Optimization of FDM parameters for improving part quality, productivity and sustainability of the process using Taguchi methodology and desirability approach. *Prog. Addit. Manuf.* **2020**, *5*, 59–65. [CrossRef]
38. Hikmat, M.; Rostam, S.; Ahmed, Y.M. Investigation of tensile property-based Taguchi method of PLA parts fabricated by FDM 3D printing technology. *Results Eng.* **2021**, *11*, 100264. [CrossRef]
39. Heidari-Rarani, M.; Ezati, N.; Sadeghi, P. Optimization of FDM process parameters for tensile properties of polylactic acid specimens using Taguchi design of experiment method. *J. Thermoplast. Compos. Mater.* **2022**, *35*, 2435–2452. [CrossRef]
40. Heidari-Rarani, M.; Rafiee-Afarani, M.; Zahedi, A.M. Mechanical characterization of FDM 3D printing of continuous carbon fiber reinforced PLA composites. *Compos. Part B Eng.* **2019**, *175*, 107147. [CrossRef]
41. Radhwan, H.; Shayfull, Z.; Abdellah, A.E.H. Optimization parameter effects on the strength of 3D-printing process using Taguchi method. In *AIP Conference Proceedings*; AIP Publishing: Melville, NY, USA, 2019; Volume 2129.
42. Khan, I.; Tariq, M.; Abas, M.; Shakeel, M.; Hira, F.; Al Rashid, A.; Koç, M. Parametric investigation and optimisation of mechanical properties of thick tri-material based composite of PLA-PETG-ABS 3D-printed using fused filament fabrication. *Compos. Part C Open Access* **2023**, *12*, 100392. [CrossRef]
43. Tamburrino, F.; Graziosi, S.; Bordegoni, M. The influence of slicing parameters on the multi-material adhesion mechanisms of FDM printed parts: An exploratory study. *Virtual Phys. Prototyp.* **2019**, *14*, 316–332. [CrossRef]
44. Lopes, L.R.; Silva, A.F.; Carneiro, O.S. Multi-material 3D printing: The relevance of materials affinity on the boundary interface performance. *Addit. Manuf.* **2018**, *23*, 45–52. [CrossRef]
45. Ahmed, S.W.; Hussain, G.; Altaf, K. On the effects of process parameters and optimization of interlaminate bond strength in 3D printed ABS/CF-PLA composite. *Polymers* **2020**, *12*, 2155. [CrossRef] [PubMed]

polymers

MDPI

Article

Design and Optimization of 3D-Printed Variable Cross-Section I-Beams Reinforced with Continuous and Short Fibers

Xin Zhang [1,2], Peijie Sun [3], Yu Zhang [3], Fei Wang [4], Yun Tu [5], Yunsheng Ma [6,*] and Chun Zhang [1,2,*]

1 School of Aeronautics, Northwestern Polytechnical University, Xi'an 710072, China; zhangxin504@mail.nwpu.edu.cn
2 Institute of Aircraft Composite Structures, Northwestern Polytechnical University, Xi'an 710072, China
3 School of Mechanics, Civil Engineering and Architecture, Northwestern Polytechnical University, Xi'an 710072, China
4 School of Aircraft, Xi'an Aeronautical Institute, Xi'an 710077, China
5 Key Laboratory of Pressure Systems and Safety, School of Mechanical and Power Engineering, Ministry of Education, East China University of Science and Technology, Shanghai 200237, China
6 Shandong Chambroad Holding Group Co., Ltd., Binzhou 256599, China
* Correspondence: yunsheng.ma@chambroad.com (Y.M.); c.zhang@nwpu.edu.cn (C.Z.)

Abstract: By integrating fiber-reinforced composites (FRCs) with Three-dimensional (3D) printing, the flexibility of lightweight structures was promoted while eliminating the mold's limitations. The design of the I-beam configuration was performed according to the equal-strength philosophy. Then, a multi-objective optimization analysis was conducted based on the NSGA-II algorithm. 3D printing was utilized to fabricate I-beams in three kinds of configurations and seven distinct materials. The flexural properties of the primitive (P-type), the designed (D-type), and the optimized (O-type) configurations were verified via three-point bending testing at a speed of 2 mm/min. Further, by combining different reinforcements, including continuous carbon fibers (CCFs), short carbon fibers (SCFs), and short glass fibers (SGFs) and distinct matrices, including polyamides (PAs), and polylactides (PLAs), the 3D-printed I-beams were studied experimentally. The results indicate that designed and optimized I-beams exhibit a 14.46% and 30.05% increase in the stiffness-to-mass ratio and a 7.83% and 40.59% increment in the load-to-mass ratio, respectively. The CCFs and SCFs result in an outstanding accretion in the flexural properties of 3D-printed I-beams, while the accretion is 2926% and 1070% in the stiffness-to-mass ratio and 656.7% and 344.4% in the load-to-mass ratio, respectively. For the matrix, PAs are a superior choice compared to PLAs for enhancing the positive impact of reinforcements.

Keywords: 3D printing; optimization; continuous carbon fiber; I-beam; equal strength

1. Introduction

Fiber-reinforced composites (FRCs) are replacing metal materials and are emerging as the primary structural material for lightweight transportation and aviation, such as drones, new energy vehicles, and high-speed rails [1,2]. Due to their exceptional strength, stiffness, fatigue resistance, and corrosion resistance, the applications of FRCs in structures are steadily growing, especially in main load-bearing structures [3,4]. For example, the utilization of pultruded carbon fiber-reinforced beams on large wind turbine blades has become a consensus in the energy industry [5]. Molds, however, have historically restricted the design and production of conventional FRCs, resulting in high costs and drawn-out production cycles [6]. Three-dimensional (3D) printing, also known as additive manufacturing, usually refers to the process of freely depositing patterns in the XY plane and stacking them layer by layer in the z direction to create a structure [7,8]. The emergence of this technology greatly reduces material waste in the manufacturing process and makes rapid manufacturing possible [9]. Moreover, it is a promising technology for manufacturing composite materials, which has the potential to free FRCs from the limitations of molds [10,11]. When producing porous structures using conventional composite materials, it is typically required to utilize compound molds. By combining 3D printing with FRCs, it is feasible to achieve quick one-shot manufacturing without the use of molds [9].

In recent years, 3D printing of composite materials based on fused deposition modeling (FDM) has been developed due to the affordable equipment and consumables [12,13]. The addition of fibrous reinforcements, such as continuous and short fibers, has demonstrated significant enhancements in mechanical properties of bending-resistant structures [14–16]. By employing continuous carbon fibers (CCFs), the tensile and bending strengths of 3D-printed structures can be increased from 28 MPa and 53 MPa to 80 MPa and 59 MPa, respectively, compared to the structures made of polylactides (PLAs) [14]. Moreover, if a sizing agent is used for pretreatment, the tensile and flexural strengths will be increased by 3.25 and 2.94 times [14], respectively. Adding short carbon fibers (SCFs) to the 3D-printed polycarbonate structure has resulted in a 127% increase in the modulus and an 18.9% increase in the flexural strength [15]. Furthermore, when 15% and 30% of SGFs were added to the acrylonitrile butadiene styrene (ABS) matrix, the flexural properties enhanced by 44% and 59%, respectively [16].

With the introduction of 3D printing, the designability of CCFs has been significantly improved due to the removal of mold restrictions. Multiple types of beams with diverse shapes were fabricated using 3D printing and were subsequently evaluated by mechanical investigation [17–21]. Distinct infill patterns, including rhombus, square, and classic hexagonal, were utilized to manufacture 3D-printed cellular-cored sandwich beams made of CCF composite materials [17]. The three-point bending experimental results indicate that the flexural modulus and bending strength of these beams are comparable with those of competing structures made of aluminum or glass fiber-reinforced composites. The 3D-printed PLA-based T-beam exhibits a flexural strength of 85 MPa and a flexural modulus of 3.5 GPa with an 8 wt.% of SCFs, while the flexural strength increased to 255 MPa and the flexural modulus could reach 16 GPa with an 8 wt.% of CCFs [18].

Despite the significance of I-beams in several industries such as machining [19], architecture [20], and energy [21], there is currently an absence of research on 3D-printed fibrous I-beams. The pultrusion process is currently the most efficient method for manufacturing composite I-beams. However, it is not capable of producing structures with variable cross-sections [22]. Advancements in 3D-printing technology enable the flexible design and manufacturing of I-beams with customizable cross-sections. Nevertheless, the existing research regarding I-beam structural design and the optimization of 3D-printed fiber-reinforced composite materials remains restricted, as well as material selection.

In the present work, the primary objective is to enhance the flexural properties of 3D-printed I-beams through structural design and optimization. The 3D printing of I-beams was conducted with three unique configurations and seven diverse materials. Three-point bending testing was performed to validate the flexural properties of various configurations, reinforcements, and matrix materials. Beams are typical and vital components that withstand bending loads and are extensively utilized in the manufacturing, building, energy, and other industries. This study aims to offer a reference for the design, optimization, and material selection and to guide the utilization of 3D-printed I-beams in lightweight, bending-resistant structures, which might improve the implementation of 3D-printed composite beam structures in commercial applications.

2. Materials and Methods

2.1. 3D-Printing Process

Two types of 3D printers were utilized to fabricate continuous and chopped fiber-reinforced composite I-beams. The scheme of the Mark Two desktop 3D printer (from Markforged Inc., Waltham, MA, USA) is illustrated in Figure 1a. There are two printing nozzles in the Mark Two 3D printer, one for printing continuous carbon fiber (CCF)-reinforced polyamide prepregs and the other for printing Onyx filaments. The adoption of CCF/PA prepregs could enhance the mechanical properties of the structures, whereas the

Onyx materials are applied to smooth surfaces and fill corners to ensure the form accuracy of parts. The CCF/PA prepregs and Onyx filaments were both purchased from Markforged Inc., Waltham, MA, USA.

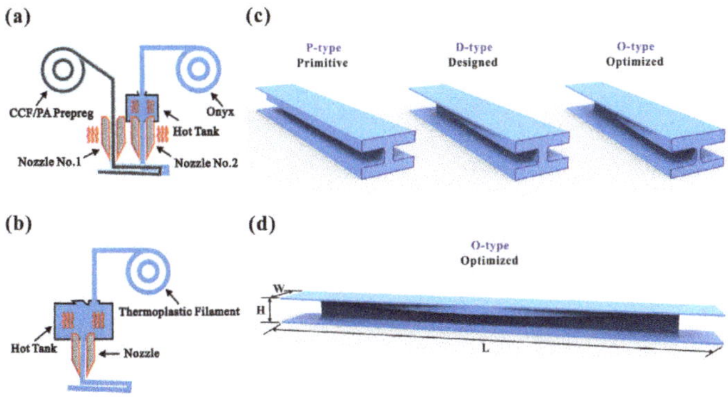

Figure 1. Schematic diagrams of the 3D-printing process and structural configurations: (**a**) a scheme of the Mark Two desktop 3D printer, (**b**) a scheme of the HB250 desktop 3D printer, (**c**) the three configurations of 3D-printed I-beams, and (**d**) the geometry parameters of the 3D-printed I-beams.

Figure 1b shows an illustration of the HB250 desktop 3D printer from Dongguan Xianglong 3D Technology Co., Ltd., Dongguan, China. The nozzle extruded the thermoplastic filament after it melted inside the hot tank. It is capable of printing a range of thermoplastic materials that have extrusion temperatures below 350 degrees, including PAs and PLAs with short carbon fibers (SCFs) or short glass fibers (SGFs). The grades and basic mechanical properties of filaments used in 3D printing are summarized in Table 1, whereas the printing parameters are listed in Table 2. The filaments with discontinuous or no reinforcements were both purchased from the eSUN 3D printing department, Shenzhen Guanghua Weiye Co., Ltd., Shenzhen, China.

Table 1. The type and mechanical properties of printing materials.

Suppliers	Type	Tensile Strength/MPa	Modulus/GPa	Elongation at Break/%	Density/g·cm^{-3}
Markforged	PA-CCF	800	51	1.5	1.4
	Onyx	35.71	3.59	4	1.2
eSUN	ePAHT-CF	173.37	5.61	8.93	1.4
	ePA-GF	76.93	1.72	20.07	1.35
	ePA	57	1.5	196	1.12
	ePLA-CF	28	3.55	4.27	1.21
	ePLA-GF	59.27	4.41	7.99	1.31
	ePLA-Lite	61.34	3.82	29.92	1.23

Carbon fibers and carbon fiber-reinforced polymers (CFRPs) have had significant advancements in recent years, with a compound annual growth rate (CAGR) of around 12.5% globally over the past two decades [2]. An important reason for this is that carbon fibers offer several benefits, including an exceptional strength-to-weight ratio, a superior stiffness-to-weight ratio, and an excellent resistance to chemicals and heat. In this study, the tensile strength of adopted T300 carbon fibers achieves 3530 MPa, while the tensile modulus could reach up to 230 GPa [2]. These values significantly exceed the tensile strength (2400 MPa) and tensile modulus (72 GPa) of adopted E-glass fibers [23].

It should be noted that three configurations of I-beams were 3D printed, as presented in Figure 1c. In order to clearly illustrate the distinctions between the configurations, the beams are dissected and displayed in a bisected form. The primitive configuration (P-type) has a constant cross-section throughout the entire design interval length (210 mm). The designed configuration (D-type) modifies the dimensions of the I-beam flanges in terms of width and thickness, while maintaining uniformity of the top and bottom flanges. The optimized configuration (O-type) not only adjusts the dimensions of the I-beam flanges in terms of width and thickness of both top and bottom flanges but also varies the thickness of different cross-sections across the whole design interval length. Figure 1d. displays the entire O-type beam with a length (*L*), width (*W*), and height (*H*) of 250 mm, 25 mm, and 15 mm, respectively. The process of obtaining these arrangements will be further elucidated in Sections 2.3–2.5. Here, I-beams with seven different materials and three different configurations were 3D printed to compare the flexural properties of different configurations, reinforcements, and matrices experimentally.

Table 2. Parameters of 3D printing.

Type	Layer Thickness/mm	Nozzle Temperature/°C	Hatch Space/mm	Printing Speed/mm·s^{-1}
PA-CCF	0.125	275	1	15
Onyx	0.2	260	0.5	15
ePAHT-CF	0.2	240	0.5	50
ePA-GF	0.2	240	0.5	50
ePA	0.2	240	0.5	50
ePLA-CF	0.2	210	0.5	50
ePLA-GF	0.2	210	0.5	50
ePLA-Lite	0.2	210	0.5	50

2.2. *Three-Point Bending Testing*

The three-point bending test loading method is considered a conventional experimental technique for bending-performance evaluation due to its straightforwardness and the relative simplicity of the fixture and test parameters [9]. Therefore, in this study, the three-point bending test was adopted instead of the four-point bending test. The DNS200 universal electronic testing machine from the Changchun Research Institute for Mechanical Science Co., Ltd., (Jilin, China) was adopted to perform three-point bending testing at a speed of 2 mm/min. The testing machine automatically collects the experimental load and displacement data. The schematic diagram of the testing is shown in Figure 2a, and the corresponding snapshot is presented in Figure 2b. The radius of the loading head and the supporting roller is 5mm, and the span of the supporting roller is 200 mm.

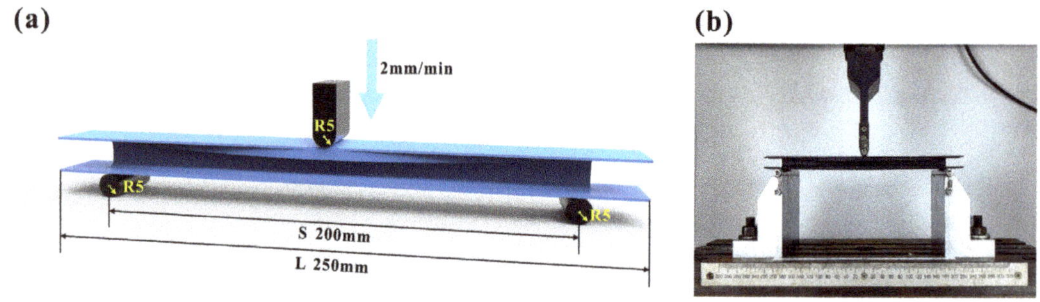

Figure 2. A (**a**) schematic diagram and (**b**) snapshot of the three-point bending testing.

The load-to-mass ratio (*LMR*) and the stiffness-to-mass ratio (*SMR*) were chosen as indicators to quantify the bending performance of the 3D-printed I-beams. The load-to-mass ratio (*LMR*) was calculated by dividing a maximum bending load by the mass of the beams:

$$LMR = \frac{F_{max}}{m} \quad (1)$$

where F_{max} is the maximum bending load, while m is the mass of the beams.

The stiffness-to-mass ratio (*SMR*) was calculated by dividing the stiffness of the linear stage of the load-to-displacement curve by the mass of the beams:

$$SMR = \frac{E}{m} \quad (2)$$

where E is the stiffness of the linear stage, while m is the mass of the beams.

2.3. Design of Configuration

The primary determinant of strength for beams in engineering is the bending stress in the normal direction. In general, the shear stress strength requirements could also be accommodated when the beam is designed according to the normal stress strength criteria, as follows:

$$\sigma_{max} = \frac{M_{max}}{W_z} \leq [\sigma] \quad (3)$$

where σ_{max} is the maximum bending stress in the normal direction, M_{max} is the maximum bending moment, W_z is the flexural modulus of the section, and $[\sigma]$ is the allowable normal stress of the material.

When the beam is subjected to three-point bending loading with a span (S) and a load (F), as shown in Figure 2a, the bending moment M (x) on the section is a function of the distance x between the section and the nearest supporting roller:

$$M(x) = \frac{F}{2}x(0 \leq x \leq \frac{S}{2}) \quad (4)$$

This indicates that the bending moment decreases as the distance from the center increases. According to the equal-strength philosophy [24,25], it is advisable to design a series of suitable cross-sections and to reduce material in areas where the bending moment is relatively low. For the I-beam, the flexural modulus of the section is the moment of inertia relative to the neutral layer (I), as follows:

$$I = \frac{t_w h_w^3}{12} + 2 \times (\frac{bt^3}{12} + bt \times (\frac{h_w + t}{2})^2) \quad (5)$$

where t is the thickness of the upper and lower beam flanges, t_w is the thickness of the web, h is the height of the I-beam, h_w is the height between the upper and lower beam flanges, and b is the width of the upper and lower e beam flanges.

Material arrangement can be optimized by designing the width (b) and thickness (t) of the upper and lower beam flanges and the height between them at different locations. The design of simultaneous variable widths and thicknesses of beam flanges in the form of a hyperbola was adopted in this study, and their specific geometry parameters of configuration are shown in Section 2.5.

2.4. Optimization of Configuration

2.4.1. Finite Element Analysis

Catia (version V5-2016) was utilized for modeling and implementing parameterization. Subsequently, the commercial finite element software Abaqus (version 2022) was introduced for a finite element analysis (FEA). Due to the structural symmetry of the 3D-printed beams, a half-beam model is created by establishing symmetry along the positive X-axis direction around the YZ plane, as seen in Figure 3. Consequently, symmetric boundary constraints are then applied to the model. The static analysis step was adopted. A rigid supporting roller and a loading head with a radius of R5 were established. The global size of meshes was 1.2 mm, and the type of meshes was C3D4. The number of meshes was around 4600, while the number of nodes was about 11,000.

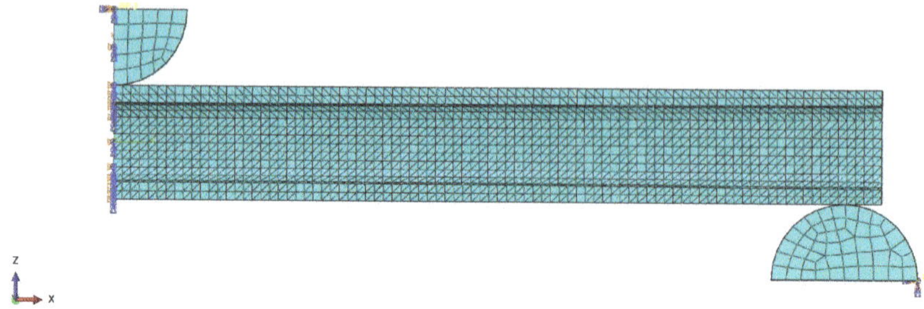

Figure 3. The finite element mode for three-point bending loading conditions.

Contact constraints were established between the supporting roller, the loading head, and the beam. The tangential friction coefficient was set to 0.2, whereas the normal behavior was defined as hard contact. Fixed boundary conditions ($U_1 = U_2 = U_3 = UR_1 = UR_2 = UR_3 = 0$) were assigned at the referenced point on the supporting roller, while displacement boundary conditions ($U_1 = U_2 = 0$, $U_3 = 5$) were defined at the referenced point on the loading head.

2.4.2. Indexes and Objective Functions

As shown in Figure 4, six optimizable parameters were constructed for the 3D-printed I-beams, including B_1, B_2, B_3, H_1, H_2, and R, and R = B_4/B_3. The aim of structural optimization is to realize the design of variable cross-sections by manipulating these optimizable parameters.

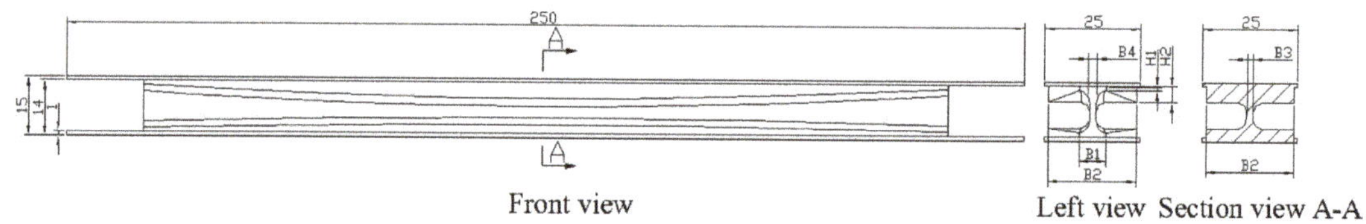

Figure 4. Optimizable parameters of the 3D-printed I-beams.

To achieve the variable cross-section design of the 3D-printed I-beams while ensuring their structural integrity and improving their structural efficiency, the following conditions need to be fulfilled: (1) The structure exhibits minimal deflection deformation when subjected to the identical boundary condition, indicating a high level of bending stiffness. (2) To increase the structure's desirable load-bearing capacity, the stress level under the assigned boundary condition needs to be decreased; here, the first principal stress has been chosen as the index. (3) The mass of the structure ought to be minimized for the highest load-to-mass ratio. Since the density of the same material is identical, the volume can also be utilized to denote structural mass.

The bending stiffness, the first principal stress, and the volume were utilized as indexes to evaluate the 3D-printed I-beams. The constructed objective functions are as follows:

$$\begin{cases} \max f_1 = W(B_1,\ B_2,\ B_3,\ H_1, H_2,\ R) \\ \max f_2 = \sigma(B_1,\ B_2,\ B_3,\ H_1, H_2,\ R) \\ \max f_3 = V(B_1,\ B_2,\ B_3,\ H_1, H_2,\ R) \end{cases}$$
$$\begin{gathered} s.t. \quad 0 < B_1 \leq B_2 \\ 0 < B_1 \leq 25 \\ 0 < RB_3 + 2 \leq B_1 \\ 0 < H_1 \leq 7.5 \\ 0 < H_2 \leq 7.5 \end{gathered} \tag{6}$$

where W (x_1, x_2, ..., x_6) is the function of the bending stiffness; σ (x_1, x_2, ..., x_6) is the function of the first principal stress; and V (x_1, x_2, ..., x_6) is the function of the volume.

Due to the diverse units of the three objective functions, there are significant numerical disparities that impact the outcomes of an algorithmic analysis. Therefore, it is necessary to normalize the functions. The Latin hypercube sampling technique is employed to sample inside the feasible space and to generate a dataset. This dataset is then used to construct a surrogate model, which serves as a substitute for the time-consuming finite element analysis. The mean and variance of the dataset were computed, followed by normalization using the z-score standardization procedure. The final objective function is as follows:

$$f_i^* = \frac{f_i - \mu_i}{\sigma_i}(i = 1,\ 2,\ 3) \tag{7}$$

where μ is the mean of the dataset, and σ is the variance of the dataset.

2.4.3. Multi-Objective Optimization Analysis

The simultaneous optimization of a 3D-printed I-beam necessitates the consideration of three objective functions, and the optimal solutions of these functions are contradictory. Hence, it is necessary to coordinate the goal function in order to achieve the Pareto optimal solution within the feasible space. The final multi-objective optimization model for the I-beam design is obtained by incorporating weight values to regulate the influence of the objective function. The final multi-objective is as follows:

$$\min f = (-w_1 f_1^*, w_2 f_2^*, w_3 f_3^*) \tag{8}$$

where w_1 = 1.16, w_1 = 1.2, and w_1 = 1.

Once the configuration of the beams is established, some geometric parameters need to be optimized for increased structural efficiency. Hence, the NSGA-II algorithm was introduced to optimize the geometric parameters of the I-beams, with the objective function being to minimize mass, reduce stress, and maximize stiffness. The NSGA-II multi-objective optimization algorithm was adopted, in which each objective parameter is processed separately. The algorithm performs standard genetic operations of mutation and crossover on design variables. The selection procedure relies on two primary mechanisms: non-dominant sorting and crowding distance sorting, and its optimization flow chart is shown in Figure 5.

Multi-objective optimization aims to identify a collection of diverse solutions that together represent the optimal trade-off surface for several objectives. The surface generated in space is referred to as the Pareto front, specifically denoting the Pareto optimum solutions. Figure 6 displays the collection of Pareto optimum solutions that were obtained. And the following parameters were selected: B_1 = 7, B_2 = 25, B_3 = 1.5, H_1 = 2.1, H_2 = 4.2, and R = 2.01. Due to the uneven tensile and compressive strength of 3D-printing materials, with the tensile strength being superior to the compressive strength, a thickness ratio of 1.5 was employed. The thickness of the upper and lower edge strips was determined according to the following formula:

$$\begin{aligned} H_u &= H_2 + (R_{ub} - 1) \times H_2/(1 + R_{ub}) \\ H_b &= H_2 - (R_{ub} - 1) \times H_2/(1 + R_{ub}) \end{aligned} \tag{9}$$

here H_u is the thickness of the upper edge strip, H_b is the thickness of the lower edge strip, and R_ub is the ratio of the thickness of the upper and lower edge strips, which is 1.5.

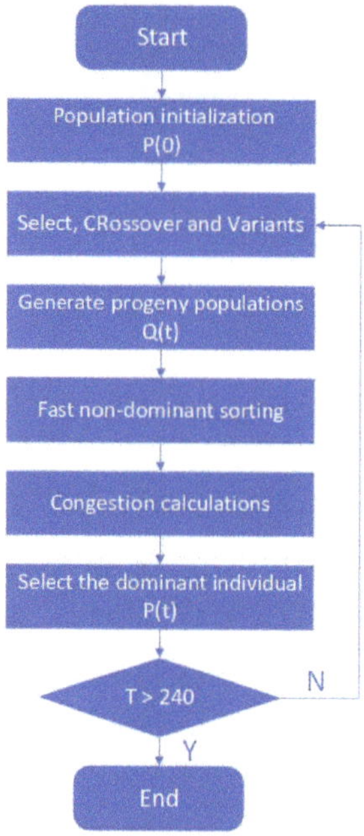

Figure 5. A flow chart of NSGA-II multi-objective optimization algorithm.

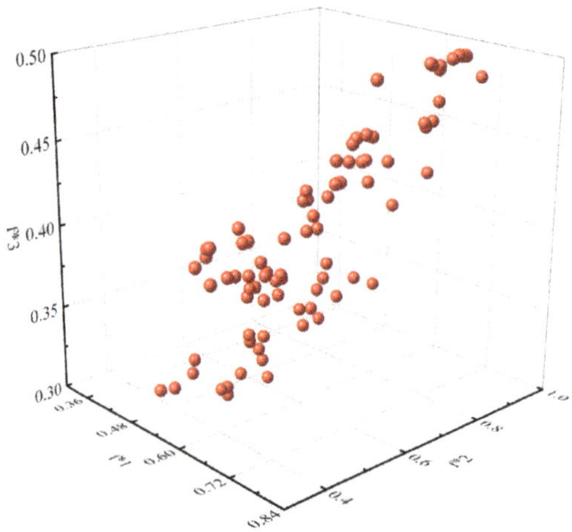

Figure 6. The collection of Pareto optimum solutions.

2.5. Geometry Parameters of Configuration

The geometry of variable cross-sections is presented in Figure 7, while the portions were divided into intervals of 21 mm and the sections were labeled sequentially from A-A to F-F. The corresponding parameters are listed in Table 3, in which W_b is the width of the beam, W_w is the width of the web, H_u is the thickness of the upper edge strip, and H_b is the thickness of the lower edge strip.

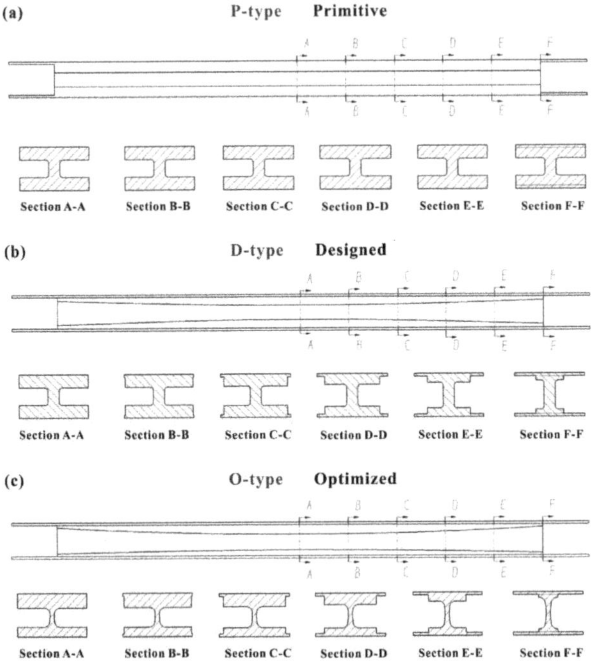

Figure 7. The geometry of variable cross-sections.

Table 3. Geometry parameters of configuration.

Type	Section Lable	W_b/mm	W_w/mm	H_u/mm	H_b/mm
P	A-A	25	4	4.5	4.5
	B-B	25	4	4.5	4.5
	C-C	25	4	4.5	4.5
	D-D	25	4	4.5	4.5
	E-E	25	4	4.5	4.5
	F-F	25	4	4.5	4.5
D	A-A	25	4	4.5	4.5
	B-B	24.4	4	4.41	4.41
	C-C	22.6	4	4.12	4.12
	D-D	19.6	4	3.66	3.66
	E-E	15.4	4	3.09	3.09
	F-F	10	4	2.5	2.5
O	A-A	25	1.5	5.04	3.36
	B-B	24.28	1.5	4.92	3.31
	C-C	22.12	1.5	4.57	3.16
	D-D	18.52	1.5	3.98	2.91
	E-E	13.48	1.5	3.16	2.55
	F-F	7	1.5	2.1	2.1

3. Results and Discussion

3.1. Flexural Properties of Different Configurations

Figure 8a illustrates the force-to-displacement curve of the primitive, designed, and optimized configurations of the SCFs-reinforced PAs. The P-type, D-type, and O-type I-beams exhibit linear characteristics at the beginning. Despite achieving a 30.63% reduction in mass through optimization (Figure 8b), the O-type I-beam still maintains a stiffness comparable to that of the P-type I-beam. However, the D-type I-beam demonstrates a significantly lower stiffness compared to both the O-type and P-type I-beams, while the mass is reduced by 14.46% compared to the P-type I-beams. Conversely, both the design and optimization resulted in an increase in the displacement of the 3D-printed I-beam at its maximum load as compared to the primitive one. This indicates that the 3D-printed I-beams' ability to maintain structural integrity was improved. Furthermore, the enhancement of structural efficiency and the achievement of lightweight effects in the designed and optimized 3D-printed I-beams are prominently demonstrated in Figure 8c,d. Compared with the primitive one, the designed and optimized structures also exhibit a 14.46% and 30.05% increase in the stiffness-to-mass ratio and a 7.83% and 40.59% increase in the load-to-mass ratio, respectively. These increases mean that D-type and O-type I-beams have smaller deflections than P-type I-beams under identical load conditions. Moreover, when the design standard requires the same structural failure load, D-type and O-type I-beams have less weight compared to P-type I-beams.

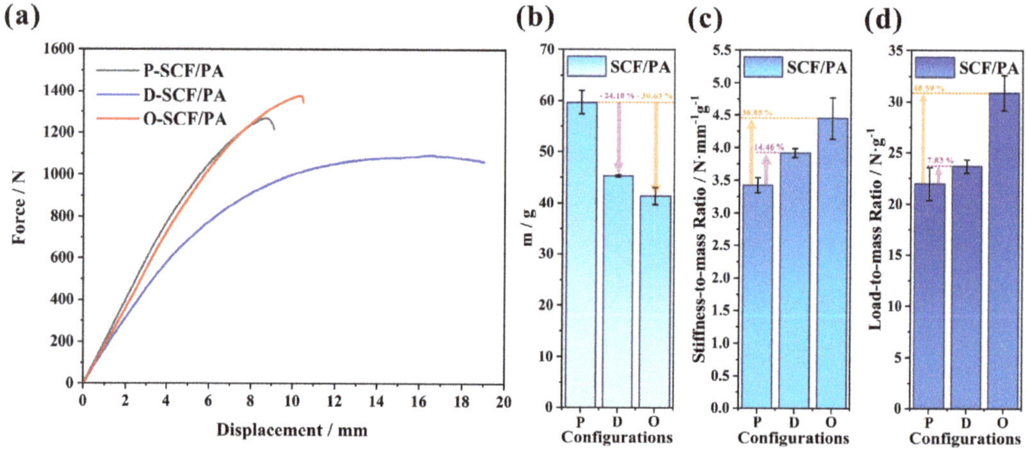

Figure 8. The flexural properties of the primitive, designed, and optimized configurations: (**a**) the force-to-displacement curve, (**b**) the mass, (**c**) the stiffness-to-mass ratio, and (**d**) the load-to-mass ratio.

A comparison of the failed specimens of the primitive, designed, and optimized configurations of the SCFs-reinforced PAs is presented in Figure 9. In all three configurations of the 3D-printed I-beams, considerable residual plastic deformation was revealed after the three-point bending testing. Additionally, marginal circular cross-section indents remain as the result of the compression of the loading head. These indents are visible in the zoom view (iv) of the top surface in Figure 9a–c. The primary mode of failure for the P-type I-beam is delamination, where the separations between the 3D-printed layers propagate and extend dramatically along the length direction of the beams (Figure 9a). Furthermore, this delamination failure mode appears on both the top and bottom surfaces of the P-type I-beam. The occurrence of delamination leads to rapid structural failures, swift load drops, and the inadequate utilization of material properties. Because at this point, structural damage is dominated by weak interfaces between 3D-printed layers, which is an undesirable form of damage.

The predominant cause of failure for the D-type and O-type I-beams is a combination of delamination and tensile failure. The fractures caused by delamination failure exhibit elongated and narrow apertures, dispersed longitudinally over the top surface of the D-type I-beam, while the top surface of the O-type I-beam only has indents attributable to the loading head. The fracture, resulting from tensile failure, exhibits an irregular serrated shape that is perpendicular to the direction of length and is observed on the bottom surface of the D-type and O-type I-beams. This is consistent with the three-point bending conditions in which the top surface of the I-beams is subject to the tensile load, and the bottom surface is under compression. Improvements in failure modes impact the structural flexural properties of 3D-printed I-beams. As presented in Figure 9c, the appearance of tensile fractures and the disappearance of delamination fractures indicate that the properties of the material were utilized to a greater extent. This is further supported by the substantial enhancement of the O-type I-beam in comparison to its primitive configuration, specifically in terms of its structural flexural properties (Figure 8).

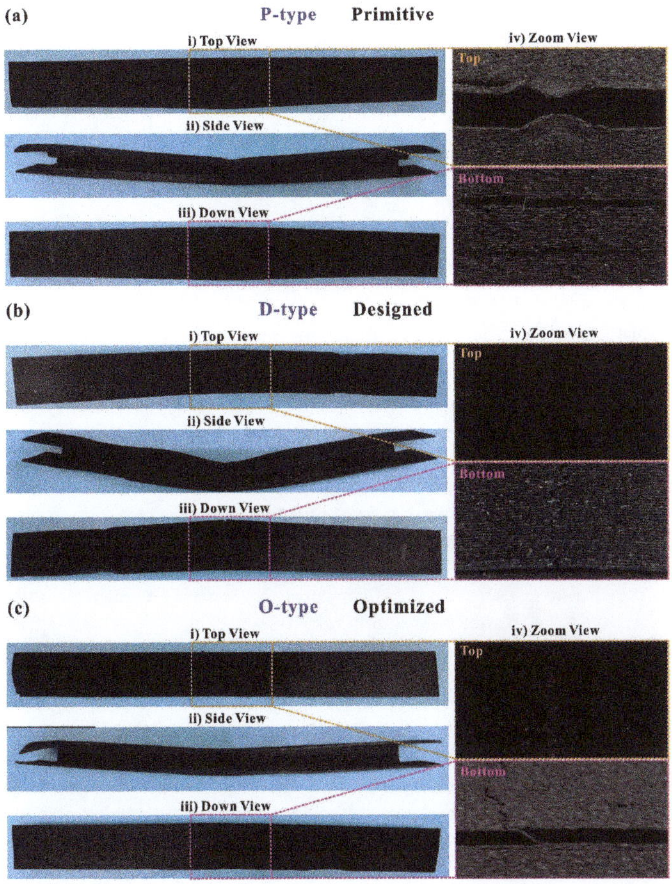

Figure 9. A comparison of the failed specimens of (**a**) the primitive, (**b**) designed, and (**c**) optimized configurations of SCFs-reinforced PAs in (**i**) top, (**ii**) side, (**iii**) down, and (**iv**) zoom views.

3.2. *Flexural Properties of Different Reinforcements*

Figure 10a presents the force-to-displacement curve of the CCFs-, SCFs-, and SGFs-reinforced PAs and PAs in the configuration of the O-type I-beam. The 3D-printed I-beam of PAs with no fiber reinforcements demonstrates an extremely brief linear stage and exhibits obvious nonlinear characteristics in advance of attaining the peak load. However, fibers have the capacity to substantially reduce the nonlinearity exhibited by 3D-printed I-beams. Among them, the reduction effect of short carbon fibers has a more pronounced reduction impact compared to short glass fibers. Additionally, continuous carbon fibers have an even greater reduction effect than short carbon fibers. Nevertheless, the displacement of the peak load also decreases when fibers are added. This indicates that the addition of continuous fibers leads to a greater decrease compared to short fibers. Furthermore, the addition of short carbon fibers results in a greater decrease than short glass fibers. As the displacement at the peak load decreases, brittleness supersedes ductility as the prevailing property of the 3D-printed beams.

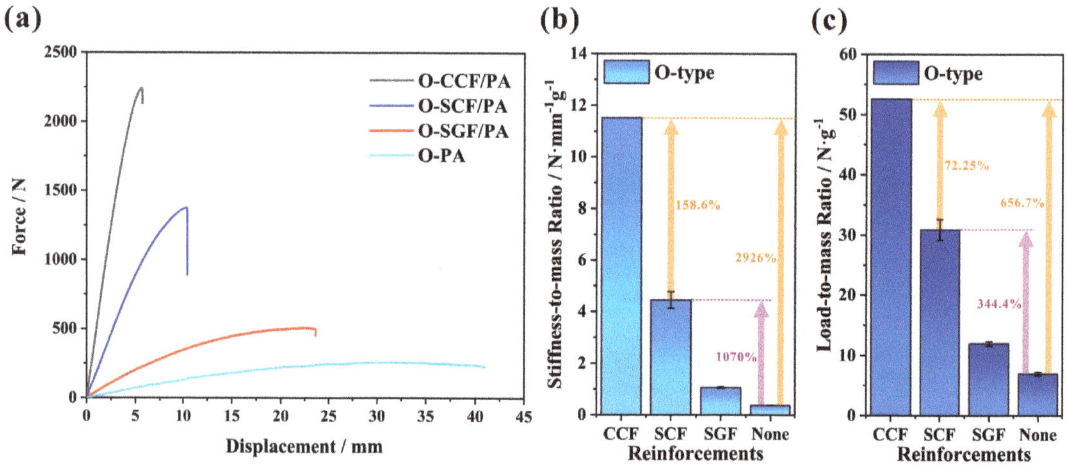

Figure 10. The flexural properties of CCFs-, SCFs-, and SGFs-reinforced PAs and PAs in the configuration of the O-type I-beam: (**a**) the force-to-displacement curve, (**b**) the stiffness-to-mass ratio, and (**c**) the load-to-mass ratio.

As shown in Figure 10b,c, due to the lower density of carbon fibers, the improvement in flexural properties is superior after taking into account the reduction in mass. Compared to the corresponding structure without reinforcements, the addition of CCFs increases the stiffness-to-mass ratio of the 3D-printed I-beam by 2926%. When the structure is subjected to the same design constraints for structural deformation, the improvement in the ratio of stiffness to mass allows for a substantial reduction in the weight of the structure. The continuous carbon fibers, which are evenly distributed along the length of the I-beam, serve as the framework for the beam structure. Working in conjunction with the matrix, they facilitate the smooth and uninterrupted transmission of loads. And the increase is 158.6% when compared to the corresponding structure with short carbon fibers. The addition of continuous carbon fibers greatly improves structural efficiency and significantly facilitates achieving lightweight structures. However, currently, printers and consumables for 3D printing with continuous carbon fibers are relatively expensive. Based on the supplier's quotation, the price for 100 g of CCF/PA prepreg yarn is around USD 337.5, whereas SCF/PA consumables of equivalent quality are priced at only USD 5.5. Consequently, the utilization of short carbon fibers to enhance structures is a more cost-effective option in certain industrial application scenarios, such as the structural components of drones, battery packs of new energy vehicles, and medical device casings, etc. The addition of short carbon fibers increases the stiffness-to-mass ratio of the 3D-printed I-beams by 1070% in comparison to the I-beams 3D-printed by PAs, representing a highly notable improvement. Similarly, the addition of fibers also significantly improves the load-to-mass ratio, which indicates the load-bearing capacity per unit mass. Compared with the structure without reinforcements, continuous and short carbon fibers increase the load-to-mass ratio of 3D-printed I-beams by 656.7% and 344.4%, respectively.

The comparison of the failed specimens of CCFs-, SGFs-, and non-reinforced PAs in the configuration of the O-type I-beam is presented in Figure 11. Both the CCFs- and non-reinforced 3D-printed I-beams exhibit little plastic deformation, while the underlying causes differ. The CCF-reinforced 3D-printed I-beam comprises over 50% of fiber content, while the content of the short glass fibers is 25% in the structures. Regarding the CCF-reinforced 3D-printed I-beam, it is important to note that the low resin content diminishes the structural plasticity. Conversely, the addition of CCFs leads to the structure attaining its maximum displacement at 5.61 mm and subsequently fracturing rapidly (Figure 10a). The limited structural displacement is insufficient to induce significant residual plastic deformation. For the 3D-printed I-beam made from PAs, the structural stiffness is relatively low (14.06 N·mm^{-1}), resulting in a certain level of overall flexibility in the structure. Upon eliminating the loading boundary conditions, the entire structure rebounds, leading to negligible residual plastic deformation. As shown in Figure 11, considering the mode of failure, delamination is the primary factor affecting CCFs-, SGFs-, and non-reinforced 3D-printed I-beams. There is slight open cracking along the length direction on both their upper and bottom surfaces, and the cracking on the lower surface is relatively obvious.

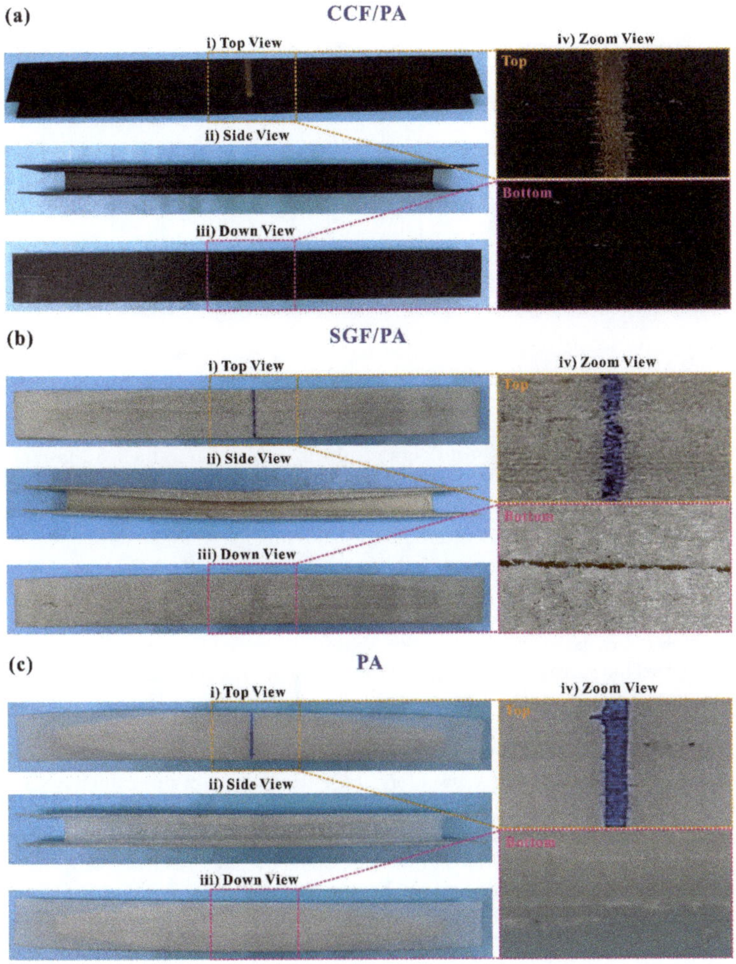

Figure 11. A comparison of the failed specimens of (**a**) CCFs-, (**b**) SGFs-, and (**c**) non-reinforced PAs in the configuration of the O-type I-beam in (**i**) top, (**ii**) side, (**iii**) down, and (**iv**) zoom views.

3.3. *Flexural Properties of Different Matrices*

Apart from the PA matrix, PLA is also utilized with SCFs, SGFs, and no reinforcements for 3D-printed O-type I-beams. And their flexural properties are illustrated in Figure 12. The force-to-displacement curve is present in Figure 12a, in which the addition of SCFs and

SGFs both increase the stiffness of the structure to 96.69 and 124.55 N·mm^{-1}, respectively. As depicted in Figure 12b, the mass of the O-type I-beams is reduced by 7.53% through the combination with SCFs. On the contrary, the mass of the structure is increased by 5.29% upon the addition of SGFs. Nevertheless, when stiffness and mass are both taken into account, SGFs prove to be a superior option. Both SGFs and SCFs contribute to an increase of 32.75% and 13.64% in the stiffness-to-mass ratio of the 3D-printed I-beams, respectively.

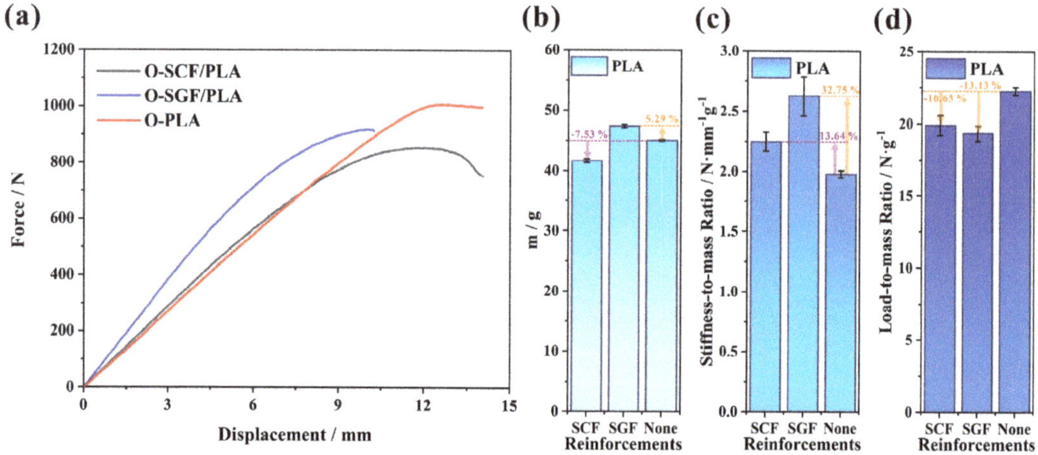

Figure 12. The flexural properties of SCFs- and SGFs-reinforced PLAs, and PLAs in the configuration of the O-type I-beam: (**a**) the force-to-displacement curve, (**b**) the mass, (**c**) the stiffness-to-mass ratio, and (**d**) the load-to-mass ratio.

Nevertheless, the addition of short fibers results in a decrease in the load-to-mass ratio of the PLA I-beams. Both SGFs and SCFs contribute to a decrease of 10.63% and 13.13% in the stiffness-to-mass ratio, respectively (Figure 12d). As listed in Table 1, the PLA matrix has an elongation at break of 29.92%, which is much lower than the elongation of the PA matrix (196%). And the E-glass fiber has a 3% elongation at break, whereas the T300-carbon fiber only has 1.5% elongation at break. By incorporating short fibers, the elongation at break of the entire structure will decrease, particularly affecting PLA, which has an elongation at break that is less than one-sixth of PAs. The force-to-displacement curve also illustrates the substantial decrease in the peak load of SCF- and SGF-reinforced PLA I-beams in comparison to the structure with a pure matrix (Figure 12a).

The comparison of the failed specimens of SCFs, SGFs, and non-reinforced PLA in the configuration of the O-type I-beam is presented in Figure 13. Due to their comparatively high stiffness and load displacement at peak force, SGF-reinforced I-beams exhibit negligible residual plastic deformation, while the failed specimens of SCFs- and non-reinforced I-beams demonstrate substantial residual bending distortion. The SGFs- and non-reinforced I-beams exhibit extensive structural damage, accompanied by fractures with uneven and jagged edges that permeate the whole structure. During the experiment, initial observations revealed the presence of white creases on the surfaces, which subsequently progressed into the material's fracture. The material's fracture dominates, and the delamination cracking caused by 3D printing is not observed, which indicates that the materials were effectively utilized. Consequently, the SGFs- and non-reinforced I-beams demonstrate prominent peak loads (Figure 12a). The initial failure of the SCFs-reinforced I-beams primarily manifests on the web (Figure 13a), resulting in a reduction in load capacity, which is an undesirable mode of failure.

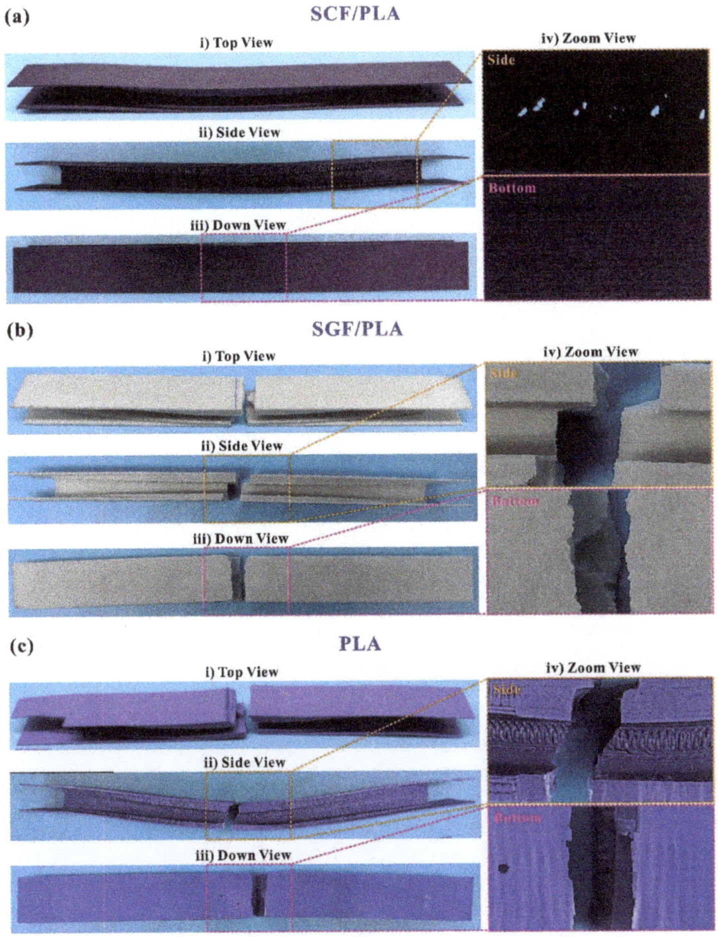

Figure 13. A comparison of the failed specimens of (**a**) SCFs-, (**b**) SGFs-, and (**c**) non-reinforced PLAs in the configuration of the O-type I-beam in (**i**) top, (**ii**) side, (**iii**) down, and (**iv**) zoom views.

3.4. Material Selection

The experimental results of the indexes of flexural properties, including the stiffness-to-mass ratio and load-to-mass ratio, are summarized in the Ashby-type materials selection charts (Figure 14). The CCFs-reinforced PAs have the most remarkable structural lightweight effect compared to other materials tested in this study. This is noticeable in their significantly greater stiffness-to-mass ratio (11.52 $N{\cdot}mm^{-1}g^{-1}$) and load-to-mass ratio (52.59 $N{\cdot}g^{-1}$). The SCFs-reinforced PAs have the potential to serve as economical substitutes for the CCFs-reinforced PAs. The 3D-prined I-beam of the O-type configuration produced by it has a relatively high stiffness-to-mass ratio (4.45 $N{\cdot}mm^{-1}g^{-1}$) and load-to-mass ratio (30.89 $N{\cdot}g^{-1}$). The application of SGFs is not recommended, due to the lack of significant enhancement in mechanical properties and the subsequent increase in structure weight, unless it is employed in specific industrial scenarios where lightweighting is not required.

For the matrix of the 3D-printed composites, PAs and PLAs have extremely distinct properties. The 3D-printed I-beams with pure PA resin demonstrate the lowest bending characteristics among all specimens. They possess the average stiffness-to-mass ratio of 0.38 $N{\cdot}mm^{-1}g^{-1}$ and the average load-to-mass ratio of 6.95 $N{\cdot}g^{-1}$. However, whether the addition of SGFs, SCFs, or CCFs indeed enhances the mechanical properties of the overall structure significantly remains to be determined. It is noted that reinforcements always play a crucial and beneficial function in PA composite I-beams. With regard to PLA resin, the mechanical performances are completely different. While the incorporation of SGFs and SCFs enhances the stiffness-to-mass ratio of the structure, it leads to premature structural failure and diminishes the load-to-mass ratio.

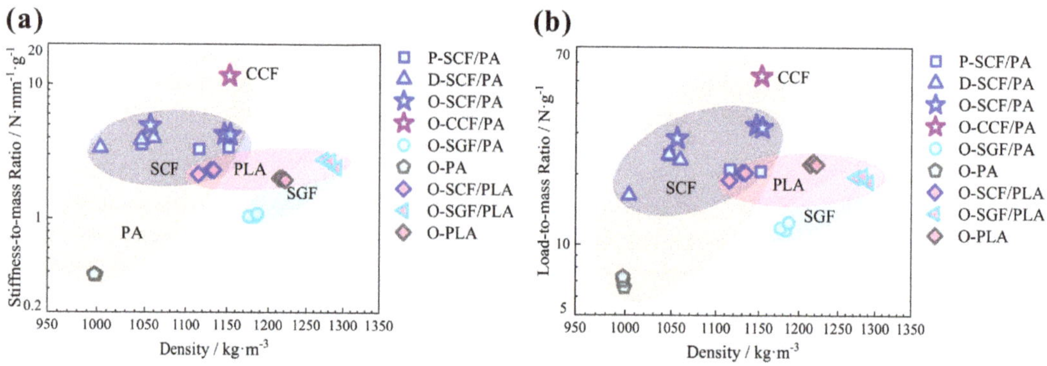

Figure 14. Ashby-type material selection maps: (**a**) stiffness-to-mass ratio and (**b**) load-to-mass ratio.

A comparison of the stiffness-to-mass ratio and load-to-mass ratio between this work and the literature is listed in Table 4. It should be noted that the flexural properties of beams are greatly affected by the span of supporting rollers. In general, the flexural properties tend to diminish as the span increases. Despite the fact that the span chosen for the present study is 200 mm, which is greater than the spans used in previous studies [9,26] (141 mm and 80 mm, respectively), the load-to-mass ratio demonstrates a reasonably similar level when identical material is used. As for the stiffness-to-mass ratio, this is more significantly influenced by the span than the load-to-mass ratio and becomes non-comparable across different spans. For example, the stiffness-to-mass ratio of [26] is only 4.53% of that of [9] when the span changes from 80 to 141 mm. Hence, it is suggested that the Ashby-type material selection maps should be used for the purpose of comparing the flexural properties of beams when the span and the design envelope (length, width, and height) are the same.

Table 4. Comparison between this work and the literature.

Configuration	Materials	Span/mm	H/mm	Stiffness-to-Mass Ratio/mm	Load-to-Mass Ratio/mm	Reference
O-type	CCF/PA	200	15	11.52	52.59	This work
O-type	SCF/PA	200	15	4.45	30.89	
D-type	SCF/PA	200	15	3.92	23.69	
Corrugated core	CCF/PA	141	23.2	28.49	57.98	[26]
	SCF/PA	141	23.2	10.6	23.44	
Diamond core	CCF/PA	80	10	627.7	68.33	[9]

4. Conclusions

In this study, the design and optimization of I-beams were performed to obtain improved configurations. 3D printing was utilized to manufacture I-beams with three distinct configurations and seven kinds of materials. Three-point bending testing was conducted to verify the flexural properties of different configurations, reinforcements, and matrices. The following conclusions can be derived:

1. The flexural properties of 3D-printed I-beams were significantly improved via structural design and multi-objective optimization, which were based on the equal-strength

philosophy and the NSGA-II algorithm, respectively. Design and optimization reduced the mass of I-beams made of SCFs-reinforced PAs by 24.10% and 30.63%, respectively. The structures also exhibited a 14.46% and 30.05% increase in the stiffness-to-mass ratio and a 7.83% and 40.59% increase in the load-to-mass ratio, respectively. The design and optimization method proved to be effective for 3D-printed I-beams.

2. Compared with pure PA structures, although the CCFs-reinforced PA I-beam exhibited a substantial improvement (2926%) in its stiffness-to-mass ratio, the addition of SCFs also increased the stiffness-to-mass ratio and the load-to-mass ratio by 1070% and 344.4%, respectively. Considering that the price of CCFs-reinforced PAs (USD 337.5) is much higher than SCFs-reinforced PAs (USD 5.5), SCFs-reinforced PAs have the potential to serve as an economical substitute for CCFs-reinforced PAs. The 3D-prined I-beam of the O-type configuration produced by SCF-reinforced PAs had a relatively high stiffness-to-mass ratio (4.45 $N{\cdot}mm^{-1}g^{-1}$) and load-to-mass ratio (30.89 $N{\cdot}g^{-1}$), while the corresponding PA structure had a stiffness-to-mass ratio of 0.38 $N{\cdot}mm^{-1}g^{-1}$ and a load-to-mass ratio of 6.95 $N{\cdot}g^{-1}$.
3. Compared with PLAs (elongation at break of 29.92%), PAs (elongation at break of 196%) are a better choice of matrix to demonstrate the adverse effects of reinforcements. The addition of reinforcements, such as SGFs, SCFs, or CCFs, improves the mechanical properties of PA composite I-beams to a great extent. However, in PLA resin, the incorporation of SGFs and SCFs increased the stiffness-to-mass ratio but caused premature structural failure and reduced the load-to-mass ratio. Although the addition of SCFs increased the stiffness-to-mass ratio of PLA I-beams by 13.64%, the load-to-mass ratio was reduced by 10.63%.

This work provides valuable insights into the design and optimization of 3D-printed I-beams, specifically focusing on their flexural properties and material selection. Optimization using the NSGA-II algorithm and structural design based on the equal-strength philosophy were both demonstrated to be effective for 3D-printed I-beams. Simultaneously, it is worth noting that the mechanical performances of 3D-printed materials are diminished during compression as opposed to tension. When financial considerations permit, CCFs are the optimal material for enhancing mechanical properties. SCFs, however, are comparatively economical alternatives. It was discovered that the highly ductile resin matrix served as a successful medium for adding reinforcements. These findings could facilitate the lightweight design and optimization of bending-resistant structures.

Author Contributions: Conceptualization, X.Z. and F.W.; funding acquisition, Y.T., Y.M. and C.Z.; investigation, Y.Z.; methodology, P.S.; project administration, Y.M.; resources, F.W. and Y.M.; supervision, Y.T.; validation, P.S. and Y.Z.; writing—original draft, X.Z.; writing—review and editing, C.Z. All authors have read and agreed to the published version of the manuscript.

Funding: This research was funded by the National Natural Science Foundation of China with grant number of 5207517, the Natural Science Foundation of Chongqing, China with grant number of cstc2021jcyj-msxmX0194, and the Shaanxi Provincial Natural Science Basic Research Program—General Project with grant number of 2023-JC-YB-019. And The APC was funded by the National Natural Science Foundation of China with grant number of 5207517.

Institutional Review Board Statement: Institutional Review Board Statement is not applicable for this study because there is no humans or animals involved.

Data Availability Statement: Data can be obtained from the authors on request.

Conflicts of Interest: Author Yunsheng Ma was employed by the company Shandong Chambroad Holding Group Co., Ltd. The remaining authors declare that the research was conducted in the absence of any commercial or financial relationships that could be construed as a potential conflict of interest.

References

1. Hsissou, R.; Seghiri, R.; Benzekri, Z.; Hilali, M.; Rafik, M.; Elharfi, A. Polymer composite materials A comprehensive review. *Compos. Struct.* **2021**, *262*, 113640. [CrossRef]
2. Zhang, J.; Lin, G.; Vaidya, U.; Wang, H. Past, present and future prospective of global carbon fibre composite developments and applications. *Compos. Part B Eng.* **2023**, *250*, 110463. [CrossRef]
3. Gemi, L.; Madenci, E.; Özkılıç, Y.O.; Yazman, Ş.; Safonov, A. Effect of fiber wrapping on bending behavior of reinforced concrete filled pultruded GFRP composite hybrid beams. *Polymers* **2022**, *14*, 3740. [CrossRef]
4. Zhao, F.; Bao, H. Shape sensing approach for composite and sandwich beam with generic cross-sections: Application to fiber-reinforced polymer composite airfoil. *Aerosp. Sci. Technol.* **2023**, *138*, 108314. [CrossRef]
5. Li, J.; Wang, J.; Yu, L.; Liu, W.; Zhang, L. Study on failure mechanisms of composite box beams on wind turbine blades under extreme combined loading. *Mech. Adv. Mater. Struct.* **2022**, *29*, 2176–2185. [CrossRef]
6. Li, Y.; Xiao, Y.; Yu, L.; Ji, K.; Li, D. A review on the tooling technologies for composites manufacturing of aerospace structures: Materials, structures and processes. *Compos. Part A Appl. Sci. Manuf.* **2022**, *154*, 106762. [CrossRef]
7. Chen, J.; Liu, X.; Tian, Y.; Zhu, W.; Yan, C.; Shi, Y.; Kong, L.B.; Qi, H.J.; Zhou, K. 3D-Printed anisotropic polymer materials for functional applications. *Adv. Mater.* **2022**, *34*, 2102877. [CrossRef]
8. Wang, Y.; Li, L.; Hofmann, D.; Andrade, J.E.; Daraio, C. Structured fabrics with tunable mechanical properties. *Nature* **2021**, *596*, 238–243. [CrossRef]
9. Wang, K.; Wang, D.; Liu, Y.; Gao, H.; Yang, C.; Peng, Y. Path Planning and Bending Behaviors of 3D Printed Continuous Carbon Fiber Reinforced Polymer Honeycomb Structures. *Polymers* **2023**, *15*, 4485. [CrossRef] [PubMed]
10. Tian, Y.; Chen, K.; Zheng, H.; Kripalani, D.R.; Zeng, Z.; Jarlöv, A.; Chen, J.; Bai, L.; Ong, A.; Du, H.; et al. Additively Manufactured Dual-Faced Structured Fabric for Shape-Adaptive Protection. *Adv. Sci.* **2023**, *10*, 2301567. [CrossRef] [PubMed]
11. Zhang, D.; Tian, X.; Zhou, Y.; Wang, Q.; Yan, W.; Zia, A.A.; Wu, L.; Li, D. Spatial 3D Printing of Continuous Fiber-Reinforced Composite Multilayer Truss Structures with Controllable Structural Performance. *Polymers* **2023**, *15*, 4333. [CrossRef]
12. Parandoush, P.; Lin, D. A review on additive manufacturing of polymer-fiber composites. *Compos. Struct.* **2017**, *182*, 36–53. [CrossRef]
13. Zhang, X.; Zheng, X.; Song, L.; Tian, Y.; Zhang, D.; Yan, L. Compressive properties and failure mechanisms of 3D-printed continuous carbon fiber-reinforced auxetic structures. *Compos. Commun.* **2023**, *43*, 101744. [CrossRef]
14. Li, N.; Li, Y.; Liu, S. Rapid prototyping of continuous carbon fiber reinforced polylactic acid composites by 3D printing. *J. Mater. Process. Technol.* **2016**, *238*, 218–225. [CrossRef]
15. Gupta, A.; Fidan, I.; Hasanov, S.; Nasirov, A. Processing, mechanical characterization, and micrography of 3D-printed short carbon fiber reinforced polycarbonate polymer matrix composite material. *Int. J. Adv. Manuf. Technol.* **2020**, *107*, 3185–3205. [CrossRef]
16. HR, M.; Benal, M.G.M.; GS, P.; Tambrallimath, V.; Ramaiah, K.; Khan, T.Y.; Bhutto, J.K.; Ali, M.A. Effect of Short Glass Fiber Addition on Flexural and Impact Behavior of 3D Printed Polymer Composites. *ACS Omega* **2023**, *8*, 9212–9220. [CrossRef] [PubMed]
17. Zeng, C.; Liu, L.; Bian, W.; Leng, J.; Liu, Y. Bending performance and failure behavior of 3D printed continuous fiber reinforced composite corrugated sandwich structures with shape memory capability. *Compos. Struct.* **2021**, *262*, 113626. [CrossRef]
18. Wu, S.; Shan, Z.; Chen, K.; Wang, S.; Zou, A.; Sun, Q. Bending properties and failure behavior of 3D printed fiber reinforced resin T-beam. *Polym. Compos.* **2022**, *43*, 4556–4568. [CrossRef]
19. Long, S.; Zong, H.; Liu, T.; Yao, X. Soft impact behavior of composite I-beams. *Compos. Struct.* **2023**, *304*, 116477. [CrossRef]
20. Chen, S.; Wei, Y.; Zhao, K.; Dong, F.; Huang, L. Experimental investigation on the flexural behavior of laminated bamboo-timber I-beams. *J. Build. Eng.* **2022**, *46*, 103651. [CrossRef]
21. Wang, J.; Zhang, L.; Li, J.; Huang, X.; Li, C. Investigation on torsional performance of wind turbine blades under pure torsion through subcomponent test. *Compos. Struct.* **2023**, *304*, 116466. [CrossRef]
22. Ibrahim, T.H.; Allawi, A.A.; El-Zohairy, A. Experimental and FE analysis of composite RC beams with encased pultruded GFRP I-beam under static loads. *Adv. Struct. Eng.* **2023**, *26*, 516–532. [CrossRef]
23. Ruzuqi, R. Tensile strength analysis of polymer composite materials fiber reinforced in the fiber boat application. *J. Res. Opin.* **2020**, *7*, 2763–2769.
24. Craig, R.R., Jr.; Taleff, E.M. *Mechanics of Materials*; John Wiley & Sons: Hoboken, NJ, USA, 2020.
25. Polilov, A.N.; Tatus, N.A.; Tian, X. Analyzing the correctness of equal strength composite profiled beam bending problems. *J. Appl. Mech. Tech. Phys.* **2019**, *60*, 144–155. [CrossRef]
26. Um, H.J.; Lee, J.S.; Shin, J.H.; Kim, H.S. 3D printed continuous carbon fiber reinforced thermoplastic composite sandwich structure with corrugated core for high stiffness/load capability. *Compos. Struct.* **2022**, *291*, 115590. [CrossRef]

polymers

MDPI

Article

The Development of Biocomposite Filaments for 3D Printing by Utilizing a Polylactic Acid (PLA) Polymer Matrix Reinforced with Cocoa Husk Cellulose Fibers

Victor Hugo Martins de Almeida [1,*], Raildo Mota de Jesus [2], Gregório Mateus Santana [2], Sabir Khan [2], Erickson Fabiano Moura Sousa Silva [1], Iago Silva da Cruz [1], Ian de Souza Santos [1] and Paulo Neilson Marques dos Anjos [2]

[1] Department of Engineering and Computing, State University of Santa Cruz (UESC), Jorge Amado Highway, Km 16, Ilhéus 45662-900, Bahia, Brazil; efmssilva@uesc.br (E.F.M.S.S.); iagocruz.engmec@gmail.com (I.S.d.C.); iansouzasantos18@gmail.com (I.d.S.S.)

[2] Department of Exact Sciences, State University of Santa Cruz (UESC), Jorge Amado Highway, Km 16, Ilhéus 45662-900, Bahia, Brazil; rmota@uesc.br (R.M.d.J.); gregorioengflorestal@gmail.com (G.M.S.); sabir_chemist@yahoo.com (S.K.); pauloneilson@uesc.br (P.N.M.d.A.)

* Correspondence: vhmalmeida@uesc.br

Abstract: Vegetable fibers are increasingly used in biocomposites, but there is a need for further development in utilizing by-products like cocoa husks. Three-dimensional printing, through Fused Filament Fabrication (FFF), is advancing rapidly and may be of great interest for applying biocomposite materials. This study focuses on developing innovative and fully biodegradable filaments for the FFF process. PLA filaments were prepared using cellulose fibers derived from cocoa husks (5% mass ratio). One set of filaments incorporated fibers from untreated husks (UCFFs), while another set utilized fibers from chemically treated husks (TCFFs). The fabricated materials were analyzed using scanning electron microscopy (SEM), thermogravimetric analysis (TGA), and Fourier transform infrared (FTIR) techniques, and they were also tested for tensile strength. ANOVA reveals that both UCFFs and TCFFs significantly predict tensile strength, with the UCFFs demonstrating an impressive R^2 value of 0.9981. The optimal tensile strength for the filament test specimens was 16.05 MPa for TCFF8 and 13.58 MPa for UCFF8, utilizing the same printing parameters: 70% infill and a layer thickness of 0.10 mm. Additionally, there was an 18% improvement in the tensile strength of the printed specimens using the filaments filled with chemically treated cocoa husk fibers compared to the filaments with untreated fibers.

Keywords: additive manufacturing; bioplastics; lignocellulosic materials; natural fiber; composites

1. Introduction

Biocomposite materials of thermoplastic polymers reinforced with vegetable fibers have attracted the attention of researchers and companies seeking more sustainable alternatives with superior mechanical performance [1]. The mechanical strength properties of these biocomposites primarily depend on the type of incorporated vegetable fiber, the aspect ratio (length/diameter), the quality of adhesion at the interface between the embedded fiber and the polymer matrix, and the volume fraction of fiber added to the mixture [2]. While incorporating vegetable fibers into a polymer composite can improve its material properties, it can also worsen the properties of the material, depending on the interaction of the abovementioned factors. Beyond the mechanical properties, incorporating vegetable fibers can help reduce the consumption of pure polymer and make these materials more environmentally friendly. Research and development on and the application of these bio-based materials can benefit the ecosystem and contribute to local socio-economic development [3].

Several studies have shown the application of PLA together with a natural filler, such as bamboo, jute, flax, pineapple, sisal, or kenaf, to produce green composites [3–5]. In Kumar's study [6], a tensile strength value of 61.07 MPa was reported for a PLA biocomposite with cellulose nanocrystals at proportions of 99/1%, respectively. Another notable study was presented by Liu [7] with a PLA biocomposite with sugarcane bagasse at proportions of 94/6%, respectively, which exhibited a tensile strength of 57.1 MPa. These studies have demonstrated that it is possible to produce biocomposites with optimized mechanical strength compared to pure PLA, serving as a reference to this study.

Cocoa husks are highlighted in this research due to their local abundance and frequent disposal as waste from cocoa processing, making them a significant focus among lignocellulosic raw materials. Their utilization could offer sustainable solutions while addressing environmental concerns and contributing to local economic (*Theobroma cacao* L.) production; the husks represent approximately 80% of the total weight of the fruit. Cocoa fruit husks are produced in significant amounts within cocoa bean industries, yet they are often discarded due to their limited market value, despite being generated as waste in cocoa plantations [8]. Cocoa husks are a source of natural vegetable fiber, with lignocellulosic material comprising approximately 87% of the dry weight of the husks. Its chemical composition on a dry weight basis (%*w*/*w*) is approximately 35.4% cellulose, 37% hemicellulose, and 14.7% lignin, and the remainder consists of ash and other components, such as wax, pectin, moisture, impurities, etc. [9]. Therefore, pretreatments are essential in the cellulose extraction process and the removal of impurities.

This study outlines the development of biocomposite materials intended for use as raw materials (filament) in a 3D printer. Cocoa husk fibers were crushed, and cellulose microcrystals were chemically isolated. Two types of biocomposites (in the form of 3D filaments) were produced: one with untreated cocoa husk fibers and PLA and another with chemically isolated cellulose microcrystals and PLA. Alkaline and bleaching treatments were employed due to their recognized effectiveness in removing impurities from cellulose fibers, as well as them being more accessible and cost-effective techniques [10].

The amount of fiber added significantly impacts the mechanical properties of the composite. Studies indicate that the tensile strength tends to decrease, with a more pronounced reduction as the fiber content increases compared to the polymer. Overall, a lower fiber content tends to yield superior mechanical performance in biocomposites [8]. The fiber weight percentage added was 5% relative to PLA in both cases.

The untreated cocoa husk fibers and chemically treated microfibrils were analyzed using TGA and FTIR tests, while the mechanical properties of the printed parts with the filaments were tested using tensile tests. The mechanical tests followed international ASTM D638-14 [11].

2. Materials and Methods

2.1. Materials

PLA granules (Ingeo Biopolymer 3051D) suitable for 3D-printing the filaments, with a density of 1.24 g.cm ^{3}, were used, sourced from NatureWorks© (Minnetonka, MN, USA). Cocoa husks were purchased from a local market in Ilhéus, southern Bahia, Brazil. Sodium hydroxide (NaOH), sodium hypochlorite (NaClO), and glacial acetic acid from Sigma-Aldrich (Barueri, São Paulo, Brazil) were also used, available at the Laboratory of Materials and Environment (LAMMA) of the State University of Santa Cruz (UESC).

2.2. Processing of Cellulose Fibers

The processing of the cocoa husks is summarized in Figure 1. The cocoa husks were washed to remove dirt and impurities and then dried at 105 °C in an oven until they reached a constant weight. Next, the pieces were ground using a knife mill. After grinding, the fiber was sieved through a mesh #325 (particles smaller than or equal to 44 μm). Part of the ground fibers, referred to as untreated cocoa husk fibers (UCF), was stored in a hermetic container for later use and analysis (chemical and morphological). The other part of the fibers was treated in a beaker with a 4% w/v sodium hydroxide (NaOH) solution at 80 °C for 4 h with mechanical stirring. After this treatment, the fibers underwent vacuum filtration, followed by washing with distilled water and then vacuum filtration again. Next, the fibers were treated in a 1:1 aqueous dilution of sodium hypochlorite (NaClO) for 30 min at 80 °C. Acetic acid was added until the solution reached a pH of 4. The fibers were then vacuum-filtered, washed with distilled water, dried, and finally ground in a ball mill for 60 min. The chemically treated fibers, referred to as treated cocoa husk fibers (TCFs), were stored in hermetic containers for later use and analysis.

Figure 1. Processing steps for cocoa husk fibers.

2.3. Preparation of the Filaments and Printing of the Test Specimens

Two types of filaments were created (Figure 2), one using untreated cocoa husk fiber, referred to as UCFF, and another with chemically treated cocoa husk fiber, called TCFF. In both cases, the cellulose microfibers and PLA pellets were weighed at a mass ratio of 5% fiber to 95% PLA and then dehumidified for 12 h at 60 °C. The filaments were manufactured using a single-screw extruder, where the fiber–PLA mixture was fed through a funnel to the extruder, and the heater was adjusted to temperatures between 185 °C and 205 °C. The filament diameter was 1.75 ± 0.05 mm.

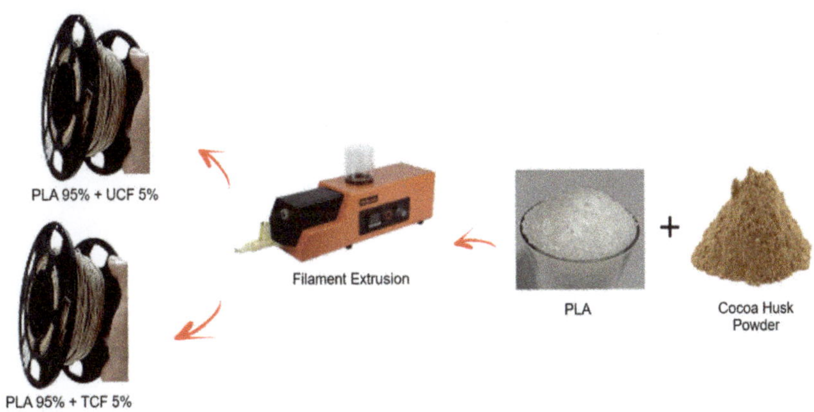

Figure 2. Extrusion steps for PLA with untreated cocoa husk fiber filament (UCFFs) and PLA with chemically treated cocoa husk fiber filaments (TCFFs).

The test specimens (TSs) printed for the tensile test were sized according to the guidelines of the international standard ASTM D638-14 [11]. The Type I TS used as a reference is illustrated in Figure 3.

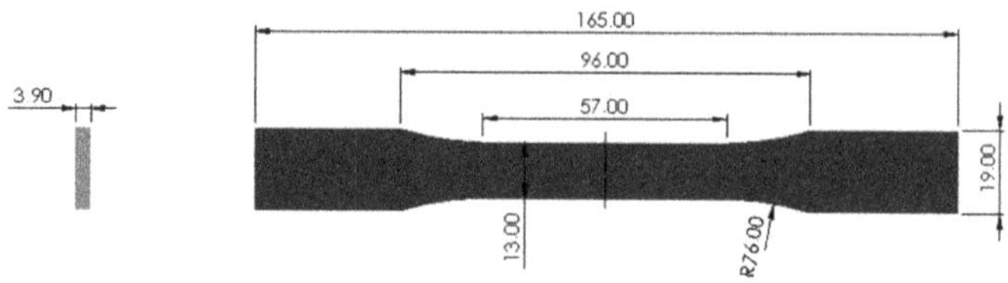

Figure 3. Type I test specimen for tensile strength testing according to ASTM D638-14.

The planning of the experiments, as well as the quantity of TSs manufactured by the 3D printing process, was defined using face-centered central composite design with two factors, two blocks, and ten experiments (Table 1).

Table 1. Levels of adjustment for input variables: infill (%) and layer height (mm) for central composite design, face-centered.

Factor	Lower Limit (−1)	Central Point (0)	Upper Limit (+1)
Infill (%)	30	50	70
Layer height (mm)	0.10	0.15	0.20

The input variables chosen were infill percentage and layer height. For the infill percentage factor, a lower limit of 30%, a central point of 50%, and an upper limit of 70% were adopted. For the layer height factor, a lower limit of 0.1 mm, a central point of 0.15 mm, and an upper limit of 0.20 mm were adopted. These two variables of the 3D printing process were identified as critical for the analysis and will be the input parameters studied for comparison of the results in subsequent tests. The combinations of these parameters are detailed in Table 2.

Table 2. Central composite design—central face, experiments for the input variables infill and layer height.

Test Specimen (TS)	Infill (%)	Layer Height (mm)
01	30	0.10
02	30	0.15
03	30	0.20
04	50	0.10
05	50	0.15
06	50	0.15
07	50	0.20
08	70	0.10
09	70	0.15
10	70	0.20

The mechanical designs of the test specimens (TSs) were created using SolidWorks 2023© software, where they were also later exported into STL file format. These files were imported into the slicing software PrusaSlicer© 2.7.4, where the G-codes used in the 3D printer to manufacture each model were generated. The Fused Filament Fabrication (FFF) process parameters used in the Ender-3 Pro 3D printer (Creality, Shenzhen, China), at Laboratory of Mechanical Projects and Tribology (LAPMET) of UESC and configured in the PrusaSlicer© software for the manufacturing of the CPs are detailed in Table 3. These parameters were kept constant for all the test specimens to ensure standardization. The tensile tests were carried out in triplicate, totaling 30 printed test specimens for each of the two types of filaments.

Table 3. FFF process parameters used in the printing of the TSs.

Parameter	Infill (%)
Extruder Diameter (mm)	0.40
Extrusion Multiplier	1.10
Layer Height	As per Table 2
Number of Top Layers	4
Number of Bottom Layers	4
Number of Perimeter Lines	3
Internal Infill Pattern	Hexagonal
External Infill Pattern	Rectilinear
Infill (%)	As per Table 2
Internal Infill Angle (°)	45
Extrusion Temperature (°C)	180
Bed Temperature (°C)	60
Extruder Travel Speed (mm/s)	40
Filament Diameter (mm)	1.75

2.4. Characterization Techniques for the Fibers and the Biocomposites

A thermogravimetric (TGA) analyzer model DTG-60H (Shimadzu©, Barueri, São Paulo, Brazil) at the Laboratory of Research and Innovation in Advanced Materials (LAPIMA) of UESC was used to perform the thermogravimetric analysis on the untreated cocoa husk fibers (UCFs) and chemically treated cocoa husk fibers (TCFs). Approximately 6 mg of powder was placed in an alumina crucible and heated to 1000 °C at an incremental rate of 10 $°C \cdot min^{-1}$ under a nitrogen atmosphere with a flow rate of 100 $mL \cdot min^{-1}$.

The Fourier transform infrared (FT-IR) spectra of the untreated cocoa husk fibers (UCFs) and chemically treated cocoa husk fibers (TCFs) were recorded using an Attenuated Total Reflectance (ATR) spectrometer, model IRPrestige-21 (Shimadzu©), at the Laboratory of Research and Innovation in Advanced Materials (LAPIMA) of UESC. The spectra were obtained in the wavenumber range between 4000 and 450 cm^{-1}.

The morphology of the untreated cocoa husk fibers (UCFs) and chemically treated cocoa husk fibers (TCFs) was visualized using a Quanta 250 Scanning Electron Microscope (SEM), model (FEI Company©, Hillsboro, OR, USA), at the Center for Electron Microscopy (CME) of UESC. The powdered materials were placed on double-sided carbon adhesive tape in the sample holder. The sample was gently pressed with paper, and the excess material was removed. Subsequently, the morphological characteristics were visualized.

The tensile strength mechanical tests were conducted according to the guidelines of the international ASTM standards. For the tensile test, the reference was ASTM standard D638-14 [11]. The Instron 5982 universal testing machine (Instron©, Norwood, MA, USA) was used to perform the tensile test, with a crosshead speed of 0.0000254 $m.s^{-1}$ (1.524 $mm.min^{-1}$) at the Laboratory of Experimental Measurements and Uncertainty Assessment (LAMEAI) of UESC.

2.5. Statistical Analyses

Response Surface Methodology (RSM) was employed to investigate the relationship between tensile strength (σt) and the parameters of internal infill and layer height. The experimental data were analyzed using Excel 365© and Minitab 21.3.0 © software. Analysis of variance (ANOVA) was conducted to determine the significance level of each factor (linear, quadratic, and a linear combination of these factors).

A significance level of 95% was adopted, corresponding to a p-value of ≤ 0.05. According to [12], there are five indices for validating a model: the coefficient of determination (R^2), mean absolute percentage error (MAPE), root mean square error (RMSE), relative error (%), and multi-objective error function (MO).

The R^2 and RMSE indices were used. There are no specific values for each of these indices; however, for R^2, the higher the percentage value, the greater the influence of the analyzed variables on the response. RMSE evaluates the performance of prediction models by providing a measure of how well the predictions match the observed data. A lower RMSE value indicates a better fit of the model to the observed data, with a value near zero indicating a closer match between the observed and predicted values [12].

3. Results

3.1. Characterizations Analysis

3.1.1. Morphological Characterization

A detailed comparative analysis of the surface morphologies of both the untreated cocoa husk fiber samples and those chemically treated is depicted in detail in Figure 4 using scanning electron microscopy (SEM). Figure 4a shows a SEM micrograph of the untreated cocoa husk fibers (UCFs), while Figure 4b displays a SEM micrograph of the cocoa husk fiber samples subjected to chemical treatment (TCFs). Analyzing the surface morphology of the fibers was pivotal to discern the surface charge of these samples, especially in the case of the untreated fibers. In Figure 4a, it is evident that the surface of the untreated cocoa husk fiber (UCF) exhibits greater uniformity, forming a denser and more filled material due to the presence of waxes, hemicellulose, pectin, lignin, and other impurities [13]. Conversely, in Figure 4b, the surface of the chemically treated cocoa husk fiber (TCF) displays increased porosity, with more empty spaces between the material, attributed to the partial or complete removal of waxes, hemicelluloses, pectin, and lignin.

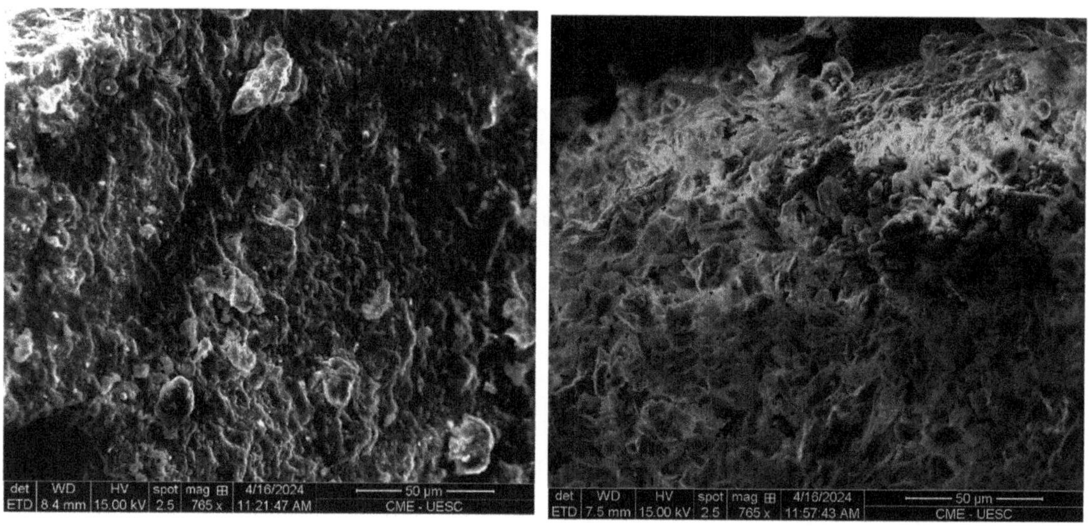

(**a**) (**b**)

Figure 4. SEM images of (**a**) the cellulose fibers from untreated cocoa husk (UCFs); and (**b**) of the cellulose fibers from the chemically treated cocoa husk (TCFs).

The SEM micrographs of the fractured test specimens of pure PLA, the test specimens created with filament using untreated cocoa husk fiber (UCFF), and test specimens created with filament using cocoa husk fiber subjected to chemical treatment (TCFF) are presented in Figure 5. The basic goal was to determine whether there were any observable fiber aggregates on the fractured surfaces. In comparison to pure PLA (Figure 5a), the surfaces of the UCFF (Figure 5b) and TCFF (Figure 5c) biocomposites exhibited rougher textures, with protrusions, small voids (red arrows), and well-dispersed fibers (yellow circles). In the TCFF specimen (Figure 5c), areas attributed to small fiber aggregates (red circles) can be observed, which are only visible in the TCFF sample. However, overall, a good dispersion of fibers was achieved in the UCFF and TCFF biocomposites.

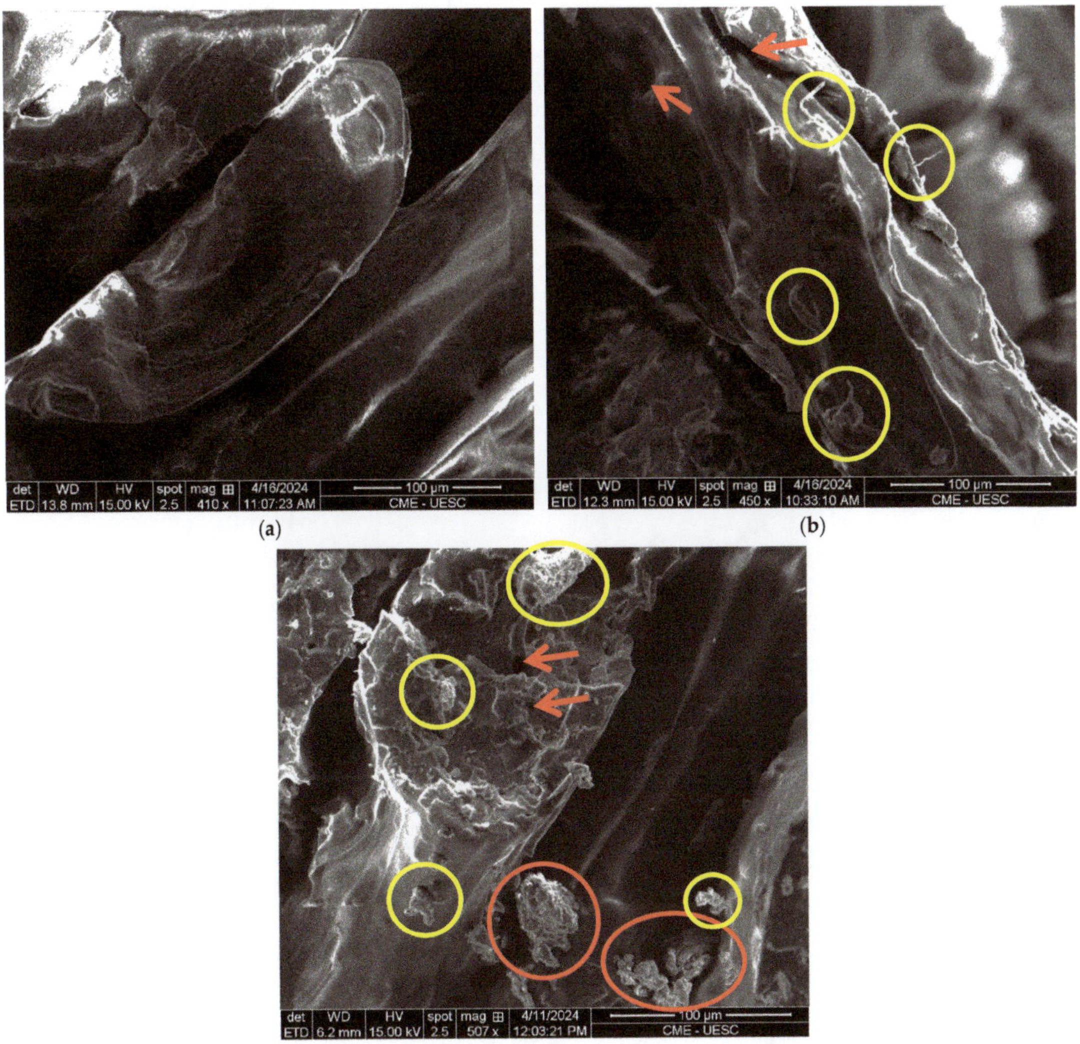

(c)

Figure 5. SEM images of fractured specimens after tensile testing: (**a**) pure PLA specimen; (**b**) PLA specimen with untreated cocoa husk cellulose fibers (UCFFs); (**c**) PLA specimen with chemically treated cocoa husk cellulose fibers (TCFFs).

3.1.2. Thermogravimetric Analysis

The thermogravimetric analysis of untreated cocoa husk fiber (UCF) and chemically treated cocoa husk fiber (TCF) is presented in Figure 6a. The TGA plot illustrates the percentage of mass loss concerning the temperature increase. Both samples exhibit distinct stages of mass loss in the thermogravimetric analysis, indicating the thermal degradation behavior of untreated and chemically treated cocoa husk fibers. A marginal decrease in mass percentage is observed in the samples from 20 °C to 110 °C, with the UCF displaying the highest mass loss in this range, around 14%, and the TCF showing the lowest loss, approximately 6%. The decline in mass percentage observed in this range is attributed to the evaporation of moisture adsorbed on the sample surfaces, along with chemisorbed water molecules bound to the hydrogen within the samples [13,14]. The second region was identified between 130 and 380 °C in the TCF sample and between 210 and 380 °C in the UCF sample. This range corresponds to the degradation primarily of hemicellulose and cellulose, along with a fraction of the lignin. The third region of thermal degradation occurs at temperatures between 380 and 600 °C. Temperatures above 380 °C are associated with the pyrolysis of lignin [15] and the oxidation of the fixed carbon present in cocoa husks [13].

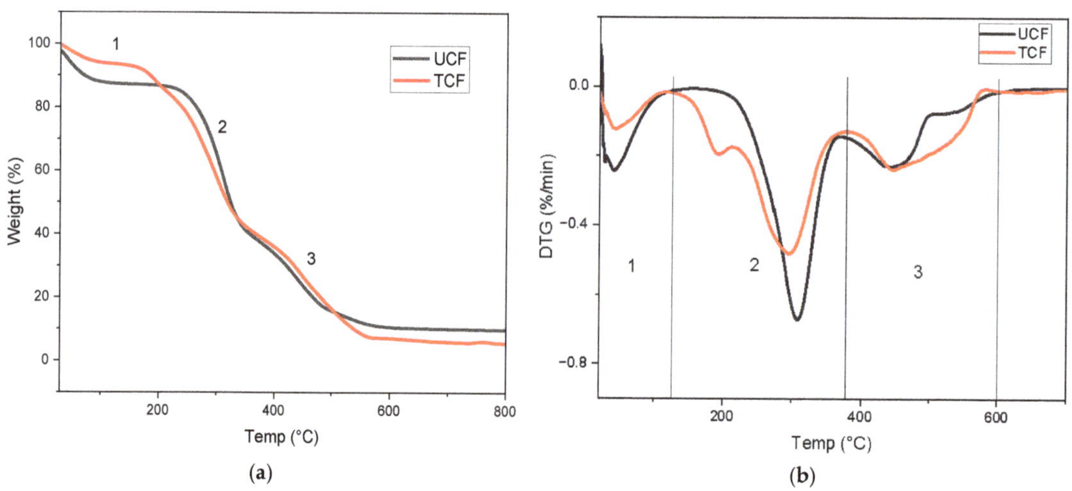

Figure 6. TGA (**a**) and DTG (**b**) graphs of cellulose fibers from untreated cocoa husks (UCF) and cellulose fibers from chemically treated cocoa husks (TCF).

Beyond 600 °C, ash formation constitutes the final residue, comprising inorganic elements. The untreated cocoa husk fiber (UCF) and chemically treated fiber (TCF) left residues of 9.8% and 5.5% of the initial mass, respectively.

The DTG curves (Figure 6b) of TCF and UCF revealed cellulose degradation peaks at 295 and 309 °C, respectively. TCF exhibited a maximum thermal stability temperature of 295 °C, lower than UCF due to the reduced cellulose crystallinity [9], as confirmed by the FTIR spectral analyses. TCF exhibited a peak at 190 °C, attributed to the residual hemicellulose [12]. Above 380 °C, a continuous and slow weight loss is attributed to lignin degradation [13], with a degradation peak at 440 °C for both UCF and TCF.

3.1.3. Fourier Transform Infrared Spectroscopy (FTIR)

The change in the chemical composition of chemically treated cocoa husk fiber (TCF) relative to untreated cocoa husk fiber (UCF) is illustrated in the FTIR spectrum according to the absorbance in Figure 7. Through this analysis, the components of the raw material were verified due to the existence of the principal bands related to cellulose, hemicellulose, and lignin. While it is a qualitative analysis, the peak amplitudes can indicate the substance concentration or the number of molecules with specific vibrations. Thus, the differences in the amplitudes of the spectra between UCF and TCF are indicative of the removal of lignin, hemicellulose, and extractives.

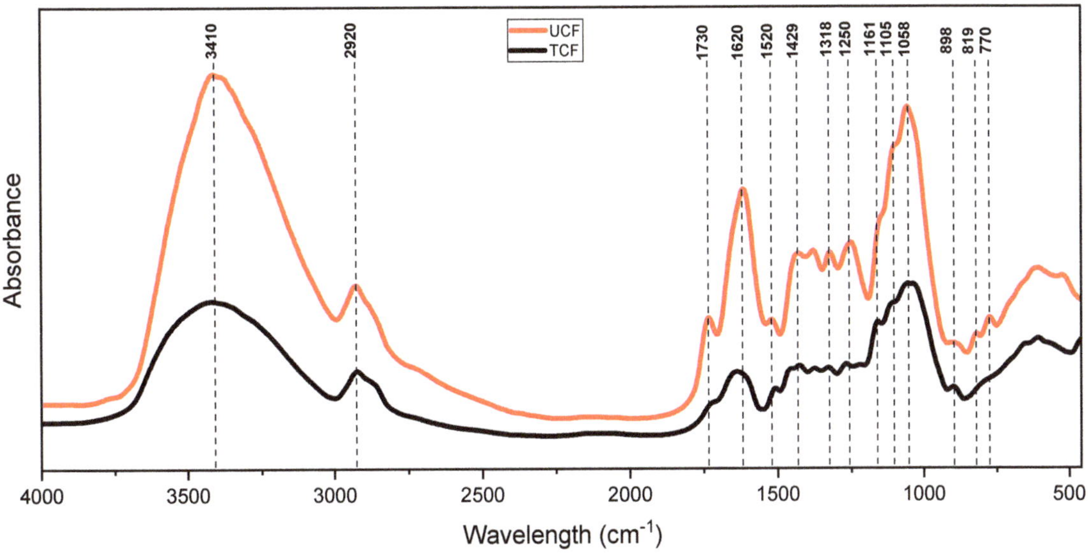

Figure 7. FTIR spectra of untreated cocoa husk fiber (UCF) and chemically treated cocoa fiber (TCF).

The spectral region between 3200 and 3600 cm^{-1} indeed corresponds to the stretching vibration of the hydroxyl (OH) group band, which is commonly found in biopolymers such as cellulose, lignin, and hemicellulose. Additionally, this peak includes inter- and intramolecular hydrogen bonding vibrations in cellulose [13].

With the removal of part of the lignin and hemicellulose, this peak tended to become narrower and sharper [16] since the TCF was also milled in the ball mill, which caused some crystalline cellulose structures to break, resulting in a more amorphous structure; a reduction in signal intensity at 3410 cm^{-1} can be observed. These results are consistent with literature reports on the effects of mechanical milling on the structure of cellulose [17].

The peak at 2920 cm^{-1}, more prominent in the UCF spectrum, corresponds to the stretching vibrations of C-H from the methyl and methylene groups of lignin and hemicellulose [16,18]. The reduction observed in the TCF indicates the partial removal of these components. The peak at 1730 cm^{-1} corresponds to the C=O stretching vibrations of the acetyl and ester groups found in hemicellulose and fiber extractives [19]. It is solely present in UCF, suggesting the removal of these components in TCF. The peak at 1620 cm^{-1}, attributed to the aromatic ring of lignin [20], can be observed in both spectra, with a lower intensity in TCF, indicating once again the partial removal of lignin. Similarly, the reduction in the bands at 1520 cm^{-1} and 1250 cm^{-1}, related to C=C group and C-H bond vibrations in lignin's methylene and methyl groups [21,22], also corroborates these results.

The absorption band at 1429 cm^{-1} corresponds to the asymmetric stretching of CH_2, indicating the crystallinity of the cellulose material. This peak becomes broader if the cellulose material has lower crystallinity [23]. In the case of TCF, this signal is broader, which corroborates the decrease in the crystallinity of the cellulose present, as the TCF samples were milled in a ball mill.

A band at 1318 cm^{-1} corresponds to the deformation of O-H and the vibration of CH2 in cellulose [24]. The vibration at 1161 cm^{-1} is attributed to the anti-symmetric stretching vibration of the C-O-C bridge, while the vibration at 1105 cm^{-1} is designated for the anti-symmetric stretching band of the in-plane ring. The intensity of these two peaks also decreases as the crystallinity of cellulose decreases [23]. The band at 1058 cm^{-1} has been associated with the stretching of C-C and C-O in polysaccharides and lignin [16,17,25]; there was a reduction in signal intensity in TCF, corroborating the removal of hemicellulose and lignin during the chemical treatment.

The vibration at 898 cm^{-1} results from the stretching of C-O-C in the β-(1-4) glycosidic linkage. This band is identified as the absorption band of amorphous cellulose, whose relative intensity increases with a reduction in crystallinity [17,23,26]. There was an increase in this peak in TCF due to the increase in the amorphous phase of cellulose after milling in the ball mill.

In UCF, there are peaks in the region near 819 cm^{-1} and 770 cm^{-1}, which have been associated with the functional groups of the deformed C-H bonds present in lignin [27,28]; there was a significant reduction in these bands in TCF, confirming the removal of part of the lignin after the chemical treatments applied to the raw fibers.

3.2. Results of Mechanical Tensile Tests

The results from the mechanical tensile test for the PLA/cocoa husk fiber biocomposite filaments are presented in Table 4. The plot shows the average values of tensile strength (σt), modulus of elasticity (Et), and strain (ε%) for the PLA filament with untreated cocoa husk fibers (UCFF) and the PLA filament with chemically treated cocoa husk fiber (TCFF). The best tensile strength (σt) result for the printed specimen using TCFF filaments was 16.05 MPa (TCFF8), while for the UCFF filament specimen, it was 13.58 MPa (UCFF8), both printed with the same settings: 70% infill and a 0.10 mm layer thickness.

Table 4. Tensile strength values (σt), elastic modulus (Et), and deformation (ε%) for test specimens printed with PLA filaments containing untreated cocoa husk fibers (UCFFs) and PLA filaments containing chemically treated cocoa husk fibers (TCFFs).

Test Specimen (TS)	Infill (%)	Layer Height (mm)	Tensile Strength (MPa)	Elastic Modulus (MPa)	Strain (ε%)
UCFF1	30	0.10	10.08	358.6	2.82
UCFF2	30	0.15	9.62	322.9	2.99
UCFF3	30	0.20	8.67	279.8	3.11
UCFF4	50	0.10	12.95	466.5	2.81
UCFF5	50	0.15	12.77	442.7	2.89
UCFF6	50	0.15	12.69	436.3	2.92
UCFF7	50	0.20	11.54	379.9	3.05
UCFF8	70	0.10	13.58	474.8	2.88
UCFF9	70	0.15	13.33	449.5	2.99
UCFF10	70	0.20	12.43	389.8	3.19
TCFF1	30	0.10	11.56	382.5	3.02
TCFF2	30	0.15	10.98	336.7	3.27
TCFF3	30	0.20	10.36	306.5	3.39
TCFF4	50	0.10	14.34	485.3	2.96
TCFF5	50	0.15	13.56	446.4	3.04
TCFF6	50	0.15	13.70	446.1	3.08
TCFF7	50	0.20	12.89	408.0	3.16
TCFF8	70	0.10	16.05	542.4	2,96
TCFF9	70	0.15	15.70	511.8	3.07
TCFF10	70	0.20	14.96	474.1	3.16

This comparison highlights a remarkable 18% enhancement in tensile strength for the TCFF filaments, suggesting that the chemical treatment applied to the fibers had a significant impact on the mechanical property of tensile strength. However, when comparing the tensile strength of the TCFF8 filaments at 16.05 MPa with the pure PLA filaments reported by [29] with σt = 29.5 MPa, we observe a decrease of approximately 45% in tensile strength for the TCFFs.

The highest elastic modulus (Et) values recorded for the test specimens printed with the filaments were TCFF8 = 542.4 MPa and FFCN8 = 474.8 MPa, respectively. These figures indicate that the FFCQ filaments demonstrated greater rigidity compared to FFCN under the specific conditions of the tensile test, emphasizing the positive effect of chemical treatment on the material's elastic properties.

The analysis of variance (ANOVA) for tensile strength yielded an $R^2 = 0.9981$ and an RMSE = 0.07079 for UCFF and an $R^2 = 0.9979$ and an RMSE = 0.08482 for TCFF. With an $R^2 = 0.9981$ for the variable (UCFF), this suggests that roughly 99.81% of the variation in tensile strength can be explained by the statistical model used, indicating a strong and significant relationship between the independent and dependent variables in this context. The $R^2 = 0.9979$ for (TCFF) suggests that approximately 99.79% of the variability in tensile strength is explained by the corresponding model for this variable. Both results indicate a strong relationship between the independent variables (infill and layer height) and tensile strength. Considering the stress data scale in MPa, the RMSE is relatively small in both cases, suggesting that the prediction errors are proportionate to the inherent variability of the data. These findings imply that the model holds significance and statistical significance.

Based on the obtained data and the results of the ANOVA for tensile strength, it was possible to establish a model of tensile strength (σt) as a function of the parameters percentage of infill (%) and layer height (mm). Figure 8a illustrates the response surface plot for the UCFF material, whereas Figure 8b shows the response surface plot for the TCFF material.

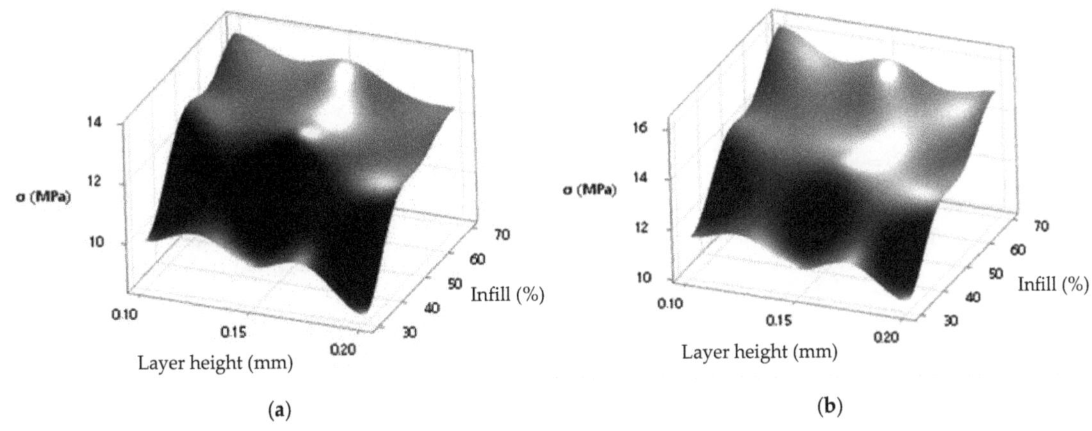

Figure 8. Surface response plot for tensile strength: σt (MPa) × layer height (mm)/infill (%) for (**a**) UCFF filament specimens and (**b**) TCFF filament specimens.

4. Discussion

Biocomposites with PLA granules and cellulose fibers from cocoa husks were premixed and compounded by extrusion in an attempt to create filaments for 3D printing. The results obtained in this study demonstrated the feasibility of using this agricultural waste as reinforcement in PLA biocomposites, applied as raw material in the filament production industry for 3D printers.

FTIR and TGA analyses revealed significant differences in the structural and thermal characteristics of the cocoa husk fiber samples. The presence of bands related to cellulose was confirmed in both samples. Additionally, reductions were observed in bands related to water molecules, low-molecular-weight organic compounds, metallic oxides, lignin, and hemicellulose in the chemically treated cocoa husk fibers (TCF). Thermal degradation changes between UCF and TCF were evident, corroborating the structural changes indicated in the FTIR analyses. The analysis highlights the effectiveness of chemical treatment in modifying the thermal and structural properties of cocoa husk fibers.

Morphological studies of cocoa husk fiber and the generated biocomposites showed good fiber dispersion in PLA; however, small aggregates occurred in TCFF, indicating that the fibers were not as well dispersed in this composite.

The mechanical tests demonstrated a superior performance by 18% in terms of the tensile strength of the PLA filament specimens with cellulose fibers from cocoa husks subjected to chemical treatment (TCFFs) compared to filaments containing untreated fibers (UCFFs). This analysis highlights the significant influence of chemical treatment on fibers for enhancing biocomposite performance. However, when comparing the tensile strength result for the pure PLA filament specimens with that of the TCFF filament specimens, it was found that the incorporation of fibers resulted in a reduction of approximately 45% in tensile strength properties. Although there was a decrease in tensile strength, the study highlights the importance of investigating variations in the manufacturing process to understand their impact on composite performance better. Additionally, a significant improvement in quality was observed when the fibers were treated. However, other aspects of the manufacturing process, such as the bonding between the fibers and PLA, the optimal fiber quantity, and the possibility of adding additives to enhance the composite properties, also require investigation to optimize performance and applicability.

In all cases, printing settings with a 70% infill density and a layer height of 0.10 mm yielded the best results. This configuration demonstrated an effective combination of high strength, material efficiency, and printing speed, making it a favorable choice. It is worth noting that balancing the desired strength with practical considerations such as print time and overall process efficiency is important for optimizing part performance in real-world contexts. These results provide valuable insights for the optimization of 3D printing processes, as they indicate which specific settings can result in the desired mechanical properties.

This study demonstrates that the process of creating biocomposites with cocoa husk fibers is a promising method; however, further research is needed to improve the dispersion and adhesion of fibers in PLA.

Author Contributions: Conceptualization, V.H.M.d.A., R.M.d.J., and G.M.S.; methodology, V.H.M.d.A., R.M.d.J., and G.M.S.; software, E.F.M.S.S., I.S.d.C., I.d.S.S., and P.N.M.d.A.; validation, V.H.M.d.A., R.M.d.J., G.M.S., and S.K.; formal analysis, V.H.M.d.A., R.M.d.J., and G.M.S.; investigation, V.H.M.d.A., R.M.d.J., and G.M.S.; resources, V.H.M.d.A., R.M.d.J., G.M.S., S.K., E.F.M.S.S., I.S.d.C., I.d.S.S., and P.N.M.d.A.; data curation, V.H.M.d.A., R.M.d.J., and G.M.S.; writing—original draft preparation, V.H.M.d.A., R.M.d.J., G.M.S., and S.K.; writing—review and editing, V.H.M.d.A., R.M.d.J., G.M.S., S.K., E.F.M.S.S., I.S.d.C., I.d.S.S., and P.N.M.d.A.; visualization, V.H.M.d.A., R.M.d.J., G.M.S., S.K., E.F.M.S.S., I.S.d.C., I.d.S.S., and P.N.M.d.A.; supervision, R.M.d.J. and G.M.S.; project administration, V.H.M.d.A., R.M.d.J., and G.M.S. All authors have read and agreed to the published version of the manuscript.

Funding: This research received no external funding. This project was supported by UESC through the DEC and DCEX.

Institutional Review Board Statement: Not applicable.

Data Availability Statement: The original contributions presented in the study are included in the article, further inquiries can be directed to the corresponding author.

Conflicts of Interest: The authors declare no conflicts of interest.

References

1. Inseemeesak, B.; Siripaiboon, C.; Somkeattikul, K.; Attasophonwattana, P.; Kiatiwat, T.; Punsuvon, V.; Areeprasert, C. Biocomposite Fabrication from Pilot-Scale Steam-Exploded Coconut Fiber and PLA/PBS with Mechanical and Thermal Characterizations. *J. Clean. Prod.* **2022**, *379*, 134517. [CrossRef]
2. Das, A.K.; Agar, D.A.; Rudolfsson, M.; Larsson, S.H. A Review on Wood Powders in 3D Printing: Processes, Properties and Potential Applications. *J. Mater. Res. Technol.* **2021**, *15*, 241–255. [CrossRef]
3. Dai, L.; Cheng, T.; Duan, C.; Zhao, W.; Zhang, W.; Zou, X.; Aspler, J.; Ni, Y. 3D Printing Using Plant-Derived Cellulose and Its Derivatives: A Review. *Carbohydr. Polym.* **2019**, *203*, 71–86. [CrossRef] [PubMed]

4. Kopparthy, S.D.S.; Netravali, A.N. Review: Green Composites for Structural Applications. *Compos. Part C Open Access* **2021**, *6*, 100169. [CrossRef]
5. Shekar, H.S.S.; Ramachandra, M. Green Composites: A Review. *Mater. Today Proc.* **2018**, *5*, 2518–2526. [CrossRef]
6. Kumar, S.D.; Venkadeshwaran, K.; Aravindan, M.K. Fused Deposition Modelling of PLA Reinforced with Cellulose Nano-Crystals. *Mater. Today Proc.* **2020**, *33*, 868–875. [CrossRef]
7. Liu, H.; He, H.; Peng, X.; Huang, B.; Li, J. Three-dimensional Printing of Poly(Lactic Acid) Bio-based Composites with Sugarcane Bagasse Fiber: Effect of Printing Orientation on Tensile Performance. *Polym. Adv. Technol.* **2019**, *30*, 910–922. [CrossRef]
8. Vásquez, Z.S.; De Carvalho Neto, D.P.; Pereira, G.V.M.; Vandenberghe, L.P.S.; De Oliveira, P.Z.; Tiburcio, P.B.; Rogez, H.L.G.; Neto, A.G.; Soccol, C.R. Biotechnological Approaches for Cocoa Waste Management: A Review. *Waste Manag.* **2019**, *90*, 72–83. [CrossRef]
9. Daud, Z.; Awang, H.; Mohd Kassim, A.S.; Mohd Hatta, M.Z.; Mohd Aripin, A. Cocoa Pod Husk and Corn Stalk: Alternative Paper Fibres Study on Chemical Characterization and Morphological Structures. *Adv. Mater. Res.* **2014**, *911*, 331–335. [CrossRef]
10. Almeida, V.H.M.; Jesus, R.M.; Santana, G.M.; Pereira, T.B. Polylactic Acid Polymer Matrix (Pla) Biocomposites with Plant Fibers for Manufacturing 3D Printing Filaments: A Review. *J. Compos. Sci.* **2024**, *8*, 67. [CrossRef]
11. *ASTM D638-14*; Standard Test Method for Tensile Properties of Plastics. ASTM International: West Conshohocken, PA, USA, 2022.
12. Myers, R.H.; Montgomery, D.C.; Anderson-Cook, C.M. *Response Surface Methodology: Process and Product Optimization Using Designed Experiments*, 4th ed.; Wiley Series in Probability and Statistics; Wiley: Hoboken, NJ, USA, 2016; ISBN 978-1-118-91601-8.
13. Akinjokun, A.I.; Petrik, L.F.; Ogunfowokan, A.O.; Ajao, J.; Ojumu, T.V. Isolation and Characterization of Nanocrystalline Cellulose from Cocoa Pod Husk (CPH) Biomass Wastes. *Heliyon* **2021**, *7*, e06680. [CrossRef]
14. Yu, W.; Dong, L.; Lei, W.; Zhou, Y.; Pu, Y.; Zhang, X. Effects of Rice Straw Powder (RSP) Size and Pretreatment on Properties of FDM 3D-Printed RSP/Poly(Lactic Acid) Biocomposites. *Molecules* **2021**, *26*, 3234. [CrossRef] [PubMed]
15. Díaz-Oviedo, A.F.; Ramón-Valencia, B.A.; Moreno-Contreras, G.G. Caracterización Físico-Química de La Cáscara de Mazorca de Cacao Como Posible Uso En La Elaboración de Tableros Aglomerados. *Rev. Investig. Desarro. E Innov.* **2022**, *12*, 97–106. [CrossRef]
16. Hozman-Manrique, A.S.; Garcia-Brand, A.J.; Hernández-Carrión, M.; Porras, A. Isolation and Characterization of Cellulose Microfibers from Colombian Cocoa Pod Husk via Chemical Treatment with Pressure Effects. *Polymers* **2023**, *15*, 664. [CrossRef]
17. Agarwal, U.P.; Ralph, S.A.; Baez, C.; Reiner, R.S. Contributions of Crystalline and Noncrystalline Cellulose Can Occur in the Same Spectral Regions: Evidence Based on Raman and IR and Its Implication for Crystallinity Measurements. *Biomacromolecules* **2021**, *22*, 1357–1373. [CrossRef]
18. Gao, X.; Jia, Y.; Chen, Z.; Santhanam, R.K.; Zhang, M.; He, C.; Chen, H. Synthesis of Hydrogels Based on Nanocellulose from Garlic Straw and Regulating the Release of Allicin and Its Cytotoxicity. *Food Sci. Technol.* **2022**, *42*, e43422. [CrossRef]
19. Garcia-Brand, A.J.; Morales, M.A.; Hozman, A.S.; Ramirez, A.C.; Cruz, L.J.; Maranon, A.; Muñoz-Camargo, C.; Cruz, J.C.; Porras, A. Bioactive Poly(Lactic Acid)–Cocoa Bean Shell Composites for Biomaterial Formulation: Preparation and Preliminary In Vitro Characterization. *Polymers* **2021**, *13*, 3707. [CrossRef] [PubMed]
20. Kaur, V.; Dash, B.P.; Vermani, S.; Devi, A. Extraction, Quantification and Characterization of Lignin Extracted from Bamboo Biomass. *Mater. Today Proc.* **2022**. [CrossRef]
21. Kłosowski, G.; Mikulski, D. Changes in Various Lignocellulose Biomasses Structure after Microwave-Assisted Hydrotropic Pretreatment. *Renew. Energy* **2023**, *219*, 119387. [CrossRef]
22. Veber, A.; Zancajo, V.M.R.; Puskar, L.; Schade, U.; Kneipp, J. In Situ Infrared Imaging of the Local Orientation of Cellulose Fibrils in Plant Secondary Cell Walls. *Analyst* **2023**, *148*, 4138–4147. [CrossRef]
23. Dassanayake, R.S.; Fierro, J.S.; Abidi, N.; Quitevis, E.L.; Boggavarappu, K.; Thalangamaarachchige, V.D. Characterization of Cellulose Nanocrystals by Current Spectroscopic Techniques. *Appl. Spectrosc. Rev.* **2023**, *58*, 180–205. [CrossRef]
24. Popescu, C.-M.; Larsson, P.T.; Olaru, N.; Vasile, C. Spectroscopic Study of Acetylated Kraft Pulp Fibers. *Carbohydr. Polym.* **2012**, *88*, 530–536. [CrossRef]
25. Shi, J.; Xing, D.; Lia, J. FTIR Studies of the Changes in Wood Chemistry from Wood Forming Tissue under Inclined Treatment. *Energy Procedia* **2012**, *16*, 758–762. [CrossRef]
26. Oh, S.Y.; Yoo, D.I.; Shin, Y.; Seo, G. FTIR Analysis of Cellulose Treated with Sodium Hydroxide and Carbon Dioxide. *Carbohydr. Res.* **2005**, *340*, 417–428. [CrossRef] [PubMed]
27. Eugenio, M.E.; Martín-Sampedro, R.; Santos, J.I.; Wicklein, B.; Ibarra, D. Chemical, Thermal and Antioxidant Properties of Lignins Solubilized during Soda/AQ Pulping of Orange and Olive Tree Pruning Residues. *Molecules* **2021**, *26*, 3819. [CrossRef] [PubMed]
28. Bhagia, S.; Ďurkovič, J.; Lagaňa, R.; Kardošová, M.; Kačík, F.; Cernescu, A.; Schäfer, P.; Yoo, C.G.; Ragauskas, A.J. Nanoscale FTIR and Mechanical Mapping of Plant Cell Walls for Understanding Biomass Deconstruction. *ACS Sustain. Chem. Eng.* **2022**, *10*, 3016–3026. [CrossRef]
29. Mansingh, B.B.; Binoj, J.S.; Tan, Z.Q.; Eugene, W.W.L.; Amornsakchai, T.; Hassan, S.A.; Goh, K.L. Comprehensive Characterization of Raw and Treated Pineapple Leaf Fiber/Polylactic Acid Green Composites Manufactured by 3D Printing Technique. *Polym. Compos.* **2022**, *43*, 6051–6061. [CrossRef]

polymers

MDPI

Article

Low-Velocity Impact of Clamped Rectangular Sandwich Tubes with Fiber Metal Laminated Tubes

Yao Wang [1,2], Jianxun Zhang [1,2,3,*], Hui Guo [1] and Hui Yuan [2]

1 Shock and Vibration of Engineering Materials and Structures Key Lab of Sichuan Province, Southwest University of Science and Technology, Mianyang 621010, China; 18563226065@163.com (Y.W.); guohui56789@126.com (H.G.)

2 State Key Laboratory for Strength and Vibration of Mechanical Structures, School of Aerospace Engineering, Xi'an Jiaotong University, Xi'an 710049, China; 13032499002@163.com

3 Key Laboratory of Mechanics on Disaster and Environment in Western China Attached to the Ministry of Education of China, Lanzhou University, Lanzhou 730000, China

* Correspondence: jianxunzhang@mail.xjtu.edu.cn

Abstract: Fiber metal laminated sandwich tubes are made up of alternating fiber-reinforced composite and metal layers. Fiber metal laminated tubes have the advantages of the high strength and high stiffness of fiber and the toughness of metal, so they have become an excellent load-bearing and energy-absorbing, lightweight structure. Due to the complexity of the fiber layup, it is difficult to establish an analytical model of the relevant structural properties. In this work, introducing the number and volume fraction of fiber layup, based on the modified rigid–plastic model, an analytical model is established for low-velocity impacts on sandwich tubes with fiber metal laminated tubes, which provided a theoretical basis for the design of fiber–metal composite tubes. In addition, a numerical simulation was conducted for low-velocity impacts on clamped rectangular sandwich tubes with fiber metal laminated (FML) tubes and a foam core. By comparing the results obtained from the theoretical analysis and numerical calculations, it is shown that the analytical results can reasonably agree with the numerical results. The influences of the metal volume fraction (MVF), the strength ratio factor of the FML metal layer to the FML composite layer, and the relative strength of the foam on the dynamic response of the rectangular sandwich tubes with FML tubes and a metal foam core (MFC) are discussed. It is shown that by increasing the fiber content and fiber strength of the FML tubes and the foam strength, the load-carrying and energy-absorbing capacity of the rectangular sandwich tubes can be effectively improved, especially by changing the fiber properties. In addition, present analytical solutions can be applied to make predictions about the dynamic response of the rectangular sandwich tubes with FML tubes and MFC during impacts with low-velocity and reasonably heavy-mass.

Keywords: dynamic response; rectangular sandwich tube; FML tube; low-velocity impact

1. Introduction

Fiber metal laminated (FML) tubes are made up of alternating fiber-reinforced composite and metal layers, and the composite layer is made of polymer materials (such as glass fiber- or carbon fiber-reinforced fibers) and metal alloy materials [1]. FML has excellent properties that combines the properties of metals and composite materials, such as good impact resistance and plasticity, good mechanical adaptability, and excellent fatigue properties, as well as low weight [2]. A sandwich tube with a metal foam core (MFC) is a typical lightweight tube consisting of rigid and strong inner and outer tubes and MFC sandwiched between the inner and outer tubes. Due to its good energy absorption and impact resistance, as well as high specific strength and specific stiffness, sandwich tubes with MFC have been used in many fields such as aviation, space, navigation, high-speed rail, and so on [3–20]. Sandwich tubes designed with FML tubes and MFC provide more choice for energy-absorbing structures. Therefore, there is a need to investigate low-velocity impacts on sandwich tubes with FML tubes and MFC.

In the past decade or so, the mechanical behavior of sandwich tubes under quasi-static loading has been extensively studied. Zhang et al. [21] established the yield criterion and developed an analytical model to predict the plastic behavior of fully clamped slender rectangular sandwich tubes with MFC. Liu et al. [22] studied the mechanics of sandwich tubes under lateral loading by a quasi-static transverse indentation test and finite element calculations, and found that the relative differences in denting characteristics between the inner and outer tubes are related to the material and geometric properties of the tubes and the interactions between the layers. Guelou et al. [23] experimentally investigated the static compression of wood-based sandwich tubes, and found the energy absorption of such structures under static conditions as well as coupling effects. Niknejad et al. [24] investigated the indentation behavior of laminated and composite tubes with agglomerated cork cores under transverse loading by a cylindrical indenter experimentally, and found that cork-filled composite tubes absorb more total energy. Baroutaji et al. [25] conducted a systematic investigation on the lateral fragmentation of sandwich tubes with circular tubes and MFC under quasi-static loading by means of experimental and numerical methods, and found that a tube with a small inner diameter and larger foam thickness would be better suited for use as an energy-absorbing component. Kim et al. [26] investigated the extrusion behavior of square composite sandwich tubes with woven fabric carbon/epoxy skins and aluminum honeycomb cores by means of quasi-static and dynamic shock tests, and concluded that the dynamic crushing modes collapsed in a more stable manner than the corresponding static tests. Fan et al. [27] conducted a systematic study of thin-walled circular tubes sandwiched with MFC under quasi-static transverse extrusion through experiments and numerical simulations, and found that the compressive strength and energy absorption of the sandwiched tubes are greater than the sum of the components, and sandwich tubes are more weight-efficient than empty tubes. Shen et al. [28] used experimental, theoretical, and numerical methods to study the transverse crushing response of sandwich tubes with shot aluminum foam, and concluded that the bonding between the tubes and the core of the sandwich tubes varies in different crushing modes. Wu et al. [29] studied the crushing behavior of a sandwich tube with a meta-honeycomb core with zero Poisson's ratio under transverse and axial compressive loadings experimentally, and concluded that it has a stable crushing curve and crushing mode, and is more stable and efficient than thin-walled composite tubes in energy absorption. Wang et al. [30] investigated the axial and transverse crushing characteristics of sandwich tubes filled with plate-lattice in square tubes by means of experimental and numerical methods, and the findings indicated that the filling of plate-lattice can significantly improve the load carrying capacity and specific energy absorption of the sandwich tubes.

In addition, the dynamic bending behavior of sandwich tubes has been investigated. Zhang et al. [31] obtained an analytical solution for dynamic response of fully clamped rectangular sandwich tubes with MFC under low-velocity impact. Zhang and Guo [32] obtained an analytical solution for the dynamical response of the clamped rectangular sandwich tube with MFC under lateral loads, which was shown to be in good agreement with the numerical results. Guo et al. [33] experimentally investigated the splitting and curling behaviors of square sandwich tubes with MFC subjected to axial low-velocity impact. The deformation modes and load-displacement curves were obtained, and the analytical solution can be used to predict the steady-state load of square sandwich tubes subjected to axial low-velocity impact. Zhang et al. [34] used experimental, theoretical and numerical methods to investigate the dynamical response and energy-absorption characteristics of metal foam-filled circular sandwich tubes under interior blast loads, and observed that circular sandwiched tubes with gradient cores of negative gradient have the optimal resistance. Guelou et al. [35] investigated the dynamic crushing of sandwich tubes with tubes made of carbon or glass fibers and a poplar ply core through experiments, and the results showed that the wood makes a significant contribution tostructural response.

Shen et al. [36] conducted a systematic investigation on the dynamic behavior of sandwich tubes under internal explosive loading by both experimental and numerical analyses, and obtained an analytical solution for the dynamic response of the sandwich tube under blast loads. Wang et al. [37] studied the dynamic response of density-graded sandwich tubes with MFC under internal explosive loading through finite element calculation, and concluded that the blast resistance decreases as the density gradient of the core increases.

With the demand for FML tubes with good load-carrying capacity, impact resistance, and energy absorption, more and more attention has been focused on the study of the quasi-static behavior and impact response of FML tubes. Mansor et al. [38] studied the crashworthiness of thin-walled FML tubes in axial compression by means of experiment and numerical calculation, and discovered that FML tubes can be considered as a suitable candidate structure for lightweight energy-absorbing applications with considerable impact resistance. Ahmad et al. [39] conducted a systematic investigation on the impact characteristics and energy absorption of a thin-walled FML tube under axial impact loads by means of an experimental and numerical calculation, and concluded that it can withstand larger impact loading, absorb higher energy, and is more suitable as an impact energy absorber. Subbaramaiah et al. [40] conducted a study of the axial crushing behavior of glass laminate aluminum reinforced epoxy (GLARE) top-hat structures experimentally and numerically, and the results showed that the crushing response of the GLARE top-hat structures is preferable to that of the bare metal equivalent. Song et al. [41] analytically investigated the dynamic response of rectangular sandwich tubes with FML tubes and MFC under low-velocity impact, considering strength and coupling effects of bending and stretching. Mansor et al. [42] investigated the energy absorption and impact properties of a tubular seamless type of FML fabricated from braided fiberglass sleeve and aluminum alloy tubes experimentally, and concluded that the mechanical properties of the FML tubes have good potential for application as an efficient energy absorber. Shiravand et al. [43] proposed an analytical solution for energy absorption of a conical FML tube consisting of any number of metal and laminate composite layers; the accuracy of analytical solutions was verified by experimental and numerical results.

As far as the authors know, little work has been carried out on theoretical research of the dynamic response of rectangular sandwich tubes with FML tubes and MFC subjected to low-velocity impacts. The main aim of this paper is to analytically and numerically investigate the low-velocity impacts on fully clamped slender rectangular sandwich tubes with FML tubes and a metal foam core. The organization of this article is as follows. The problem statement is presented in Section 2. In Section 3, an analytic solution is presented for predicting the dynamic response of a rectangular sandwich tube with FML tubes and MFC under low-velocity impacts. Section 4 presents the finite element analysis. Section 5 presents a comparison of the numerical and analytical results and discusses the influence of MVF, strength ratio factor of the two different material layers of the FML, and relative strength of the foam on the dynamic response of the rectangular sandwich tube. Section 6 contains concluding remarks.

2. Problem Formulation

Now consider a fully clamped slender rectangular sandwich tube with FML tubes and a metal foam core, as shown in Figure 1. The tube length is $2L$ in length, and thelengths of cross-section are b in width and b_1 in height. The mass per unit length of the rectangular sandwich tube is G_b, and the FML tube is struck by a heavy impactor. The impactor has a mass G_s and an initial low-velocity V_I. Assume that the impactor is rigid. The MFC between the FML inner and outer tubes is assumed to be completely filled between them with no gaps, that is, it is perfectly bonded. As shown in Figure 1a,b, let the thickness of MFC be h_1, and the thickness of FML be h, the thickness of the metal layer of FML be h_m, and the thickness of the composite layer of FML be h_f. In addition, the FML tube has n

layers of metallic materials and $n-1$ layers of composite materials in an alternating layup. Thus, the thickness of the FML tube is

$$h = nh_m + (n-1)h_f \tag{1}$$

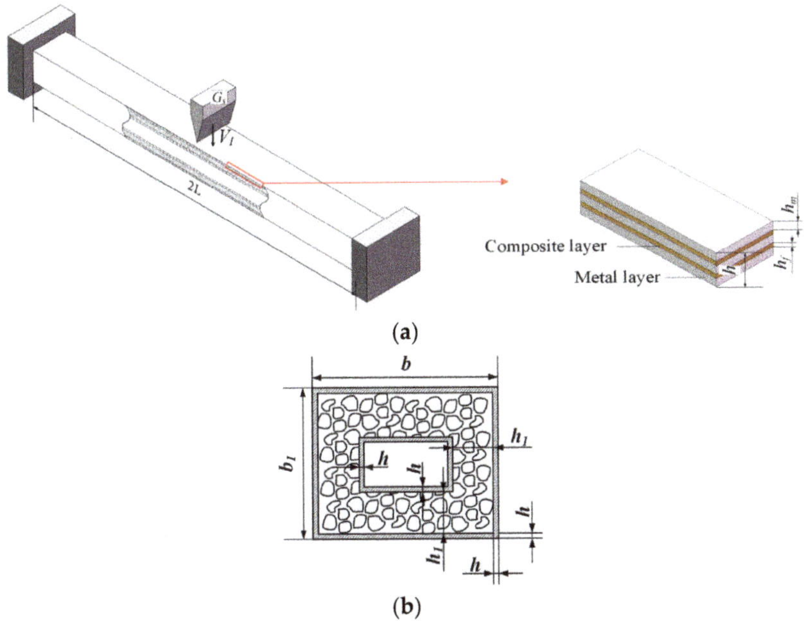

Figure 1. Sketch of a rectangular sandwich tube with FML tubes and MFC under heavy-impact with low-velocity at midspan. (**a**) Rectangular sandwich tube and (**b**) the cross-section of the rectangular sandwich tube. The red arrow indicates a partial enlarged view of the FML tube.

The density, strength, and elastic modulus of the composite layers and metal layers of FML tubes are ρ_f, ρ_m, σ_f, σ_m, E_f, and E_m, respectively. And the density, yield strength, and elastic modulus of the MFC are ρ_c, σ_c, and E_c. Define a strength ratio factor of FML metal layer to FML composite layer q, so

$$q = \frac{\sigma_m}{\sigma_f} \tag{2}$$

In order to better analyze the calculations, the different flow strengths and thicknesses of the components of the FML tubes are defined as the weighted strength σ_{fa}. The weighted strength σ_{fa} of the inner and outer rectangular tubes are supposed to follow the rigid–perfectly plastic law, while the yield strength σ_c and densification strain ε_D of the filled MFC are supposed to follow the rigid–perfectly plastic locking (*RPPL*) law.

3. Analytical Solutions

Rigid–plastic theory solutions can predict the dynamic behaviour of the FML when the plastic behaviour dominates the response of the plate [44]; the solution here is extended for predicting the dynamic response of rectangular sandwich tubes with FML tubes and MFC under low-velocity impact.

To describe the proportion of metal layers in FML, a metal volume fraction (MVF) f is defined, and can be given by

$$f = \frac{nh_m}{h} \times 100\% \tag{3}$$

So, the weighted strength σ_{fa} can be given by

$$\sigma_{fa} = f\sigma_m + (1-f)\sigma_f \tag{4}$$

It is assumed that the ratio of height to span of the rectangular tube is small enough that the overall deformation of the rectangular tube does not produce a localized denting under the impactor. Therefore, the cross-section of the rectangular sandwich tube with FML tubes and MFC maintains its original shape, and the overall deformation profile under low-velocity impact is the same as that of a fully clamped whole rectangular tube. Therefore, the equation for the balance of the mass–tube system loaded at the mid-span of the rectangular sandwich tube can be derived as

$$\left(\frac{G_b L^2}{3} + \frac{G_s L}{2}\right)\ddot{W}_0 + NW_0 + 2M_1 = 0 \tag{5}$$

where

$$G_b = \left(4\rho_f h + 2\rho_c h_1\right)(b + b_1 - 4h - 2h_1),$$

in which M_1 and N are the plastic moments and axial forces of the rectangular sandwich tube; the bending moment $M_1 = M_2$; at moderate deflections, F and N can be approximately equal; W_0 is the deflection of the rectangular sandwich tube at impact point; $\ddot{W}_0$ is the acceleration of the impactor hitting the mid-span of the rectangular sandwich tube; P is the impact force of the impactor at impact position, as shown in Figure 2.

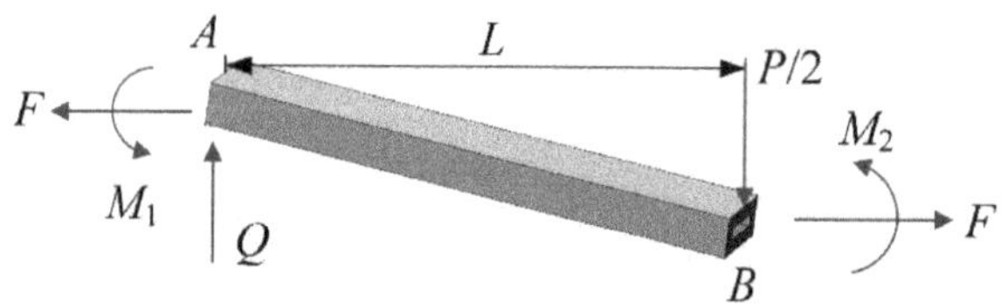

Figure 2. Global bending deformation model of the left plastic neutral surface of the rectangular sandwich tube with MFC under low-velocity impact at mid-span.

Zhang et al. [31] obtained the analytical solution of low-velocity impact analysis of a rectangular sandwich tube with MFC under low-velocity impact with a heavy-impactor. Appendix A shows the relationship between the dimensionless reaction force and the dimensionless deflection of the tube, and the relationship between the dimensionless kinetic energy of the striker and the maximum central deflection of a fully clamped rectangular sandwich tube with MFC subjected to low-velocity impact.

Inserting Equations (3) and (4) into Equations (A1) and (A3) yields the analytical solutions for the dynamic response of rectangular sandwich tubes with FML tubes and MFC subjected to low-velocity impact. Then, the relationship between the dimensionless initial impact energy U_K^* and the dimensionless maximum central deflection W_{0m}^*, and the relationship between the dimensionless reaction force P_r^* and the dimensionless deflection W_0^* for the rectangular sandwich tube with FML tubes and MFC are given by

$$
\alpha U_K^* = \begin{cases}
\frac{1}{6L_2}\left[2(n-1)\bar{h}_f T + \bar{h}_1\bar{\sigma}_c \mathrm{p}(1-f)\right]{W_{0m}^*}^3 + W_{0m}^*, \\
\qquad 0 \le W_{0m}^* \le \frac{(1-2\bar{h}_1 p)(1-f)-4(n-1)\bar{h}_f}{1-f} \\
\frac{{W_{0m}^*}^3}{12L_2}\left[2\bar{h}_1\bar{\sigma}_c p(1-f) + \left(1-2\bar{h}_1\right)p(1-f)T\right] + \frac{L_5}{4L_2}{W_{0m}^*}^2 + \frac{L_6}{L_2}W_{0m}^* + S_1 \\
\quad + \frac{{W_{0m}^*}^2}{4L_2}\left[2\bar{h}_1\bar{\sigma}_c + \left(1-2\bar{h}_1\right)T\right]\left[3(n-1)\bar{h}_f + \left(2\bar{h}_1 p - 1\right)(1-f)\right]p, \\
\qquad \frac{(1-2\bar{h}_1 p)(1-f)-4(n-1)\bar{h}_f}{1-f} \le W_{0m}^* \le \frac{(1-2\bar{h}_1 p)(1-f)-2(n-1)\bar{h}_f}{1-f} \\
-\frac{L_4}{12L_2}{W_{0m}^*}^3 + \frac{1}{2}\left\{\frac{L_4}{2L_2}\left[\frac{(n-1)\bar{h}_f}{1-f} + \bar{h}_1 p - \frac{3}{2}\right] - \frac{L_3}{2L_2}\right\}{W_{0m}^*}^2 + S_2 \\
\quad + \frac{W_{0m}^*}{L_2}(n-1)\bar{h}_f\left[1 - \frac{(n-1)\bar{h}_f}{1-f}\right]T + \frac{L_4}{2L_2}\left\{\left[\frac{(n-1)\bar{h}_f}{1-f}\right]^2 - \frac{1}{4}\right\}W_{0m}^*, \\
\qquad \frac{(1-2\bar{h}_1 p)(1-f)-2(n-1)\bar{h}_f}{1-f} \le W_{0m}^* \le \frac{1-f-2(n-1)\bar{h}_f}{1-f} \\
\frac{{W_{0m}^*}^3}{12L_2}p(1-f)T + \frac{1}{2}\left[\frac{L_1}{L_2} - \frac{1}{2L_2}p(1-f)T\right]{W_{0m}^*}^2 + \frac{W_{0m}^*}{4L_2}p(1-f)T + S_3, \\
\qquad \frac{1-f-2(n-1)\bar{h}_f}{1-f} \le W_{0m}^* \le 1 \\
\frac{L_1}{2L_2}{W_{0m}^*}^2 + S_4, \qquad W_{0m}^* \ge 1
\end{cases} \tag{6}
$$

and

$$
P_r^* = \begin{cases}
\frac{3G^*}{3G^*+1}\left\{\frac{{W_0^*}^2}{2L_2}\left[2\bar{h}_f(n-1)T + \bar{h}_1\bar{\sigma}_c p(1-f)\right] + 1\right\}, \\
\qquad 0 \le W_0^* \le \frac{(1-2\bar{h}_1 p)(1-f)-4(n-1)\bar{h}_f}{1-f} \\
\frac{3G^*}{3G^*+1}\left\{p\left[2\bar{h}_1\bar{\sigma}_c + \left(1-2\bar{h}_1\right)T\right]\left[3\bar{h}_f(n-1) + \left(2\bar{h}_1 p - 1\right)(1-f)\right]\frac{W_0^*}{2L_2}\right\} \\
\quad + \frac{3G^*}{3G^*+1}\left\{\frac{{W_0^*}^2}{4L_2}p(1-f)\left[2\bar{h}_1\bar{\sigma}_c + \left(1-2\bar{h}_1\right)T\right] + \frac{L_5}{2L_2}W_0^* + \frac{L_6}{L_2}\right\}, \\
\qquad \frac{(1-2\bar{h}_1 p)(1-f)-4(n-1)\bar{h}_f}{1-f} \le W_0^* \le \frac{(1-2\bar{h}_1 p)(1-f)-2(n-1)\bar{h}_f}{1-f} \\
\frac{3G^*}{3G^*+1}\left\{\frac{1}{L_2}\bar{h}_f(n-1)pT\left[1 - \frac{(n-1)\bar{h}_f}{1-f}\right] + \frac{L_4}{2L_2}\left[\frac{(n-1)\bar{h}_f}{1-f} - \frac{1}{2}\right]^2\right\} \\
\quad + \frac{3G^*}{3G^*+1}\left\{\frac{L_4{W_0^*}^2}{8L_2} + \left[\frac{L_4}{2L_2}\left(\frac{\mathrm{n}-1}{1-f}\bar{h}_f + \bar{h}_1 p - \frac{1}{2}\right) - \frac{L_3}{2L_2}\right]W_0^*\right\}, \\
\qquad \frac{(1-2\bar{h}_1 p)(1-f)-2(n-1)\bar{h}_f}{1-f} \le W_0^* \le \frac{1-f-2(n-1)\bar{h}_f}{1-f} \\
\frac{3G^*}{3G^*+1}\left\{\frac{{W_0^*}^2+1}{4L_2}p(1-f)T + \left[\frac{L_1}{L_2} - \frac{1}{2L_2}p(1-f)T\right]W_0^*\right\}, \\
\qquad \frac{(1-2\bar{h}_1 p)(1-f)-2(n-1)\bar{h}_f}{1-f} \le W_0^* \le 1 \\
\frac{3G^*}{3G^*+1}\frac{L_1}{L_2}W_0^*, \qquad W_0^* \ge 1
\end{cases} \tag{7}
$$

where

$$\begin{aligned}
&\bar{h}_f = \frac{h_f}{b_1},\ \bar{\rho}_c = \frac{\rho_c}{\rho_f},\ \bar{\sigma}_c = \frac{\sigma_c}{\sigma_f},\ q = \frac{\sigma_m}{\sigma_f},\ T = (q-1)f+1,\ \bar{V} = \frac{V_0}{\sqrt{\sigma_f/\rho_f}},\ \bar{t} = \frac{t}{b_1\sqrt{\rho_f/\sigma_f}},\\
&L_1 = 2\bar{h}_f p(n-1)\left(\bar{h}_1\bar{\sigma}_c + T\right) + \bar{h}_1\bar{\sigma}_c p\left[p(1-f) - 2\bar{h}_f(n-1)\right]\\
&\quad -\left[\bar{h}_1\bar{\sigma}_c p(1-f) + 2\bar{h}_f(n-1)T\right]\left[\frac{4(n-1)\bar{h}_f}{1-f} + 2\bar{h}_1 p - 1\right],\\
&L_2 = \bar{h}_1 p\left[1 - \frac{2(n-1)\bar{h}_f}{1-f} - \bar{h}_1 p\right]\left\{2\bar{h}_f(n-1)T + \left[p(1-f) - 2\bar{h}_f(n-1)\right]\bar{\sigma}_c\right\}\\
&\quad +\left[\left(1-2\bar{h}_1\right)T + 2\bar{h}_1\bar{\sigma}_c\right]\left[1 - f - 3\bar{h}_f(n-1) - 2\bar{h}_1 p(1-f)\right]p\frac{(n-1)\bar{h}_f}{1-f}\\
&\quad +\frac{1}{2}\left[2\bar{h}_f(n-1)T + p(1-f)\bar{h}_1\bar{\sigma}_c\right]\left[1 - \frac{4(n-1)\bar{h}_f}{1-f} - 2\bar{h}_1 p\right]^2\\
&\quad +\bar{h}_f pT(n-1)\left[1 - \frac{(n-1)\bar{h}_f}{1-f}\right],\\
&L_3 = -2\bar{h}_f p(n-1)\left[2\bar{h}_1\bar{\sigma}_c + \left(1-2\bar{h}_1\right)T\right]\\
&\quad -\left[4\bar{h}_f(n-1)T + 2\bar{h}_1\bar{\sigma}_c p(1-f)\right]\left[1 - \frac{4(n-1)\bar{h}_f}{1-f} - 2\bar{h}_1 p\right],\\
&L_4 = 4\bar{h}_f(n-1)T + 2\bar{\sigma}_c\left[p(1-f) - 2\bar{h}_f(n-1)\right],\\
&L_5 = p\bar{h}_f(n-1)\left[2\bar{h}_1\bar{\sigma}_c + \left(1-2\bar{h}_1\right)T\right]\\
&\quad +\left[2\bar{h}_1\bar{\sigma}_c p(1-f) + 4\bar{h}_f(n-1)T\right]\left[1 - \frac{4(n-1)\bar{h}_f}{1-f} - 2\bar{h}_1 p\right],\\
&L_6 = \bar{h}_f pT(n-1)\left[1 - \frac{(n-1)\bar{h}_f}{1-f}\right] + \frac{p}{4}(1-f)\left[\left(1-2\bar{h}_1\right)T + 2\bar{h}_1\bar{\sigma}_c\right]\left[1 - \frac{2(n-1)\bar{h}_f}{1-f} - 2\bar{h}_1 p\right]^2\\
&\quad +\bar{h}_1 p\left[1 - \frac{2(n-1)\bar{h}_f}{1-f} - \bar{h}_1 p\right]\left\{\left[2\bar{h}_f(n-1)T\right] + \left[p(1-f) - 2\bar{h}_f(n-1)\right]\bar{\sigma}_c\right\},\\
&S_1 = \left(1 - \frac{L_6}{L_2}\right)\left[1 - \frac{4(n-1)\bar{h}_f}{1-f} - 2\bar{h}_1 p\right] - \frac{L_5}{4L_2}\left[1 - \frac{4(n-1)\bar{h}_f}{1-f} - 2\bar{h}_1 p\right]^2\\
&\quad +\frac{T}{6L_2}\left[2\bar{h}_f(n-1) + p(1-f)\left(\bar{h}_1 - \frac{1}{2}\right)\right]\left[1 - \frac{4(n-1)\bar{h}_f}{1-f} - 2\bar{h}_1 p\right]^3\\
&\quad -\frac{p}{4L_2}\left[3\bar{h}_f(n-1) + \left(2\bar{h}_1 p - 1\right)(1-f)\right]\left[2\bar{h}_1\bar{\sigma}_c + \left(1-2\bar{h}_1\right)T\right]\left[1 - \frac{4(n-1)\bar{h}_f}{1-f} - 2\bar{h}_1 p\right]^2,\\
&S_2 = \frac{1}{24L_2}\left[4\bar{h}_1 p(1-f)\bar{\sigma}_c - 4\bar{h}_1 Tp(1-f) - L_4 + 2p(1-f)T\right]\left[1 - \frac{2(n-1)\bar{h}_f}{1-f} - 2\bar{h}_1 p\right]^3\\
&\quad +\frac{p}{4L_2}\left[3\bar{h}_f(n-1) + \left(2\bar{h}_1 p - 1\right)(1-f)\right]\left[2\bar{h}_1\bar{\sigma}_c + \left(1-2\bar{h}_1\right)T\right]\left[1 - \frac{2(n-1)\bar{h}_f}{1-f} - 2\bar{h}_1 p\right]^2\\
&\quad -\frac{1}{4L_2}\left\{\left[L_4\left(\frac{n-1}{1-f}\bar{h}_f + \bar{h}_1 p - \frac{1}{2}\right) - L_3 - L_5\right]\left[1 - \frac{2(n-1)\bar{h}_f}{1-f} - 2\bar{h}_1 p\right]^2\right\} + S_1\\
&\quad +\frac{1}{L_2}\left[1 - \frac{2(n-1)\bar{h}_f}{1-f} - 2\bar{h}_1 p\right]\left\{L_6 - \bar{h}_f pT(n-1)\left[1 - \frac{(n-1)\bar{h}_f}{1-f}\right] - \frac{L_4}{2}\left[\frac{(n-1)\bar{h}_f}{1-f} - \frac{1}{2}\right]^2\right\},\\
&S_3 = -\frac{1}{12L_2}\left[L_4 + p(1-f)T\right]\left[1 - \frac{2(n-1)\bar{h}_f}{1-f}\right]^3\\
&\quad +\frac{1}{4L_2}\left[L_4\left(\frac{n-1}{1-f}\bar{h}_f + \bar{h}_1 p - \frac{3}{2}\right) - L_3 - 2L_1 + p(1-f)T\right]\left[1 - \frac{2(n-1)\bar{h}_f}{1-f}\right]^2\\
&\quad +\frac{1}{L_2}\left\{\bar{h}_f pT(n-1)\left[1 - \frac{(n-1)\bar{h}_f}{1-f}\right]\left[1 - \frac{2(n-1)\bar{h}_f}{1-f}\right]\right\}\\
&\quad +\frac{L_4}{2L_2}\left[\left(\frac{n-1}{1-f}\bar{h}_f\right)^2 - \frac{1}{4}\right]\left[1 - \frac{2(n-1)\bar{h}_f}{1-f}\right] - \frac{pT}{4L_2}(1-f)\left[1 - \frac{2(n-1)\bar{h}_f}{1-f}\right] + S_2,\\
&S_4 = \frac{1}{12L_2}pT(1-f) + S_3.
\end{aligned}$$

4. Finite Element Analysis

This section presents a numerical study of low-velocity impact of a clamped rectangular sandwich tube with FML tubes and MFC using ABAQUS/Explicit software. In this model, both the FML tubes and the MFC are modeled with 3D eight-node linear hexahedral brick elements (type C3D8R). A rigid roller is modeled to simulate the impactor, defining a point at its center as a point of concentrated mass and predetermining an initial low-velocity. The degrees of freedom of the nodes at both ends of the rectangular sandwich tube are set to zero to simulate a fixed support, that is, all vertical, horizontal, and rotational displacements are zero. In addition, the contact surface of the rigid roller with the outer face of the FML outer tube is set to be frictionless contact.

Take the half-span length of the rectangular sandwich tube as L = 200 mm, and the width and height of the cross-section of the outer FML tube are b = 40 mm and b_1 = 16 mm. The thickness of the metal foam is h_1 = 3 mm, and the thicknesses of the metal layer and the composite layer of the FML tube are h_m = 0.2 mm and h_f = 0.1 mm, respectively. For the convenience of this study, there are three metal layers, that is, n = 3, and then there are two composite layers. Thus, the total thickness of the FML tube is h = 0.8 mm, that is, $\overline{h} = 4/15$, $\overline{h}_1 = 0.075$, $\overline{h}_f = 1/160$, p = 2.5, and f = 0.75, respectively. In addition, a suitable radius is chosen for the loading roller of R = 4.

Based on experimental data [45], the metal layer of the FML is made of aluminum alloy, and the glass composite woven layer includes quasi-isotropic glass fiber fabrics infused with vinylester resin. The composite layers of the FML tube are made of composite material with strength σ_f = 220 MPa, elastic modulus E_f = 10 Gpa, elastic Poisson's ratio ν_{ef} = 0.3, and density ρ_f = 1700 kg/m^3, respectively. Due to the stretchable nature of FML tubes, composite layers can be assumed to be linearly elastic [45]. Using aluminum as the metal layers of the FML tubes, the yield strength, elastic modulus, elastic Poisson's ratio, density, and linear hardening tangential modulus are σ_m = 460 Mpa, E_m = 70 Gpa, ν_{em} = 0.3, ρ_m = 2800 kg/m^3, and $E_{mt} = 0.02E_m$ respectively. It is assumed that the layers of the metal are able to withstand deformation without fracture, that is, they have sufficient ductility.

The plastic compressible behavior of MFC Is modeled In ABAQUS software using the Deshpande–Fleck constitutive model [46]. The yield strength, elastic modulus, elastic Poisson's ratio, plastic Poisson's ratio, and density are σ_c = 10 Mpa, E_c = 10 Gpa, ν_{ec} = 0.3, ν_p = 0, and ρ_c = 405 kg/m^3, respectively, for aluminum foam as an isotropic metallic foam core. The MFC has a long yield stress platform σ_c that continues up to the densification strain ε_D = 0.5, and it is assumed that the MFC obeys a linear hardening law with tangent modulus $E_{ct} = 0.3E_m$ beyond densification.

A calculation of mesh sensitivity is performed to ensure that the mesh division does not have an impact on the results. Figure 3 shows the results of the mesh sensitivity examination of impact force versus deflection of outer surface of the rectangular sandwich tube with FML tubes and MFC under low-velocity impact at mid-span with G^* = 1000 and V_I = 2.5 m/s. In this model, the number of elements is N = 85,773. It can be seen that extra mesh refinement does not significantly change the results of the calculations.

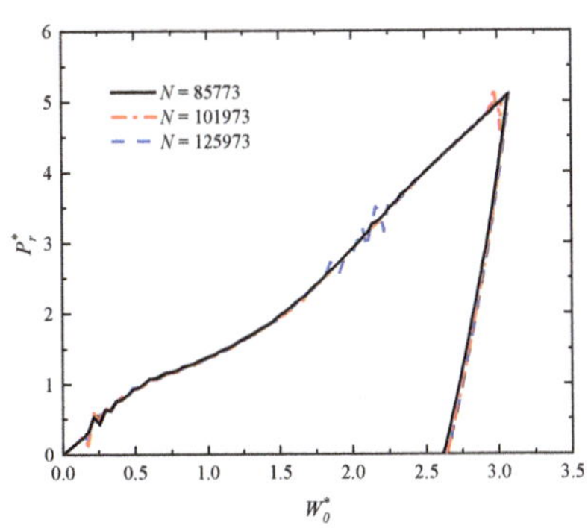

Figure 3. Results of the mesh sensitivity examination of the relationship between impact force and deflection for the rectangular sandwich tube with FML tubes and MFC under low-velocity impact with G^* = 1000 and V_I = 2.5 m/s.

5. Results and Discussion

The impact force and velocity of the impactor versus time curves are shown in Figure 4a,b. It can be seen in Figure 4a that the impact force of the impactor gradually increases to the maximum value and then decreases to zero during the impact process, indicating that the impactor causes the FML rectangular sandwich tube to deflect maximally and then is rebounded until it leaves the surface of the FML rectangular sandwich tube. As can be seen in Figure 4b, the velocity of the impactor gradually decreases to zero and then increases in the opposite direction, indicating that the impactor rebounds after impact.

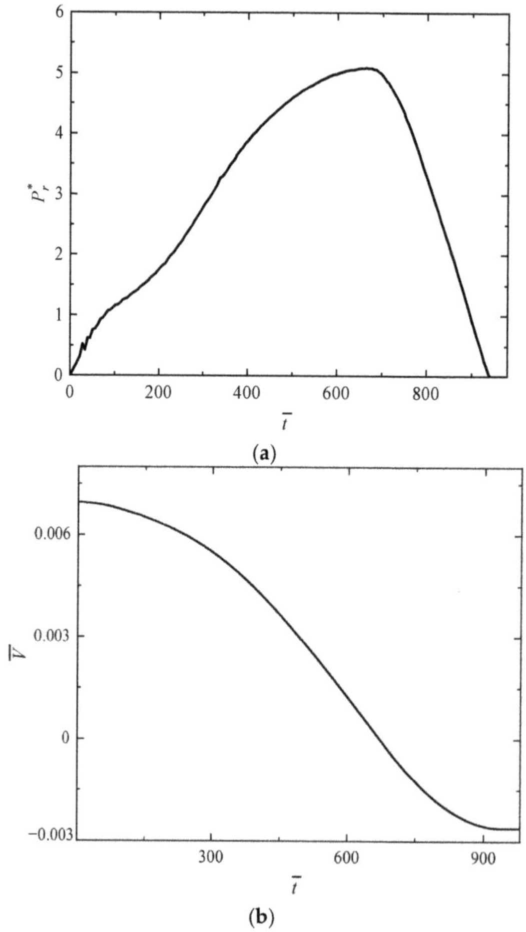

Figure 4. Numerical results of the impact response for the rectangular sandwich tube with FML tubes and MFC under low-velocity impact with $G^* = 1000$, $V_I = 2.5$ m/s. (**a**) The relationship between the impact force of the impactor and time, and (**b**) the relationship between the velocity of the impactor and time.

Figure 5a–c show the distribution of equivalent plastic strain (PEEQ) for the outer FML tube, MFC, and inner FML tube of the rectangular sandwich tubes. It can be seen that the local denting of the tube is not obvious below the impactor.

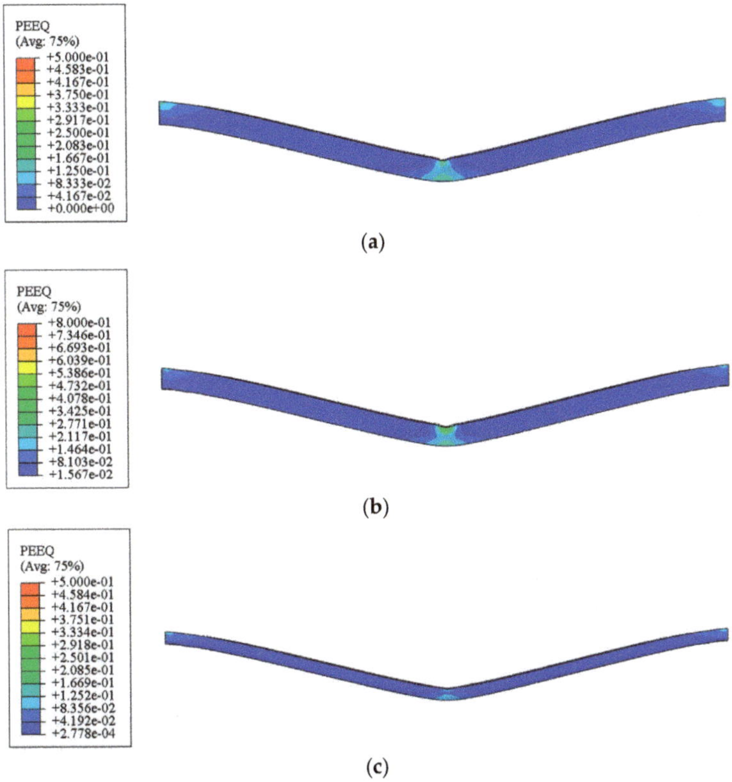

Figure 5. Numerical results of the equivalent plastic strain distribution of rectangular sandwich tubes with FML tubes and MFC impacted by a heavy mass of $G^* = 1000$ and an initial low-velocity of $V_I = 2.5$ m/s. (**a**) The outer tube of the rectangular sandwich tube, (**b**) the MFC of the rectangular sandwich tube, and (**c**) the inner tube of the rectangular sandwich tube. It should be noted that "5e-01" means "5×10^{-1}" in the color bars.

The analytical predictions and numerical results of the relationship between the dimensionless impact force P_r^* and dimensionless deflection W_0^* for the rectangular sandwich tubes with FML tubes and MFC under low-velocity heavy-mass impact at mid-span are shown in Figure 6a,b. The geometrical and material parameters are $f = 0.75$, $p = 2.5$, $q = 23/11$, $\overline{h} = 4/15$, $\overline{h}_1 = 0.075$, and $\overline{\sigma}_c = 1/22$. It is known from the figure that the curve has an upward trend and the impact force increases with the increase in deflection in analytical predictions. It can also be seen that the results obtained from the analytical predictions are reasonably well consistent with the numerical results of the post-yield phase. However, there are some differences between the analytical predictions and numerical results, which is probably caused by the effects of the shear force, elastic deformation, inertia, and strain hardening of the material that are neglected in the analytical solutions. It can be noticed that the numerical results show micro-oscillations in the initial stage, which are most likely due to the complex interaction between the impactor and the rectangular sandwich tubes.

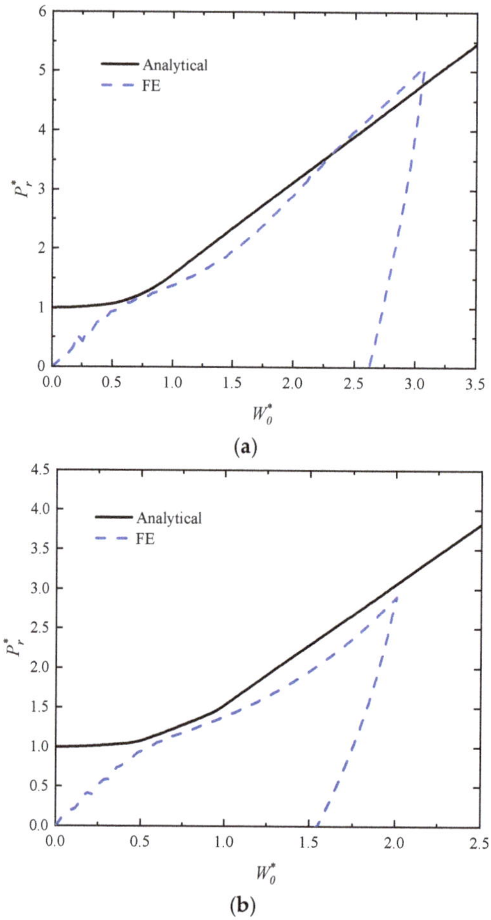

Figure 6. Analytical predictions and numerical results of the relationship between the dimensionless impact force and deflection for the rectangular sandwich tubes with FML tubes and MFC under low-velocity impact with a heavy mass impactor. (**a**) $G^* = 1000$, $V_I = 2.5$ m/s and (**b**) $G^* = 2500$, $V_I = 1$ m/s.

Figure 7a,b show the analytical results and numerical results of the dimensionless maximum deflection W_{0m}^* for the middle span of the outer surface of the rectangular sandwich tubes versus initial impact energy U_K^* of the impactor, where $f = 0.75$, $p = 2.5$, $q = 23/11$, $\overline{h} = 4/15$, $\overline{h}_1 = 0.075$, and $\overline{\sigma}_c = 1/22$. It is known from the figure that the curve has an upward trend and the maximum deflection increases with the increase ininitial impact energy in analytical predictions. In Figure 7a, it can be seen that the maximum deflection increases with the increase in impact energy when the mass ratio is constant as $G^* = 1000$. The cases for given initial velocity $V_I = 2.5$ m/s and various mass ratio G^* are considered in Figure 7b. It can be seen that the analytical predictions are in good harmony with the numerical results, while the numerical results may slightly overestimate the analytical predictions. The disagreement between the analytical predictions and numerical results is probably caused by ignoring the influence of elastic deformation and strain hardening of material in the analytical solution.

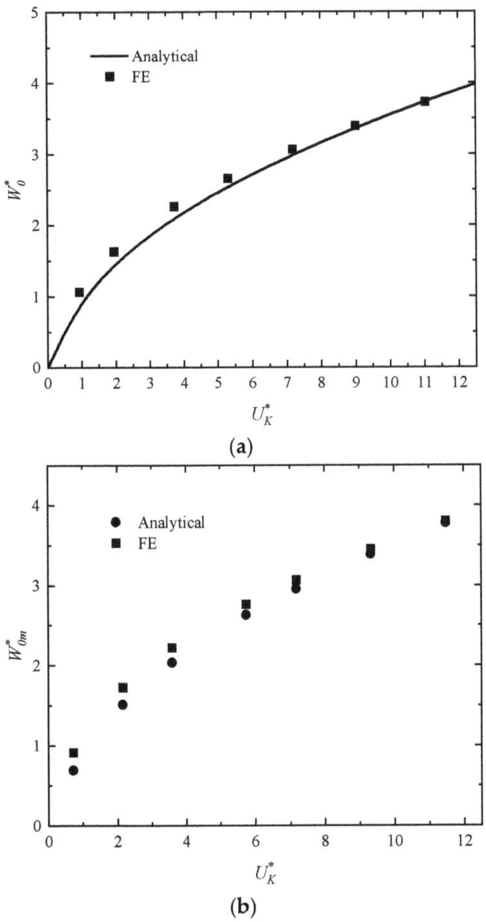

Figure 7. Comparisons of analytical and numerical results of the dimensionless maximum deflection for the middle span of the outer surface of the rectangular sandwich tubes with FML tubes and MFC versus the initial impact energy of the impactor with (**a**) a given constant mass ratio $G^* = 1000$ and (**b**) a given constant initial velocity $V_I = 2.5$ m/s.

Based on the analytical solutions, two types of curves are analyzed here, i.e., the dimensionless impact force P_r^* versus deflection W_0^* for rectangular sandwich tubes with FML tubes and MFC, and the dimensionless maximum deflection W_{0m}^* for the middle span of the outer surface versus initial impact energy U_K^* of the impactor of the rectangular sandwich tubes. The discussion here focuses on the influence of the MVF f, the strength ratio factor q, and the foam strength $\overline{\sigma}_c$ in the FML tubes on these two curves.

The Influence of the MVF f on the W_0^* versus P_r^* curves as well as the U_K^* versus W_{0m}^* curves of the rectangular sandwich tubes with FML tubes are shown in Figure 8a,b, respectively, in which $n = 3$, $p = 2.5$, $q = 23/41$, $\overline{h} = 4/15$, $\overline{h}_1 = 0.075$, $\overline{\sigma}_c = 1/82$, $G^* = 1000$. The smaller the MVF f, the greater the fiber content in the FML. As shown in Figure 8a, the impact force increases when the MVF f decreases for the constant deflection, indicating that the greater the fiber content in the FML, the better the load-bearing capacity of the rectangular sandwich tube. In Figure 8b, when the initial impact energy is constant, the maximum deflection of the rectangular sandwich tube decreases with decreasing MVF f, indicating that the greater the fiber content in the FML, the better the impact resistance of the rectangular sandwich tube.

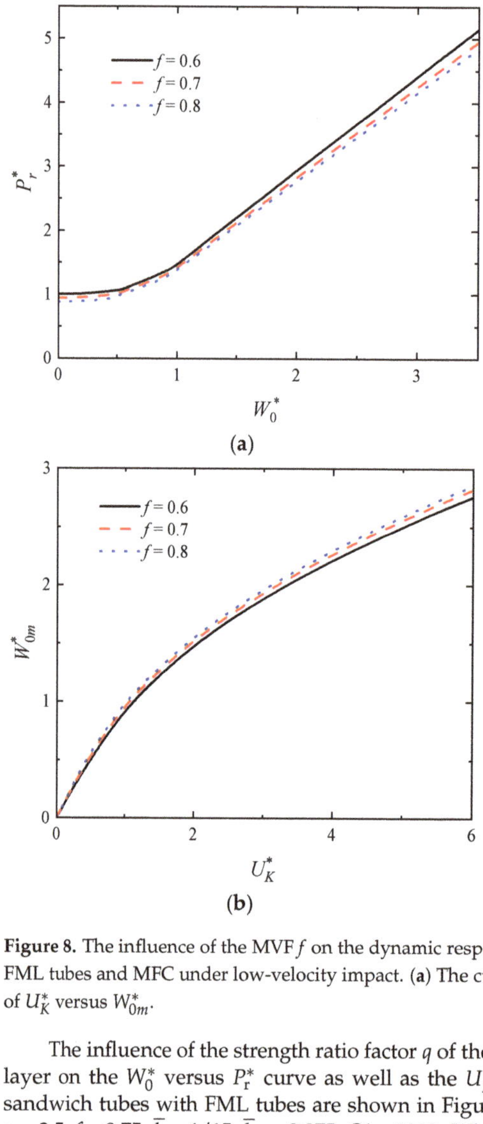

Figure 8. The influence of the MVF f on the dynamic response of rectangular sandwich tubes with FML tubes and MFC under low-velocity impact. (**a**) The curves of W_0^* versus P_r^* and (**b**) the curves of U_K^* versus W_{0m}^*.

The influence of the strength ratio factor q of the FML metal layer to FML composite layer on the W_0^* versus P_r^* curve as well as the U_K^* versus W_{0m}^* curves of rectangular sandwich tubes with FML tubes are shown in Figure 9a,b, respectively, in which $n = 3$, $p = 2.5, f = 0.75, \overline{h} = 4/15, \overline{h}_1 = 0.075, G^* = 1000$. When the strength of the metal layer in the FML is constant, as the strength of the fiber layer in the FML increases, the value of q decreases. In Figure 9a, the impact force increases obviously with the decreasing strength ratio when the deflection is constant, indicating that the higher the strength of the fiber in the FML, the better the load-carrying capacity of the rectangular sandwich tube. Also, the dimensionless maximum deflection of the rectangular sandwich tube decreases with the decreasing strength ratio for a constant initial impact energy, as shown in Figure 9b, indicating that the greater the strength of the fiber in the FML, the better the impact resistance of the rectangular sandwich tube. In addition, it is shown that the effect of q on both the load-carrying capacity and the impact resistance of the rectangular sandwich tube is evident. It can be seen that the strength ratio plays an important role in the low-velocity impact of the rectangular sandwich tube with FML tubes.

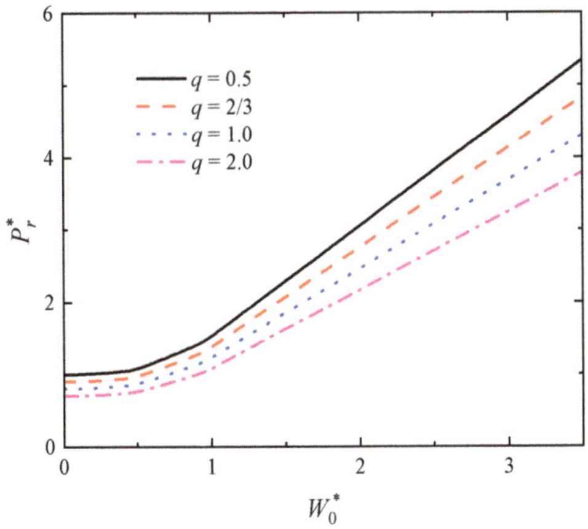

(a)

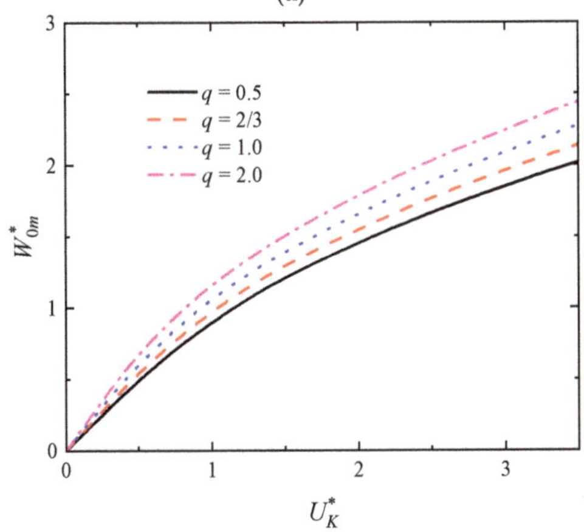

(b)

Figure 9. The influence of the strength ratio q on the dynamic response of rectangular sandwich tubes with FML tubes and MFC under low-velocity impact. (**a**) The curves of W_0^* versus P_r^* and (**b**) the curves of U_K^* versus W_{0m}^*.

In addition, the influence of the foam strength $\overline{\sigma}_c$ on the W_0^* versus P_r^* curves as well as the U_K^* versus W_{0m}^* curves are investigated as shown in Figure 10a,b, respectively, where $n = 3$, $p = 2.5$, $f = 0.75$, $q = 23/11$, $\overline{h} = 4/15$, $\overline{h}_1 = 0.075$, $G^* = 1000$. As shown in Figure 10a, the impact force increases with increasing foam strength when the deflection is constant. Moreover, as shown in Figure 10b, the dimensionless maximum deflection of the rectangular sandwich tube decreases with increasing foam strength for a constant initial impact energy. It can be concluded that the higher the strength of the foam, the higher the load-carrying capacity and the better the impact resistance of the rectangular sandwich tube.

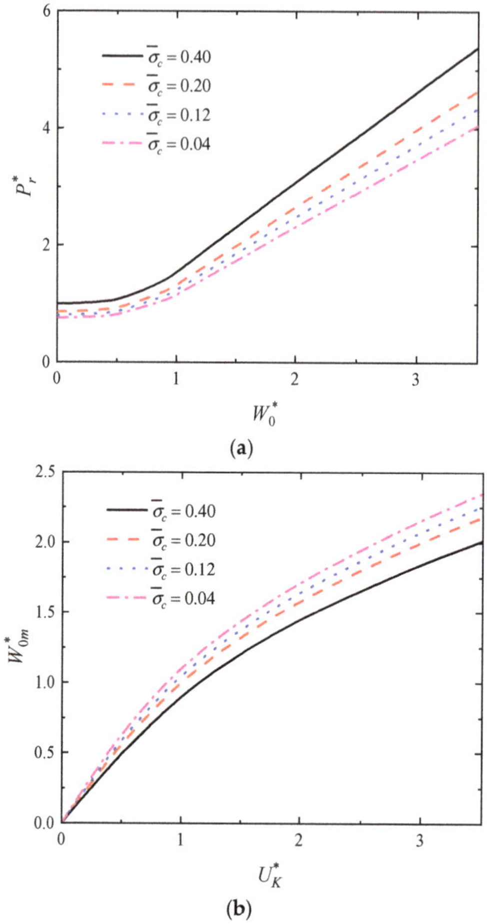

Figure 10. The influence of the foam strength $\overline{\sigma}_c$ on the dynamic response of rectangular sandwich tubes with FML tubes and MFC under low-velocity impact. (**a**) The curves of W_0^* versus P_r^* and (**b**) the curves of U_K^* versus W_{0m}^*.

6. Concluding Remarks

In this paper, the dynamic response of clamped rectangular sandwich tubes with FML tubes and MFC subjected to the impact of a low-velocity and heavy-mass impactor is theoretically and numerically investigated. As a new type of composite tube combining polymer and metal materials, FML tube plays a good role in the energy absorption and dynamic response of foam-filled tubes under low-velocity impacts. Based on modified rigid–plastic material approximation and introducing the number and volume fraction of composite layup, simple formulations for low-velocity impacts of rectangular sandwich tubes with FML tubes and MFC are obtained. The analytical solution agrees with the numerical results reasonably well. It is shown that the impact force of the rectangular sandwich tubes with FML tubes and MFC increases with increasing MVF, strength ratio factor of the FML metal layer to the FML composite layer, and relative strength of the foam. Moreover, the maximum deflection of the rectangular sandwich tubes with FML tubes and MFC decreases with increasing MVF, strength ratio factor, and foam strength. It is shown that by increasing the fiber content and fiber strength of the FML tubes and the foam strength, the load-carrying and energy-absorbing capacity of the rectangular sandwich tubes can be effectively improved, especially by changing the fiber properties. In addition, the present analytical model can be reasonably applied to make predictions about the dynamic response of rectangular sandwich tubes with FML tubes and MFC under low-velocity impacts. The analytical model also provides a theoretical basis for the design of FML tubes.

Author Contributions: Conceptualization, J.Z.; methodology, Y.W.; software, Y.W.; validation, Y.W., J.Z., H.G. and H.Y.; formal analysis, Y.W.; investigation, H.G. and H.Y.; resources, H.G.; data curation, H.Y.; writing—original draft preparation, Y.W. and J.Z.; writing—review and editing, Y.W. and J.Z.; visualization, Y.W. and J.Z.; supervision, J.Z.; project administration, J.Z.; funding acquisition, J.Z. All authors have read and agreed to the published version of the manuscript.

Funding: This research was funded by the National Natural Science Foundation of China (12272290), Fund for Shock and Vibration of Engineering Materials and Structures Key Lab of Sichuan Province (23kfgk03), and the Fundamental Research Funds for the Central Universities (lzujbky-2022-kb08).

Institutional Review Board Statement: The study did not require ethical approval.

Data Availability Statement: The authors attest that all data for this study are included in the paper.

Acknowledgments: The authors are grateful for their financial support through the National Natural Science Foundation of China (NSFC12272290), Fund for Shock and Vibration of Engineering Materials and Structures Key Lab of Sichuan Province (23kfgk03), and the Fundamental Research Fund for the Central Universities (lzujbky-2022-kb08).

Conflicts of Interest: The authors declare no conflicts of interest. The funders had no role in the design of this study; in the collection, analyses, or interpretation of data; in the writing of the manuscript; or in the decision to publish the results.

Appendix A. Analytical Solutions of a Fully Clamped Rectangular Sandwich Metal Tube Subjected to Low-Velocity Impact

This appendix contains the relationship between the dimensionless reaction force and the dimensionless deflection for a fully clamped rectangular sandwich metal tubewith MFC subjected to low-velocity impact, as well as the relationship between the dimensionless kinetic energy of the striker and maximum central deflection of the tube [31].

Rectangular sandwich tubes with MFC consist of an inner and outer rectangular tube of thickness h and a MFC of thickness h_1, in which the width, height, and length of the rectangular sandwich tube are b, b_1, $2L$ and the mass per unit length of the rectangular sandwich tube is G_b, respectively. The density of the inner and outer rectangular tube is ρ_f, and the density of isotropic MFC is ρ_c. The rectangular sandwich tube with MFC is impacted by an impactor with a heavy mass G_s and an initial velocity of V_I. The inner and outer rectangular tube are supposed to follow the RPP law of yield strength σ_f, and the filled MFC follows the RPPL material of yield strength σ_c and densification strain ε_D. Based on the flow rules associated with the fully clamped rectangular sandwich tubewith MFC at both ends, the analytical solution can be obtained using the yield condition [47] and the expressions for the elongation and angular rotation.

The relationship between the dimensionless kinetic energy of the initial impact U_K^* and the dimensionless maximum central deflection W_{0m}^* of the rectangular sandwich metal tube with MFC under low-velocity impact can be expressed as

$$\alpha U_K^* = \begin{cases} \frac{1}{6B}\left(\frac{2\bar{h}}{p}+\bar{h}_1\bar{\sigma}\right){W_{0m}^*}^3+W_{0m}^*, & 0 \le W_{0m}^* \le 1-2\bar{h}_1 p-4\bar{h} \\ \frac{{W_{0m}^*}^3}{12B}\left(2\bar{h}_1\bar{\sigma}+1-2\bar{h}_1\right)+\frac{1}{4B}\left(3\bar{h}+2\bar{h}_1 p-1\right)\left(2\bar{h}_1\bar{\sigma}-2\bar{h}_1+1\right){W_{0m}^*}^2 & \\ \quad +\frac{E}{4B}{W_{0m}^*}^2+\frac{F}{B}W_{0m}^*+K_1, & 1-2\bar{h}_1 p-4\bar{h} \le W_{0m}^* \le 1-2\bar{h}_1 p-2\bar{h} \\ -\frac{D}{12B}{W_{0m}^*}^3+\frac{1}{2}\left[\frac{D}{2B}\left(\bar{h}+\bar{h}_1 p-\frac{3}{2}\right)-\frac{C}{2B}\right]{W_{0m}^*}^2+\frac{\bar{h}}{B}\left(1-\bar{h}\right)W_{0m}^* & \\ \quad +\frac{D}{2B}\left(\bar{h}^2-\frac{1}{4}\right)W_{0m}^*+K_2, & 1-2\bar{h}_1 p-2\bar{h} \le W_{0m}^* \le 1-2\bar{h} \\ \frac{{W_{0m}^*}^3}{12B}+\frac{1}{2}\left(\frac{A}{B}-\frac{1}{2B}\right){W_{0m}^*}^2+\frac{1}{4B}W_{0m}^*+K_3, & 1-2\bar{h} \le W_{0m}^* \le 1 \\ \frac{A}{2B}{W_{0m}^*}^2+K_4, & W_{0m}^* \ge 1 \end{cases} \tag{A1}$$

where

$$\begin{aligned}
&\alpha = \frac{4G^*(1-3G^*)}{3(1+2G^*)^2},\ U_K^* = \frac{G_s {V_I}^2}{2P_c b_1},\ W_{0m}^* = \frac{W_{0m}}{b_1},\ P_c = \frac{4M_p}{L},\\
&G^* = \frac{G_s}{2G_b L},\ \bar{\sigma} = \frac{\sigma_c}{\sigma_f},\ \bar{h} = \frac{h}{b_1},\ \bar{h}_1 = \frac{h_1}{b},\ p = \frac{b}{b_1},\\
&A = 2\bar{h}\left(\bar{h}_1\bar{\sigma}+1\right)-\left(4\bar{h}+2\bar{h}_1 p-1\right)\left(\bar{h}_1\bar{\sigma}+\frac{2\bar{h}}{p}\right)+\bar{\sigma}\bar{h}_1 p\left(1-\frac{2\bar{h}}{p}\right),\\
&B = \bar{h}\left(1-\bar{h}\right)+\left[\frac{2\bar{h}}{p}+\left(1-\frac{2\bar{h}}{p}\right)\bar{\sigma}\right]\left(1-2\bar{h}-\bar{h}_1 p\right)\bar{h}_1 p\\
&\quad +\left(1-2\bar{h}_1+2\bar{h}_1\bar{\sigma}\right)\left(1-3\bar{h}-2\bar{h}_1 p\right)\bar{h}+\frac{1}{2}\left(\frac{2\bar{h}}{p}+\bar{h}_1\bar{\sigma}\right)\left(1-4\bar{h}-2\bar{h}_1 p\right)^2,\\
&C = -2\bar{h}\left(2\bar{h}_1\bar{\sigma}-2\bar{h}_1+1\right)-\left(\frac{4\bar{h}}{p}+2\bar{h}_1\bar{\sigma}\right)\left(1-4\bar{h}-2\bar{h}_1 p\right),\\
&D = \frac{4\bar{h}}{p}+2\bar{\sigma}\left(1-\frac{2\bar{h}}{p}\right),\\
&E = \bar{h}\left(2\bar{h}_1\bar{\sigma}+1-2\bar{h}_1\right)+\left(1-4\bar{h}-2\bar{h}_1 p\right)\left(2\bar{h}_1\bar{\sigma}+\frac{4\bar{h}}{p}\right),\\
&F = \bar{h}\left(1-\bar{h}\right)+\left[\frac{2\bar{h}}{p}+\left(1-\frac{2\bar{h}}{p}\right)\bar{\sigma}\right]\left(1-2\bar{h}-\bar{h}_1 p\right)\bar{h}_1 p+\frac{1}{4}\left(1-2\bar{h}_1+2\bar{h}_1\bar{\sigma}\right)\left(1-2\bar{h}-2\bar{h}_1 p\right)^2,\\
&K_1 = \left(1-\frac{F}{B}\right)\left(1-4\bar{h}-2\bar{h}_1 \mathrm{p}\right)+\frac{1}{6B}\left(\frac{2\bar{h}}{p}+\bar{h}_1-\frac{1}{2}\right)\left(1-4\bar{h}-2\bar{h}_1 p\right)^3\\
&\quad -\frac{E}{4B}\left(1-4\bar{h}-2\bar{h}_1 p\right)^2-\frac{1}{4B}\left(3\bar{h}-1+2\bar{h}_1 p\right)\left(2\bar{h}_1\bar{\sigma}+1-2\bar{h}_1\right)\left(1-4\bar{h}-2\bar{h}_1 p\right)^2,\\
&K_2 = \frac{4\bar{h}_1\bar{\sigma}-4\bar{h}_1-D+2}{24B}\left(1-2\bar{h}-2\bar{h}_1 p\right)^3+\frac{1}{4B}\left(3\bar{h}+2\bar{h}_1 p-1\right)\left(1-2\bar{h}_1+2\bar{h}_1\bar{\sigma}\right)\left(1-2\bar{h}-2\bar{h}_1 p\right)^2\\
&\quad -\frac{1}{4B}\left[D\left(\bar{h}+\bar{h}_1 p-\frac{1}{2}\right)-C-E\right]\left(1-2\bar{h}-2\bar{h}_1 p\right)^2\\
&\quad +\frac{1}{B}\left(1-2\bar{h}-2\bar{h}_1 p\right)\left[F-\bar{h}\left(1-\bar{h}\right)-\frac{D}{2}\left(\bar{h}-\frac{1}{2}\right)^2\right]+K_1,\\
&K_3 = -\frac{D+1}{12B}\left(1-2\bar{h}\right)^3+\frac{1}{4B}\left[D\left(\bar{h}+\bar{h}_1 p-\frac{3}{2}\right)-C-2A+1\right]\left(1-2\bar{h}\right)^2+\frac{\bar{h}}{B}\left(1-\bar{h}\right)\left(1-2\bar{h}\right)\\
&\quad +\frac{D}{2B}\left(\bar{h}^2-\frac{1}{4}\right)\left(1-2\bar{h}\right)-\frac{1}{4B}\left(1-2\bar{h}\right)+K_2,\\
&K_4 = \frac{1}{12B}+K_3.
\end{aligned}$$

in which the fully plastic bending moment M_p of the rectangular sandwich tube with MFC can be given by

$$M_p = \sigma_{fa} bh(b_1-h)+\sigma_3 bh_1(b_1-2h-h_1)+\sigma_2 bh(b_1-3h-2h_1)+\frac{1}{4}\sigma_1 b(b_1-4h-2h_1)^2 \tag{A2}$$

where

$$\sigma_1 = \frac{4h}{b}\sigma_{fa}+\frac{2h_1}{b}\sigma_c,\quad \sigma_2 = \frac{b-2h_1}{b}\sigma_{fa}+\frac{2h_1}{b}\sigma_c,\quad \sigma_3 = \frac{2h}{b}\sigma_{fa}+\frac{b-2h}{b}\sigma_c.$$

Meanwhile, defining the impact force between the impactor and the rectangular sandwich tube as P_r, the relationship between the dimensionless impact force P_r^* and the dimensionless deflection W_0^* of the rectangular sandwich tube with MFC under low-velocity impact can be expressed as

$$P_r^* = \begin{cases} \frac{3G^*}{3G^*+1}\left[\frac{1}{2B}\left(\frac{2\overline{h}}{p}+\overline{h}_1\overline{\sigma}\right)W_0^{*\,2}+1\right], & 0 \le W_0^* \le 1-2\overline{h}_1 p-4\overline{h} \\ \frac{3G^*}{3G^*+1}\left[\frac{W_0^{*\,2}}{4B}\left(2\overline{h}_1\overline{\sigma}+1-2\overline{h}_1\right)+\frac{1}{2B}\left(3\overline{h}+2\overline{h}_1 p-1\right)\left(2\overline{h}_1\overline{\sigma}-2\overline{h}_1+1\right)W_0^*\right] & \\ \quad +\frac{3G^*}{3G^*+1}\left(\frac{E}{2B}W_0^*+\frac{F}{B}\right), & 1-2\overline{h}_1 p-4\overline{h} \le W_0^* \le 1-2\overline{h}_1 p-2\overline{h} \\ \frac{3G^*}{3G^*+1}\left\{\frac{DW_0^{*\,2}}{8B}+\left[\frac{D}{2B}\left(\overline{h}+\overline{h}_1 p-\frac{1}{2}\right)-\frac{C}{2B}\right]W_0^*+\frac{\overline{h}}{B}\left(1-\overline{h}\right)+\frac{D}{2B}\left(\overline{h}-\frac{1}{2}\right)^2\right\}, & \\ & 1-2\overline{h}_1 p-2\overline{h} \le W_0^* \le 1-2\overline{h} \\ \frac{3G^*}{3G^*+1}\left[\frac{W_0^{*\,2}}{4B}+\left(\frac{A}{B}-\frac{1}{2B}\right)W_0^*+\frac{1}{4B}\right], & 1-2\overline{h} \le W_0^* \le 1 \\ \frac{3G^*}{3G^*+1}\frac{A}{B}W_0^*, & W_0^* \ge 1 \end{cases} \tag{A3}$$

where

$$W_0^* = \frac{W_0}{b_1},\ P_r^* = \frac{P_r}{P_c}.$$

References

1. Iriondo, J.; Aretxabaleta, L.; Aizpuru, A. Characterisation of the elastic and damping properties of traditional FML and FML based on a self-reinforced polypropylene. *Compos. Struct.* **2015**, *131*, 47–54. [CrossRef]
2. Nam, H.W.; Hwang, W.; Han, K.S. Stacking sequence design of fiber-metal laminate for maximum strength. *J. Compos. Mater.* **2001**, *35*, 1654–1683. [CrossRef]
3. Nejad, A.F.; Koloor, S.S.R.; Hamzah, S.M.S.A.S.; Yahya, M.Y. Mechanical behaviour of pin-reinforced foam core sandwich panels subjected to low impact loading. *Polymers* **2022**, *13*, 3627. [CrossRef] [PubMed]
4. Wang, S.L.; Zhang, M.; Zha, S.G.; Ding, Y.Y. In-plane crushing of a novel sinusoid-curved honeycomb under static and dynamic loadings. *Int. J. Crashworthiness* **2022**, *28*, 418–434. [CrossRef]
5. Fan, Y.S.; Yang, X.L.; He, J.; Sun, C.M.; Wang, S.K.; Gu, Y.Z.; Li, M. The variation mechanism of core pressure and its influence on the surface quality of honeycomb sandwich composite with thin facesheets. *J. Mater. Res. Technol.* **2021**, *15*, 6113–6124. [CrossRef]
6. Zhang, P.; Mo, D.H.; Ge, X.X.; Wang, H.; Zhang, C.Z.; Cheng, Y.S.; Liu, J. Experimental investigation into the synergetic damage of foam-filled and unfilled corrugated core hybrid sandwich panels under combined blast and fragment loading. *Compos. Struct.* **2022**, *299*, 116089. [CrossRef]
7. Zhang, Z.; Lei, H.S.; Yang, H.Y.; Xu, M.C.; Chen, M.J.; Wang, C.X.; Fang, D.N. Radar-stealth and load-bearing corrugated sandwich structures with superior environmental adaptability. *Compos. Sci. Technol.* **2022**, *277*, 109594. [CrossRef]
8. Guo, L.W.; Yu, J.L.; Li, Z.B. Experimental studies on the quasi-static bending behavior of double square tubes filled with aluminum foam. *Acta Mech.* **2010**, *213*, 349–358. [CrossRef]
9. Yang, S.H.; Chen, S.Y.; Liu, X.C.; Lin, Z.; Yang, L.H.; Schröder, K.U.; Schmidt, R. Dynamic crushing behavior of multi-layered hybrid foam-filled composite graded lattice sandwich panels. *Mech. Adv. Mater.* **2021**, *29*, 6694–6704. [CrossRef]
10. Wang, S.L.; Zhang, M.; Pei, W.J.; Yu, F.; Jiang, Y.H. Energy-absorbing mechanism and crashworthiness performance of thin-walled tubes diagonally filled with rib-reinforced foam blocks under axial crushing. *Compos. Struct.* **2022**, *299*, 116149. [CrossRef]
11. Wang, Z.G.; Deng, J.J.; He, K.N.; Tao, Y. Out-of-plane crushing behavior of hybrid hierarchical square honeycombs. *Thin Wall. Struct.* **2022**, *181*, 110051. [CrossRef]
12. Komorek, A.; Przybylek, P.; Szczepaniak, R.; Godzimirski, J.; Roskowicz, M.; Imilowski, S. The influence of low-energy impact loads on the properties of the sandwich composite with a foam core. *Polymers* **2022**, *14*, 1566. [CrossRef]
13. Wang, Y.Z.; Ma, L. Sound insulation performance of pyramidal truss core cylindrical sandwich structure. *Acta Mech. Solida Sin.* **2022**, *35*, 504–517. [CrossRef]
14. Li, J.Q.; Zhang, Y.; Fan, X.L.; Li, F.M. Multi bandgaps design of sandwich metamaterial plate with embedded membrane-type resonators. *J. Sandw. Struct. Mater.* **2022**, *25*, 311–329. [CrossRef]
15. Li, C.L.; Qi, J.F.; Wang, P.D.; Zhao, Z.A.; Wang, Z.; Lei, H.S.; Duan, S.Y. A novel hybrid design method of lattice structure based on failure mode. *Sci. China Phys. Mech.* **2022**, *65*, 294611. [CrossRef]
16. Wu, H.X.; Sun, F.Y.; Sun, Q.S.; Zhang, X.C.; Yang, S. Dynamic crushing behaviors of second-order hexagonal honeycombs. *Int. J. Appl. Mech.* **2022**, *14*, 2250054. [CrossRef]
17. Zhang, X.C.; Dong, S.J.; An, C.C.; Wu, H.X.; Niu, X.Y. Dynamic crushing behaviors of multi-layered gradient honeycombs with different poisson's ratios a finite element simulation. *Int. J. Appl. Mech.* **2022**, *14*, 2150127. [CrossRef]
18. Yang, J.S.; Yang, F.; Han, L.; Yang, L.H.; Wu, L.Z. Vibration response of glass fiber composite multi-layer graded corrugated sandwich panels. *J. Sandw. Struct. Mater.* **2022**, *24*, 1491–1511. [CrossRef]
19. Deng, S.N.; Wu, D.; Yang, J.S.; Luo, H.; Fu, L.L.; Schmidt, R.; Schröder, K.U. Damage recognition of glass fiber composite bi-directional corrugated sandwich cylindrical panels via non-contacted vibration method. *Mater. Today Commun.* **2022**, *32*, 103864. [CrossRef]

20. Hao, X.; Jin, T.; Ma, X.M.; Shu, X.F.; Li, X. Dynamic response of single curved fiber-metal hybrid lamina composites subject to low-velocity impact. *Int. J. Impact Eng.* **2022**, *164*, 104209.
21. Zhang, J.X.; Ye, Y.; Qin, Q.H.; Wang, T.J. Low-velocity impact of sandwich beams with fibre-metal laminate face-sheets. *Compos. Sci. Technol.* **2018**, *168*, 152–159. [CrossRef]
22. Liu, X.M.; Zhou, C.X.; Wan, F.; Tian, H.F.; Chen, W.H.; Guan, F. The difference of dent characteristics between the inner and outer tubes of sandwich pipes under lateral loading. *Ocean Eng.* **2023**, *271*, 113726. [CrossRef]
23. Guélou, R.; Eyma, F.; Cantarel, A.; Rivallant, S.; Castanié, B. Static crushing of wood based sandwich composite tubes. *Compos. Struct.* **2021**, *273*, 114317. [CrossRef]
24. Niknejad, A.; Moradi, A.; Beheshti, N. Indentation experiments on novel sandwich composite tubes. *Mater. Lett.* **2016**, *179*, 142–145. [CrossRef]
25. Baroutaji, A.; Gilchrist, M.D.; Smyth, D.; Olabi, A.G. Analysis and optimization of sandwich tubes energy absorbers under lateral loading. *Int. J. Impact Eng.* **2015**, *82*, 74–88. [CrossRef]
26. Kim, J.S.; Yoon, H.J.; Shin, K.B. Experimental investigation of composite sandwich square tubes under quasi-static and dynamic axial crushing. *Adv. Compos. Mater.* **2011**, *20*, 385–404. [CrossRef]
27. Fan, Z.; Shen, J.; Lu, G. Investigation of lateral crushing of sandwich tubes. *Procedia Eng.* **2011**, *14*, 442–449. [CrossRef]
28. Shen, J.H.; Lu, G.X.; Ruan, D.; Seah, C.C. Lateral plastic collapse of sandwich tubes with metal foam core. *Int. J. Mech. Sci.* **2015**, *91*, 99–109. [CrossRef]
29. Wu, J.Q.; Chen, X.; Zhu, H.Y.; Wang, P.; Jin, F.N.; Fan, H.L. Meta-honeycomb sandwich tubes designing, manufacturing, and crashworthiness performance. *Compos. Sci. Technol.* **2023**, *240*, 110096. [CrossRef]
30. Wang, Y.J.; Zhang, Z.J.; Xue, X.W.; Zhou, J.; Song, Z.X. Axial and lateral crushing performance of plate-lattice filled square sandwich tubes. *Compos. Struct.* **2011**, *274*, 114404. [CrossRef]
31. Zhang, J.X.; Zhu, Y.Q.; Ye, Y.; Yuan, H.; Qin, Q.H. Low-velocity impact of clamped slender rectangular sandwich tubes with metal foam core. *Int. J. Crashworthiness* **2020**, *26*, 1–14. [CrossRef]
32. Zhang, J.X.; Guo, H.Y. Dynamic response of rectangular sandwich tubes with metal foam core under blast loading. *Int. J. Appl. Mech.* **2022**, *14*, 2150126. [CrossRef]
33. Guo, H.Y.; Zhang, J.X.; Li, J.F.; Yuan, H.; Zhu, Y.Q. Splitting and curling collapse of square sandwich metal tube with aluminum foam core under axial low-velocity impact. *J. Braz. Soc. Mech. Sci.* **2022**, *44*, 514. [CrossRef]
34. Zhang, T.H.; Liu, Z.F.; Li, S.Q.; Lei, J.Y.; Wang, Z.H. Dynamic response and energy absorption performance of aluminum foam-filled sandwich circular tubes under internal blast loading. *Int. J. Impact Eng.* **2022**, *173*, 114458. [CrossRef]
35. Guélou, R.; Eyma, F.; Cantarel, A.; Rivallant, S.; Castanié, B. Dynamic crushing of wood-based sandwich composite tubes. *Mech. Adv. Mater. Struct.* **2022**, *29*, 7004–7024. [CrossRef]
36. Shen, J.H.; Lu, G.X.; Zhao, L.M.; Zhang, Q.M. Short sandwich tubes subjected to internal explosive loading. *Eng. Struct.* **2013**, *55*, 56–65. [CrossRef]
37. Wang, A.S.; Yu, X.H.; Wang, H.; Li, Y.; Zhang, J.; Fan, X.L. Dynamic response of sandwich tubes with continuously density-graded aluminum foam cores under internal explosion load. *Materials* **2022**, *15*, 6966. [CrossRef]
38. Mansor, M.A.; Ahmad, Z.; Abdullah, M.R. Crashworthiness capability of thin-walled fibre metal laminate tubes under axial crushing. *Eng. Struct.* **2022**, *252*, 113660. [CrossRef]
39. Ahmad, Z.; Abdullah, M.R.; Tamin, M.N. Experimental and numerical studies of fiber metal laminate (FML) thin-walled tubes under impact loading. In *Advanced Structured Materials Ochsner*; Ochsner, A., Altenbach, H., Eds.; Springer International Publishing: New York, NY, USA, 2015; Volume 70, pp. 433–443.
40. Subbaramaiah, R.; Prusty, B.G.; Pearce, G.M.K.; Lim, S.H.; Thomson, R.S. Crashworthy response of fibre metal laminate top hat structures. *Compos. Struct.* **2017**, *160*, 773–781. [CrossRef]
41. Song, Y.; Yuan, H.; Du, J.L.; Sun, H.; Han, Z.L.; Zhang, J.X. Analytical and numerical investigation on dynamic behavior of rectangular FML sandwich tubes with metal foam core under low-velocity impact. *Int. J. Appl. Mech.* **2022**, *14*, 2250028. [CrossRef]
42. Mansor, M.A.; Ahmad, Z.; Abdullah, M.R. Experimental studies on the impact characteristics of seamless fibre metal laminate (FML) tubes. *Mater. Today Proc.* **2021**, *39*, 1077–1081. [CrossRef]
43. Shiravand, A.; Asgari, M. Hybrid metal-composite conical tubes for energy absorption; theoretical development and numerical simulation. *Thin Wall. Struct.* **2019**, *145*, 106442. [CrossRef]
44. Jones, N. Note on the impact behaviour of fibre-metal laminates. *Int. J. Impact Eng.* **2017**, *108*, 147–152. [CrossRef]
45. Tagarielli, V.L.; Fleck, N.A.; Deshpande, V.S. Collapse of clamped and simply supported composite sandwich beams in three-point bending. *Compos. Part B Eng.* **2003**, *35*, 523–534. [CrossRef]
46. Deshpande, V.S.; Fleck, N.A. Isotropic constitutive models for metallic foams. *J. Mech. Phys. Solids* **2000**, *48*, 1253–1283. [CrossRef]
47. Qin, Q.H.; Wang, T.J. An analytical solution for the large deflections of a slender sandwich beam with a metallic foam core under transverse loading by a flat punch. *Compos. Struct.* **2008**, *88*, 509–518. [CrossRef]

polymers

MDPI

Article

Optimization of 3D Printing Parameters of High Viscosity PEEK/30GF Composites

Dmitry Yu. Stepanov [1], Yuri V. Dontsov [2], Sergey V. Panin [2,3,*], Dmitry G. Buslovich [4], Vladislav O. Alexenko [3], Svetlana A. Bochkareva [3], Andrey V. Batranin [5] and Pavel V. Kosmachev [1]

1 Microelectronics of Multispectral Quantum Introscopy Laboratory of the R&D Center "Advanced Electronic Technologies", National Research Tomsk State University, 634050 Tomsk, Russia; sdu@ispms.ru (D.Y.S.); kosmachev@mail.tsu.ru (P.V.K.)
2 Department of Materials Science, Engineering School of Advanced Manufacturing Technologies, National Research Tomsk Polytechnic University, 634050 Tomsk, Russia; doncov@mail2000.ru
3 Laboratory of Mechanics of Polymer Composite Materials, Institute of Strength Physics and Materials Science of Siberian Branch of Russian Academy of Sciences, 634055 Tomsk, Russia; aleksenkovo@ispms.ru (V.O.A.); svetlanab7@yandex.ru (S.A.B.)
4 Laboratory of Nanobioengineering, Institute of Strength Physics and Materials Science of Siberian Branch of Russian Academy of Sciences, 634055 Tomsk, Russia; buslovich@ispms.ru
5 Engineering School of Non Destructive Testing, National Research Tomsk Polytechnic University, 634050 Tomsk, Russia; batranin@gmail.com
* Correspondence: svp@ispms.ru

Abstract: The aim of this study was to optimize a set of technological parameters (travel speed, extruder temperature, and extrusion rate) for 3D printing with a PEEK-based composite reinforced with 30 wt.% glass fibers (GFs). For this purpose, both Taguchi and finite element methods (FEM) were utilized. The artificial neural networks (ANNs) were implemented for computer simulation of full-scale experiments. Computed tomography of the additively manufactured (AM) samples showed that the optimal 3D printing parameters were the extruder temperature of 460 °C, the travel speed of 20 mm/min, and the extrusion rate of 4 rpm (the microextruder screw rotation speed). These values correlated well with those obtained by computer simulation using the ANNs. In such cases, the homogeneous micro- and macro-structures were formed with minimal sample distortions and porosity levels within 10 vol.% of both structures. The most likely reason for porosity was the expansion of the molten polymer when it had been squeezed out from the microextruder nozzle. It was concluded that the mechanical properties of such samples can be improved both by changing the 3D printing strategy to ensure the preferential orientation of GFs along the building direction and by reducing porosity via post-printing treatment or ultrasonic compaction.

Keywords: ANN; PEEK; glass fiber; polymer matrix composite (PMC); FDM; optimization; Taguchi method; additive manufacturing (AM); computed micro-tomography (micro-CT)

1. Introduction

Polyetheretherketone (PEEK) is a high-performance high-temperature semi-crystalline polymer with improved physical and mechanical properties, chemical resistance, as well as bioinertness [1]. Respectively, a significant number of researchers have been investigating the features of PEEK [2,3]. However, its high cost greatly narrows the range of applications, primarily towards high-tech industries and medicine.

In the manufacturing of machine parts and friction units, PEEK is typically loaded with fillers [4,5], among which the most widely used are chopped carbon fibers (CCFs) and glass fibers (GFs) at contents of about 30 wt.% [6,7]. Antifriction composites, fabricated by injection molding, contain smaller amounts of reinforcing fibers but they are additionally filled with polytetrafluoroethylene (PTFE) and/or graphite, for instance, for improving their functional characteristics. For such cases, the term 'high-pressure velocity' (HPV) composites has been proposed [8].

Recently, 3D printing methods have become widespread for manufacturing numerous products from PEEK-based composites [9], in particular with the use of fused deposition modeling (FDM)/fused filament fabrication (FFF). Typically, neat PEEK is used as a filament matrix with the content of other fillers (like CFs, GFs, AFs, PTFE, graphite, etc.) less than $\leq$10 wt.%, since melt flow index (MFIs) is sharply reduced at greater concentrations of CCFs or GFs, preventing the formation of dense and uniform structures of additively manufactured products. An alternative approach involves the use of both granular feedstocks and 3D printing heads with microextruders [10], but few research results have been published so far on 3D printing with such high-viscosity thermoplastics [11–14].

For 3D printing, including with thermoplastic composites, it is necessary to simultaneously optimize a set of technological (input/control) parameters that ensure the high quality of additively manufactured (AM) products. The latter means the maximum precision of shapes, the minimum number of discontinuities, the uniformity of both macro- and microstructure, etc. Their quality is controlled by output (mechanical and structural) parameters. Thus, the 3D printing process can be represented as a black box, and a given set of some functional (output) parameters is achieved by varying the input ones. According to the formulation, such a problem can be solved via the design of experiments [15]. As compared to the other methods: (i) Box Behnken design [16]; (ii) Rotatable Central composite design of experiments (Box-Wilson Central Composite Designs) [17]; (iii) D-optimal design [18]; (iv) Doehlert design [19]; (v) Response surface method (RSM) [20] the Taguchi method [21] is widely applied due to a possibility of multiple reductions in the number of required results in contrast to full-scale experiments [22].

Another effective approach for solving such problems is the use of artificial neural networks (ANNs) [23]. However, their learning requires large training samples, limiting the possibilities of computer simulation of the 3D printing processes [24]. Therefore, it is important to implement methods that enable increasing the sample sizes artificially (non-experimentally) [25]. Enlarging the sample size based on experimental data was carried out by preliminary analysis of the 3D printing process and synthesis of additional preliminary knowledge for the limit values of the process parameters.

Despite numerous reported results and some achieved successes in this field of science, the challenge of 3D printing with high-viscosity thermoplastic composites remains relevant. The reasons are both specifics of AM and complex dependences of the structure formation on a set of input (technological) parameters. This issue requires in-depth research in both materials science and machine learning when ANNs are implemented to solve the problem of approximating a vector quantity in the space of multiple parameters.

The aim of this study was to optimize a set of input parameters (the V_{3D} travel speed, the T_{ext} extruder temperature, and the ν extrusion rate by varying rotation speeds of the microextruder screw) for 3D printing with the high-viscosity PEEK/30GF composite from a commercially available granulate. For this purpose, ANNs were implemented for computer simulation of full-scale experiments.

As a null hypothesis for AM of high-quality products from the PEEK/30GF composite, it was assumed that it is necessary to ensure the maximum material feeding into the printing zone by reducing the V_{3D} travel speed of the moving extruder head by increasing both the T_{ext} extruder temperature and the ν extrusion rate.

The selection of glass fibers as filler material was motivated by their: (i) availability as a commercial grade feedstock; (ii) lower price; (iii) ease of processing; (iv) large interphase adhesion, and (v) possibility of substantially improving mechanical properties of PEEK-based composites. Their input in increasing mechanical properties of PEEK-based composites is a little bit lower in contrast with carbon fibers; however, in terms of affecting 3D printability, they are quite comparable.

The article is structured as follows. Section 2 describes materials and research methods, including the design of the experiment using the Taguchi method. Section 3 presents the results of assessing the mechanical properties and analyzing the structures of the composite, including scanning electron microscopy (SEM) and IR spectroscopy. Section 4 is devoted to the optimization of the 3D printing parameters via computer simulation using ANNs. In Section 5, the obtained results are discussed and some prospects are justified for this scientific direction.

2. Materials and Methods

Initially, rectangular shape samples (templates) were fabricated by the FDM method using the 'PEEK KETASPIRE KT-880 GF30 BG20' granulate with a length of 3 mm and a diameter of 2 mm (SOLVAY, Shanghai, China) that included 30 wt.% GFs. In this way, an 'ArmPrint-2' laboratory 3D printer was deployed (Tomsk Polytechnic University, Tomsk, Russia). It was controlled by the 'LINUX CNC' OS operating system and equipped with a single-screw microextruder with a nozzle diameter of 0.5 mm. The ν extrusion rates (amounts of the fed material) were changed by varying rotation speeds of the microextruder screw. The temperature of the heated bed of the 3D printer was 180 C. The 3D printer did not include any thermal chamber, so the air convection of the manufactured samples and their cooling conditions were not instrumentally controlled. However, the temperature conditions were identical in all cases.

Digital model files were created using the 'Repetir-Host V2.1.3' software package (Hot-World GmbH & Co. KG Knickelsdorf 4247877, Willich, Germany) and the 'Slic3r' slicer (licensed under the GNU Aero General Public License, version 3). The software parameters were preset as the following:

- The layer height was 0.2 mm;
- The first layer height was 0.3 mm;
- The perimeter was 2 lines;
- The top solid layer was 0 and the bottom one was 0;
- The infill was 80%;
- The rectilinear and speed were the same, but the first layer was 50 %; the extrusion width infill was 100%.

For uniaxial static tensile testing, the type 5 dog-bone specimens were cut out according to ISO 527-2:2012 [26]. Their surfaces were processed with sandpapers (grit 240) and conditioned at room temperature for more than 24 h. The tests were carried out using a 'Gotech Al-7000M' electromechanical machine at a cross-head speed of 1 mm/min.

The structures of the samples were analyzed with a 'LEO EVO 50' SEM (Carl Zeiss, Oberkochen, Germany) at an accelerating voltage of 20 kV. Initially, the notched samples were cooled in liquid nitrogen at −197 °C for one hour and mechanically fractured then. In a vacuum, copper films ~10 nm thick were deposited on the fracture surfaces using a 'JEOL JEE-420' vacuum evaporator (JEOL USA, Inc., Peabody, MA, USA). The requirements for the thickness of the conductive films were justified by the need to preserve the morphology of the original fracture surfaces. Then, the filler distributions and the supermolecular structures of the samples were investigated.

The chemical structures of the samples were analyzed by IR spectrometry. IR spectra were recorded using both 'NICOLET 5700' (Thermo Fisher Scientific, Waltham, MA, USA) and 'FT-801' (SIMEX, Novosibirsk, Russia) Fourier-transform IR spectrometers in the diffuse reflectance range of 600–4000 cm^{-1} with a diamond (Single Reflection Diamond ATR).

Computed microtomography (micro-CT) of the samples was carried out with an 'OREL-MT' tomograph assembled at Tomsk Polytechnic University [27]. The tomograph was equipped with a 'XWT 160-TC' high-focus X-ray machine (X-RAY WorX, Garbsen, Germany), a 'PaxScan-2520V' flat-panel X-ray detector (Varian, Palo Alto, CA, USA), and a research object positioning system. Conical X-ray beam geometry and projection magnification were implemented to increase the spatial resolution of images. The 'Bruker-microCT' software package v.1.18 was used for tomographic reconstruction and visualization of the obtained data [28]. The key characteristics of the tomograph are presented in Table 1.

Table 1. The key characteristics of the 'OREL-MT' tomograph.

Characteristic	Value
X-ray machine	XWT 160-TC (X-RAY WorX, Garbsen, Germany)
Anode voltage, kV	20–160
Anode current, µA	5–1000
Focal spot, µm	1.4
Detector	PaxScan-2520V (Varian, Palo Alto, CA, USA)
Pixel size, µm	127
Detector operating area, mm	193 × 242
Detector size, pixels	1900 × 1516
Linear manipulator	MDrive Plus 42 (Intelligent Motion Systems, Inc., Marlborough, CT, USA)
Movement range, mm	0–400
Rotary system	MS208Ec (Physik Instrumente (PI) GmbH & Co., Karlsruhe, Germany)
Minimum angular step, degree	0.007

The following micro-CT mode was utilized: an anode voltage of 100 kV, an anode current of 35 µA, an angular step of scanning projections of 0.2 degrees, a number of projections of 1800, and a voxel size of 6.35 µm.

The Taguchi method was used for trial optimizing the 3D printing parameters. The input factors were the V_{3D} travel speed, the T_{ext} extruder temperature, and the ν extrusion rate. Initially, the levels of these input parameters were empirically determined for the sustainable 3D printing process within their wide ranges.

The T_{ext} extruder temperature was preset at 440 °C since this level enabled the melting of the high-viscosity PEEK/30GF composite. It was an important parameter for 3D printing at the V_{3D} travel speeds of about 40 mm/s (to ensure an acceptable production rate). Lower T_{ext} extruder temperatures significantly reduced the 3D printing rate because they did not allow a sufficient amount of the molten composite material to be extruded. So, the T_{ext} extruder temperatures were 420, 440, and 460 °C (Table 2). The V_{3D} travel speeds were preset at 20, 30 and 40 mm/s. As mentioned above, the ν extrusion rates were controlled by varying rotation speeds of the microextruder screw. Initially, a multiplier of 4.5 rpm was determined corresponding to the number of screw revolutions per unit track length upon 3D printing. According to the results of some trial experiments, the shapes of the samples additively manufactured from the PEEK/30GF composite correlated well with their digital models. So, the ν extrusion rates were preset at 4.0, 4.5, and 5.0 rpm (Table 2).

Table 2. The level numbers and the values of the Taguchi factors according to the L9 format.

Experiment Number	Levels and Values of (Technological) Factors		
	Extruder Temperature T_{ext}, °C	Travel Speed V_{3D}, mm/s	Extrusion Rate ν, Rpm
1	420 (1)	20 (1)	4.0 (1)
2	420 (1)	30 (2)	4.5 (2)
3	420 (1)	40 (3)	5.0 (3)
4	440 (2)	20 (1)	4.5 (2)
5	440 (2)	30 (2)	5.0 (3)
6	440 (2)	40 (3)	4.0 (1)
7	460 (3)	20 (1)	5.0 (3)
8	460 (3)	30 (2)	4.0 (1)
9	460 (3)	40 (3)	4.5 (2)

3. Results

3.1. Mechanical Properties

Table 3 illustrates representative values of the key mechanical properties obtained by averaging over at least four experimental points according to the Taguchi L9 design. A preliminary analysis of these data enabled the conclusion that the ranges of changes in the input parameters used when planning the experiment did not lead to multiple variations in the mechanical properties of the AM samples. In addition, the obtained values were noticeably inferior to the data reported by the feedstock manufacturer since they involved the use of injection molding or compression sintering as a fabrication method [29].

Table 3. The results of the tensile tests of the samples additively manufactured from the PEEK/30GF composite.

3D Printing Mode	Tensile Strength, MPa	Elastic Modulus, MPa	Elongation at Break, %
1	86.10 ± 6.12	2100 ± 216	4.8 ± 0.3
2	88.10 ± 6.65	2458 ± 346	4.7 ± 0.1
3	84.00 ± 2.72	2349 ± 232	4.8 ± 0.4
4	91.00 ± 6.29	2456 ± 232	4.7 ± 0.5
5	94.50 ± 4.39	2359 ± 162	5.1 ± 0.2
6	84.60 ± 2.19	2326 ± 268	4.6 ± 0.2
7	94.50 ± 1.53	2610 ± 173	4.8 ± 0.3
8	93.80 ± 2.75	2542 ± 8	5.2 ± 0.4
9	85.00 ± 1.44	2378 ± 357	4.3 ± 0.3

Then, graphs were drawn according to the Taguchi method [10], characterizing the influence of each of the (input) parameters on the mechanical properties of the AM samples. This made it possible to quantify their effect by analyzing the implemented modes (the combinations of the 3D printing parameters) and propose the rational one. For this purpose, the 'bigger is better' principle and the signal-to-noise (S/N) ratio were applied. According to this methodology, the maximum S/N ratios characterized the maximum variability of the output parameters (the mechanical properties) and corresponded to the achievement of their maximum values. In this case, such a fact reflected the greater significance of a particular level of a 3D printing parameter.

It should be noted that in contrast to the elastic modulus and the tensile strength used as the output parameters, for which the maximum values had to be achieved, a target value of elongation at break was not obvious according to the 'bigger is better' principle. Typically, particulate composites at filling degrees of 25–30% were characterized by brittle failure, corresponding to their high mechanical strength. In this approximation, just a negligible elongation at the break value reflected the preference of the 3D printing mode. On the other hand, the porosity of the PEEK/30GF composite significantly reduced its ductility, being a negative factor. Accordingly, the authors prefer not to discuss in detail the role of the elongation at break values and not to consider it when analyzing the data on the influence of technological factors.

For the 'ν extrusion rate' input factor, higher tensile strength and elastic modulus values were achieved when using its third level of 5 rpm (Figure 1a,b, respectively). At the same time, a multidirectional trend was observed for the elongation at break levels (Figure 1c), which confirmed the above reasoning on the inexpediency of taking it into account. Therefore, the maximum ν extrusion rate had to be preset according to the Taguchi method.

An analysis of the influence of the 'T_{ext} extruder temperature' input factor on the mechanical properties of the samples showed that their elastic modulus values were characterized by a pronounced rising trend with the maximum at T_{ext} = 460 °C (Figure 1b). For the tensile strength, higher values were also observed at T_{ext} = 440–460 °C (Figure 1a), while the minimum S/N ratio for the 'elongation at break' parameter corresponded to the maximum T_{ext} level of 460 °C (Figure 1c). Thereby, the high T_{ext} extruder temperatures had to be preset according to the Taguchi method.

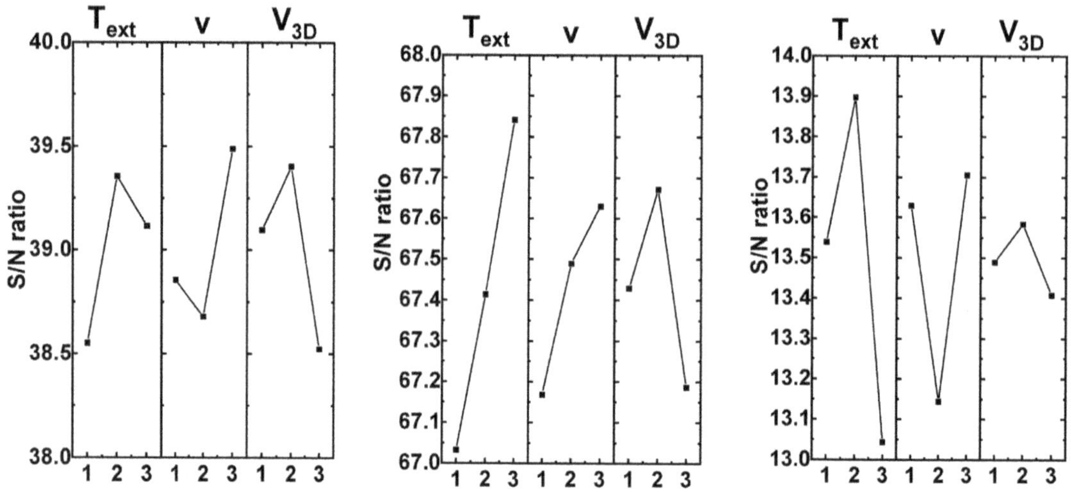

(a) (b) (c)

Figure 1. The S/N ratios for different levels of the technological parameters: (**a**) tensile strength; (**b**) elastic modulus; (**c**) elongation at break.

Varying the 'V_{3D} travel speed' input parameter from 20 up to 40 mm/s provided higher tensile strength and elastic modulus values at its average level of 30 mm/s, despite they were not significantly lower at the minimum of 20 mm/s. Elongation at break values was characterized by minimal variability at the applied V_{3D} travel speeds. For this reason, the authors again preferred to exclude this "output" parameter from the assessment. Based on the above, the low V_{3D} travel speeds had to be preset according to the Taguchi method.

Finally, the Delta (Δ) influence degrees of the input (technological) parameters were ranked according to the Taguchi principle to ensure higher mechanical properties of the samples (Table 4). From the tensile strength perspective, all three factors had comparable Δ influence degrees of 0.804–0.879. For the elastic modulus, the T_{ext} extruder temperature was two times 'more important' of a factor than the other two. In the case of elongation at break, the maximum Δ influence degree of 0.853 was exerted by the V_{3D} travel speed, while it was the minimum (Δ = 0.175) for the T_{ext} extruder temperature.

Table 4. Ranking of the input (control) parameters by their Δ influence degrees.

Property	Delta Parameter Δ (or $L_{max} - L_{min}$, i.e., the Difference in the S/N Values between the Maximum and Minimum Factor Levels)		
	Extruder Temperature T_{ext}	**Extrusion Rate v**	**Travel Speed V_{3D}**
Tensile strength	0.804	0.808	0.879
Elastic modulus	0.808	0.462	0.485
Elongation at break	0.175	0.561	0.853

As a preliminary discussion, it should be noted that the above findings were somewhat correlated with the results of Deng et al. [30], who applied the Taguchi method for optimizing the 3D printing mode for neat PEEK (its MFI was surely lower than that of the PEEK/30GF composite). In that case, the highest tensile strength of 40 MPa was achieved at the maximum T_{ext} extruder temperature of 370 °C and V_{3D} travel speed of 60 mm/s, while, the highest elastic modulus of 1563 MPa was obtained at a lower T_{ext} level of 360 °C but the same maximum V_{3D} travel speed of 60 mm/s. The greatest elongation at break value of 14.3% was observed at the minimum V_{3D} travel speed of 20 mm/s. These maximum mechanical properties were noticeably inferior to the cast polymer. In [30], it was stated that 3D printing angles, nozzle diameters, and bead widths had to be optimized as well.

So, the results of the above analysis carried out within the framework of the adopted Taguchi method did not allow both to determine the exact (but not applied in the experiments) values of the input parameters and to give an unambiguous interpretation of the reason for their Δ influence degrees. For clarifying these nuances and assessing the '3D printing mode–structure–properties' relationship, the micro- and chemical structures of the samples were examined by SEM and IR spectroscopy, respectively.

3.2. Microstructures

Figure S1 shows lower magnification SEM micrographs of the PEEK/30GF composites additively manufactured using the FDM modes presented in Table 2. Note, that no signs of agglomeration of GFs were observed. The pattern of the fracture surfaces (obtained after exposing the samples to liquid nitrogen) was not brittle. Generally, this fact was consistent with the relatively high elongation at break values of ~5%, since similar PEEK-based composites possessed minimal ductility (~1%) at comparable filling degrees. On the other hand, the presented SEM micrographs did not reveal any fundamental differences in the "macrostructures" of all investigated samples.

An analysis of the SEM micrographs"take' at higher magnification (Figure 2) enabled the conclusion that there was no predominant orientation of GFs along the direction of laying the molten filament, despite the FDM method used for 3D printing, which involved the extrusion of the molten granular feedstock. The reason could be the fact that GFs were stochastically oriented in the initial granulate.

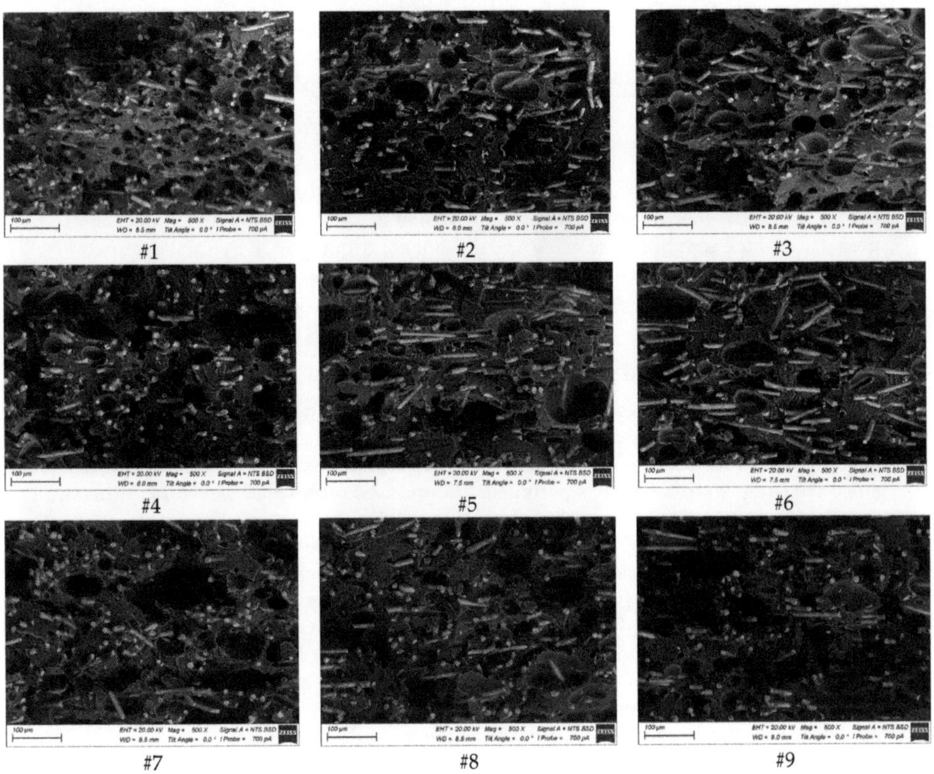

Figure 2. The SEM micrographs of the PEEK/30GF composites additively manufactured using the modes presented in Table 2.

Nevertheless, the more fundamental identified result was high porosity levels, reaching up to dozens vol.%. In the Discussion section, estimates of the effect of porosity on the elastic modulus of the PEEK/30GF composite are given, explaining its low level at the sufficiently high filling degree of reinforcing GFs.

The authors did not attempt to statistically estimate the porosity levels for each of the nine samples. The reason was the fact that the elongation at break values, which were very sensitive to porosity, did not change within noticeable limits. For this reason, one could conclude that it was inappropriate to determine a correlation between the porosity and the set of the 3D printing parameters.

In addition to the porosity statement and the approximate estimates of its levels given above, it was important to note the dimensions of pores in order to discuss the aspect of their formation, including in terms of the applied 3D printing parameters. According to Figure 2, pores were generally isolated (not through), while their characteristic sizes (with a predominantly round shape) were several tens of microns. At the same time, the pores were not predominantly formed near reinforcing GFs.

Thus, the revealed high porosity levels indicated that some thermal destruction of the polymer could occur at the elevated T_{ext} extruder temperatures of 420–460 °C (recall that the melting point of PEEK was ~343 °C). To verify this fact, the results of the IR spectral analysis of the samples additively manufactured at different T_{ext} extruder temperatures are discussed below.

Note that it was difficult to ensure uniform spreading of the high-viscosity molten composite containing fibrous inclusions, which differed significantly in rheological properties, upon 3D printing at high both ν extrusions rates and V_{3D} travel speeds. Loading PEEK with 30 wt.% GFs necessitated a higher extrusion pressure, so the 3D printer equipped with the microextruder was utilized. For this reason, it could not be expected that the formed structures were highly dense and continuous. This fact was analyzed in more detail by computer simulation using ANNs, the results of which are reported below in Section 4.

3.3. Chemical Structures

Figure S2 presents the IR spectra of the samples of the PEEK/30GF composite additively manufactured at the T_{ext} extruder temperatures of 420, 440, and 460 °C. The key task was to show that short-term heating up to 460 °C, at which the higher mechanical properties of the samples were achieved, did not contribute to the thermal destruction of the polymer matrix. Such a statement was based on the fact that the given IR spectra differed little from each other even for modes 7 and 9, characterized by the T_{ext} extruder temperature of 460 °C.

4. Application of ANNs for Simulation of the 3D Printing Process

Since the study did not allow for the explicit determination of the exact optimal values of technological parameters, partly because their preset ranges may not have been wide enough, ANNs were implemented for computer simulation of the 3D printing process. Since the sample size of nine experimental points was ultra-small, several additional experiments were carried out.

In this section, the statement of the research aim could be formulated as follows. As noted above, computer simulation was an effective tool for optimizing the 3D printing parameters, i.e., improving the mechanical properties with a minimum number of full-scale experiments. Previously, a similar problem was solved by the authors for ultrasonic welding of laminated composites [25]. The proposed methodology involved solving two key issues: (i) synthesis of a single nonlinear model of 3D printing as a process with many inputs and outputs, and (ii) the addition of a priori knowledge to ensure acceptable predictive accuracy of computer-aided simulation using ANNs.

4.1. Experimental Data Analysis

In addition to the nine modes (combinations of the 3D printing parameters) analyzed above by the Taguchi method, six additional modes No. 10–15 were tested (Table 5; Figure 3) in order to expand the training sample. As noted above, four to seven laboratory experiments were performed for each set of the input parameters (75 in total), as well as the same numbers of the output ones were determined (tensile strength, elastic modulus, and elongation at break). An illustration of the distribution of the (input) parameters in a three-dimensional space is shown in Figure 3.

Table 5. The 3D printing parameters of the laboratory experiments, in addition to those analyzed by the Taguchi method.

3D Printing Mode	Extruder Temperature T_{ext}, °C	Travel Speed V_{3D}, mm/s	Extrusion Rate ν, rpm
10	460	20	4
11	380	20	5
12	380	40	5
13	400	30	5
14	440	10	5
15	440	30	3.5

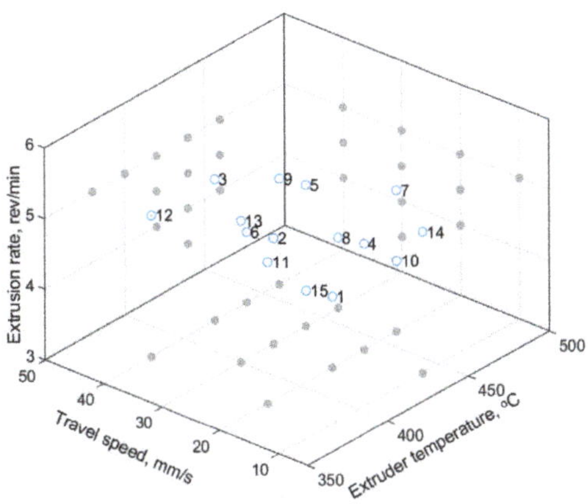

Legend.
○12—3D printing modes
●—projections of the 3D printing modes on the plane, limiting the analysis area

Figure 3. The 3D printing modes of the laboratory experiments in the space of the (input) parameters.

The obtained data were analyzed by drawing dependencies of the mechanical properties of the samples of the PEEK/30GF composite on the 3D printing parameters (Figure 4a), interdependencies of their mechanical properties (Figure 4b), and calculating regression models for each of them. Among the obtained data, six laboratory experiments were identified and excluded from a subsequent analysis, the results of which were outside the range of confidence intervals (at a significance level of 0.05).

The assumption of a significant nonlinear dependence of the mechanical properties of the samples on the 3D printing parameters was observed in Figure 4a, which was confirmed by both low coefficients of determination values and high standard errors of multiple linear regression models (Table 6).

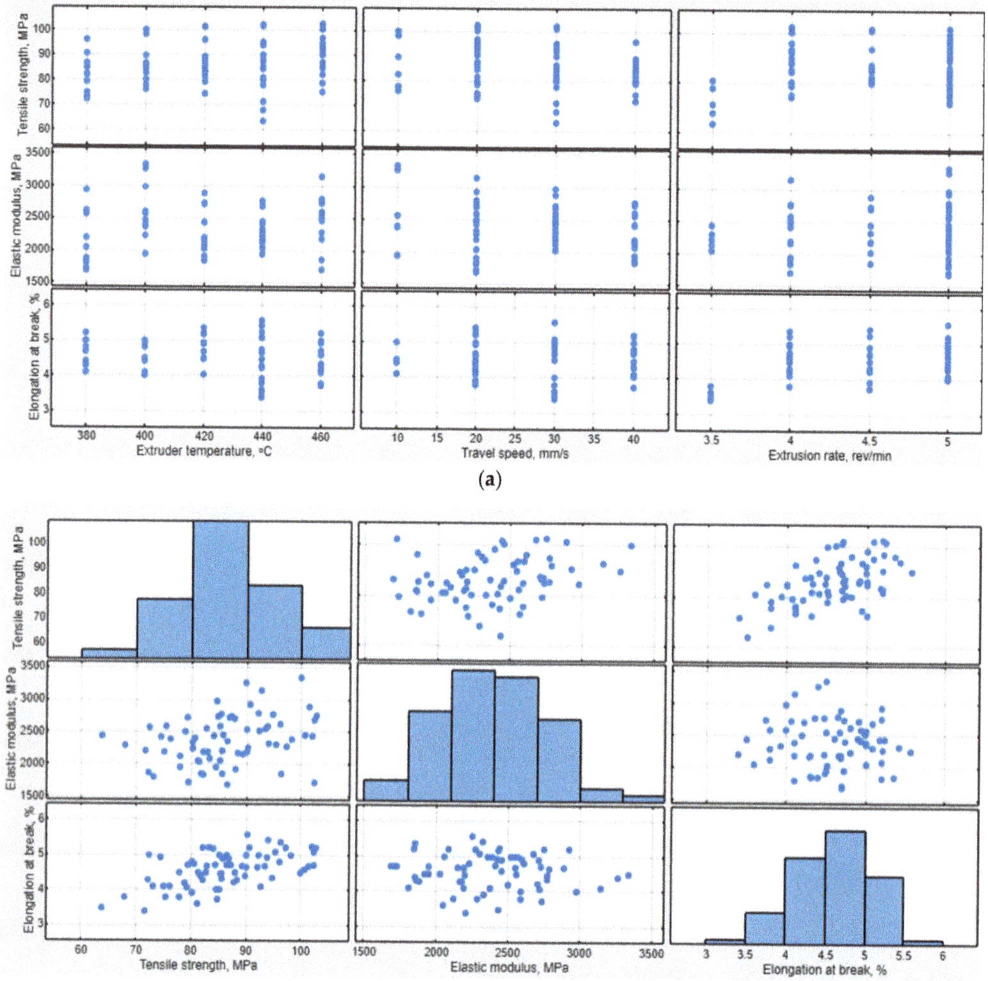

Figure 4. The dependences of the mechanical properties of the samples of the PEEK/30 GF composite on the 3D printing parameters (**a**), as well as both dependences and histograms (**b**) after verification.

Table 6. The regression statistics of the experimental data.

Statistics	Predicted Property			Average
	Tensile Strength	Elastic Modulus	Elongation at Break	
Multiple R	0.573	0.243	0.445	0.4203
R-squared	0.329	0.059	0.198	0.1953
Normalized R-squared	0.299	0.018	0.163	0.1600
Standard error	8.846 MPa	378.091 MPa	0.594%	–
MSE	0.0983	0.1260	0.0914	0.1052
Multiple R	0.573	0.243	0.445	0.4203

4.2. Analysis of Priori Knowledge

4.2.1. Prerequisites for a Priori Knowledge

A preliminary analysis of both 3D-printing parameters and predicted mechanical properties of the samples was based on previously acquired knowledge about the features of the analyzed process:

1. The 3D printing process could not be implemented under one of the following conditions: the T_{ext} extruder temperature was below the melting point of the polymer matrix (mode 16), the absence of the fed material (at $v = 0$ rpm, mode 20 or $V_{3D} = 0$ mm/s, mode 18). So, the mechanical properties of such samples were assumed to be zero.
2. It was also assumed that the mechanical properties were lower at an excessively high T_{ext} extruder temperature of 560 °C (mode 17) due to possible thermal degradation of PEEK. The same applied to high V_{3D} travel speed of 60 mm/s (mode 19) and v extrusion rate of 8 rpm (mode 21), based on the assumption that an excessive amount of the extruded polymer did not allow a homogeneous structure to form.

The mechanical properties of the samples of the PEEK/30GF composite, summarized on the basis of a priori knowledge, are presented in the Supplementary (Tables S1 and S2) section.

4.2.2. Selecting Acceptable Minimum Values

Then, the authors expertly formulated the requirements for the minimum acceptable values of the output parameters (the mechanical properties according to Table 7). Checking the results of the laboratory experiments for compliance with these acceptable values showed that there was not a single mode among the applied ones (Table 5) that would meet all the requirements. However, modes 4, 5, 7, 8, and 10 were identified, for which the acceptable minimum values were achieved in at least one experiment (Figure 5).

Table 7. The minimum acceptable property values for the samples of the PEEK/30GF composite.

Property	Minimum Acceptable Value
Tensile strength, MPa	90
Elastic modulus, MPa	2400
Elongation at break, %	4.8

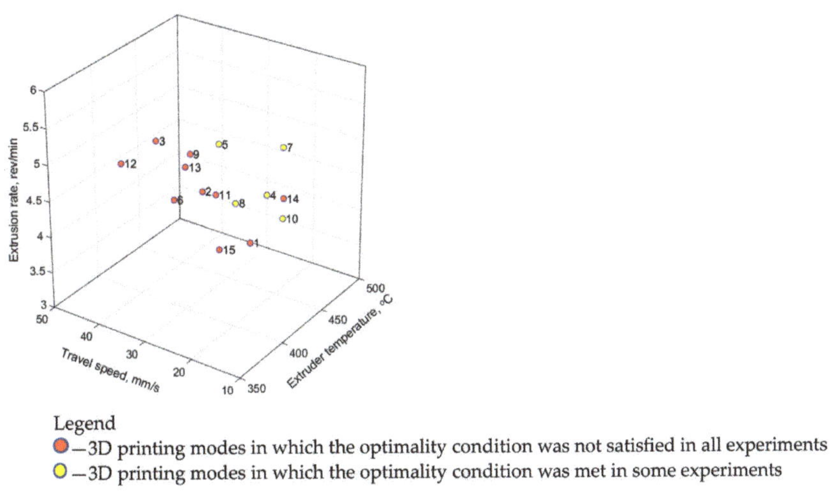

Figure 5. The parameters' space and the result of checking the 3D-printing modes for compliance with the minimum acceptable property values.

4.3. ANN-Based Computer Simulation of the 3D Printing Process

In this study, the authors used two types of simple ANNs that provided nonlinear computer simulation: feedforward networks (FFNNs) and radial basis function networks (RBFNNs). Their inputs were the 3D-printing parameters, while the outputs were the mechanical properties. The selection of both architecture and parameters was described in detail previously [25,31,32], so the applied approach corresponded to that reported in [25].

In the first stage, a sample containing the verified experimental data was used, i.e., a total of 15 vectors of the 3D printing parameters (modes) and 66 vectors of the corresponding mechanical properties. For the FFNN model, the number of hidden layer neurons (from 3 to 8) and activation functions (linear, hyperbolic tangent sigmoid, log-sigmoid) were varied. Upon their learning, synthesized property vectors formed from the experimental values with the addition of pseudo-random numbers with zero mathematical expectation and normalized standard deviation equal to 0.01 were added to the training sample. Computer simulation was carried out many times with an enumeration of the main parameters of the ANNs. For the RBFNN model, the 'spread' parameter varied from 0.1 to 0.9, while the 'goal' was changed within the 0.100–0.001 range. The results of assessing the areas of suboptimal parameters (SOPs) of the 3D printing process (Table 6) are shown in Figures 6a and 7a for the RBFNN and FFNN models, respectively.

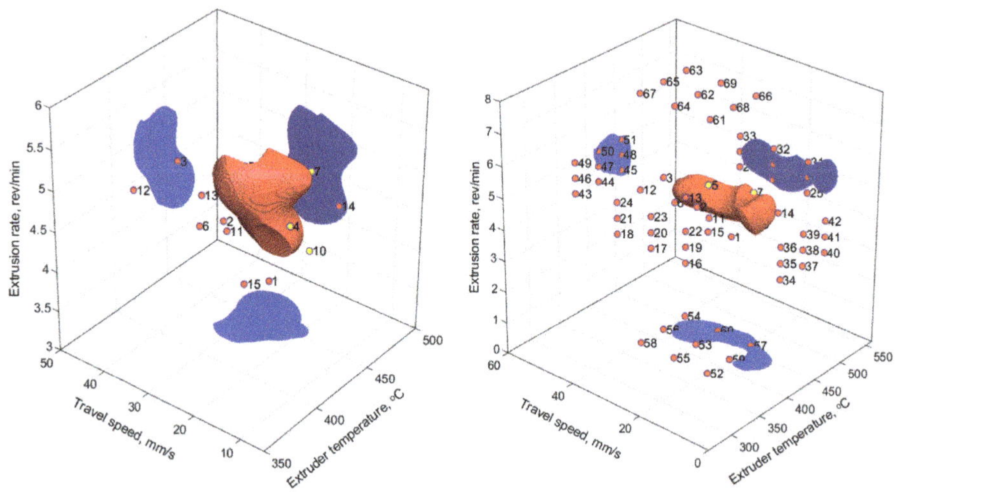

Legend
●—3D printing modes in which the optimality condition was not satisfied in all experiments
○—3D printing modes in which the optimality condition was met in some experiments

Figure 6. The 3D printing modes and a priori knowledge, as well as the SOP area, drawn using the RBFNN model: (**a**) spread = 0.3, goal = 0.001, the training sample size of 66 vectors; (**b**) spread = 0.3, goal = 0.01, the training sample size of 66 experimental + 54 a priori vectors.

In the second stage, the training sample included the experimental data and a priori knowledge. In order to estimate the influence of training sample size, its effect on the accuracy of model construction for the FFNN was carried out at variation of the number of vectors of input parameters as well as the amount of synthesized data, Figure 8. During the synthesis of the models, the mean square error (MSE) was utilized as the optimization function. It allowed for the characterization of the deviation of the model from the training set. The mean absolute deviation (MAD) of the normalized model values over priory values not taken/specified/determined in the training set was utilized for characterizing the prognosis capability of the models. The verification of the FFNN models over the data of priory knowledge planes has shown quite different dependence: the presence of MAD maximum at the sample size of 69 modes (vectors of the input parameters) and 120 vectors of the corresponding mechanical properties. At the same time, for the FFNN model, the training sample was enlarged by synthesized data with characteristics similar to those at the first stage (see above). It might be concluded that the latter has increased the prognosis capabilities of the developed models. The best characteristics of the ANNs were sought in the same ranges, so two models were selected, for which SOP areas were characterized by the simplest structures (Figures 6b and 7b).

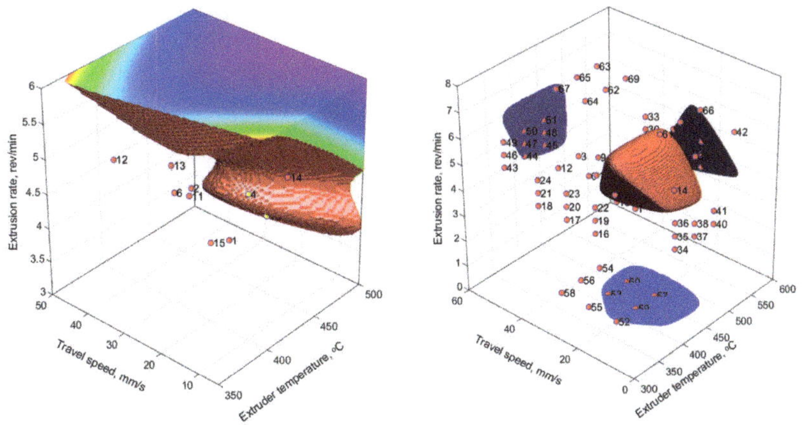

(a) (b)

Legend
●—3D printing modes in which the optimality condition was not satisfied in all experiments
○—3D printing modes in which the optimality condition was met in some experiments

Figure 7. The experimental modes and a priori knowledge, as well as the SOP area, drawn using the FFNN model: (**a**) 4 hidden layer neurons, the sample size of 66 experimental + 132 synthesized vectors; (**b**) 6 hidden layer neurons, the sample size of 66 experimental + 54 prior + 240 synthesized vectors.

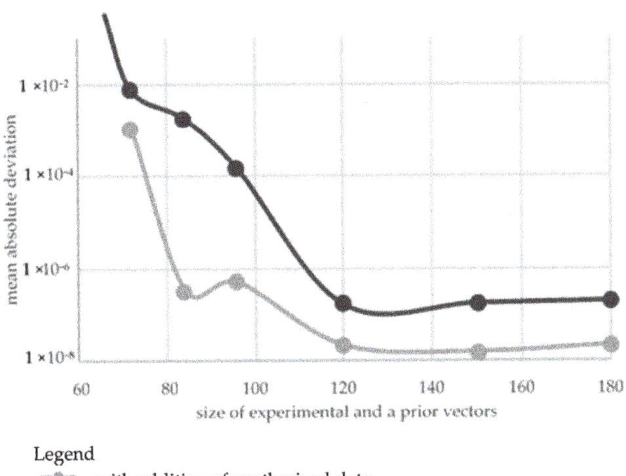

Legend
—with addition of synthesized data
—without addition of synthesized data

Figure 8. Results of models' verification within priory knowledge planes as a function of the size of experimental and prior vectors of the properties.

Table 8 presents the statistical characteristics of the learning results for the selected models. Comparing them with the data obtained using the regression model (Table 6), it has been concluded that computer simulation using the ANNs, due to its nonlinearity, was characterized by a low mean square error (MSE) but a high determination coefficient (R^2). The use of a priori knowledge made it possible to increase the determination coefficient, i.e., reduce the unexplained variance of the model. However, the root-mean-square error increased, which was a natural result of enhancing the modeling area in the extrapolation zone.

Table 8. The statistical characteristics of the selected models, the 3D printing parameters, and the corresponding predicted maximum property values for the samples of the PEEK/30GF composite.

ANN Type	Sample Size	MSE	R2	3D Printing Parameters			Mechanical Properties		
				Extruder Temperature T_{ext}, °C	Travel Speed V_{3D}, mm/s	Extrusion Rate v, rpm	Tensile Strength MPa	Elastic Modulus MPa	Elongation at Break, %
RBFNN	66	0.0235	0.4543	460.40	25.07	4.10	98.85	2559.17	5.05
	66 + 54	0.0427	0.9837	454.54	30.91	4.59	100.51	2525.50	5.58
FFNN	66 + 132	0.0255	0.4124	500	18.64	3.85	99.55	2605.99	5.31
	66 + 54 + 240	0.0560	0.9788	453.36	13.92	4.51	95.02	2645.10	5.19

For the RBFNN model (Figure 6a), the SOP area was finite and closed. The 3D printing modes, which partially satisfied the optimality condition, turned out to be outside its boundaries. The addition of a priori knowledge to the training sample led to a slight enlargement of the SOP area, compliance with the modes that partially satisfied the target mechanical properties, and significant complication of its shape (Figure 6b).

The SOP area, drawn according to the FFNN model after its learning based only on the experimental and synthesized data, was not limited to four planes: high values of both T_{ext} extruder temperature and v extrusion rate, as well as both low and high levels of the V_{3D} travel speed (Figure 7a). Such behavior was caused by the location of the satisfactory experimental results on the boundaries of the SOP area and the poor ability of the FFNN model to be extrapolated. Respectively, such a model predicted the target mechanical properties at the boundaries of the analyzed data area and could not be considered adequate. Adding a priori knowledge in the field of extrapolation to the training sample made it possible to limit the SOP area throughout the entire volume and localize the predicted optimal parameters (Figure 7b).

Within the SOP areas, the 3D printing parameters were sought, at which the mechanical properties were maximum. The weighted root-mean-square value was chosen as the maximum search measure. The search results are presented in Table 8. Excluding the "FFNN 66 + 132" model from consideration, all other ones enabled the prediction of the optimal values of both T_{ext} extruder temperatures of 450–460 °C and v extrusion rates of 4.1–4.59 rpm. A noticeable discrepancy in the predicted data was observed only for the V_{3D} travel speeds (from 13.92 up to 30.91 mm/s), which could be explained by the difference in the general simulation approaches.

The RBFNN model was based on the concept of minimizing approximation in each value of the training sample and searching for the minimum predicted values in the extrapolation zones. Therefore, it was obvious that it predicted the optimal values close to the known parameters in all cases. Conversely, the root-mean-square minimization was applied in the FFNN model to identify general trends in the training sample. This algorithm could lead (as was shown in the studied case) to the removal of the predicted optimal parameters from the region of known values.

The predicted 3D printing parameters were not duly verified for several reasons. Firstly, they differed slightly from the results of the laboratory experiments, enhancing the probability of obtaining similar data. Secondly, the results of the applied Taguchi method (Section 3) were in qualitative agreement with the data obtained by computer simulation using the ANNs. Thirdly, the achieved mechanical properties of the additively manufactured samples were lower than the values given by the manufacturer of the PEEK/30GF feedstock for injection molding, so the reasons for this phenomenon are discussed below.

5. Discussion

5.1. Computer Simulation on the Effect of Porosity on the Mechanical Properties of the PEEK/30GF Composites

Since it was convincingly shown above that the formation of porosity could significantly reduce the mechanical properties of the PEEK/30GF composite, the influence of pores (air inclusions) of various diameters (from 20 up to 100 μm) on the elastic modulus was assessed by computer simulation using the finite element method (FEM). It should be noted that the studied effect of lowering the mechanical properties of polymer composites with increasing porosity was also reported elsewhere [33,34].

The problem of determining the mechanical properties of porous neat polymers and their composites could be reduced to calculating the stress-strain state parameters of a representative volume under uniaxial tension. In this way, the equations of solid mechanics [35–37] were solved with subsequent averaging of the obtained values [38].

In this study, the problem of the plane-stress state of the porous PEEK/30GF composite was solved to obtain quantitative estimates of the changes in its mechanical properties. Square computational domains with round pores were applied. Figure 9 shows ones as finite element meshes of 500 × 500 μm in size with a pore content of 30% (the air filling degree) of different diameters (from 20 to 100 μm). The centers of pores were specified using a random number generator, so they were located as uniformly as possible over the computational domains.

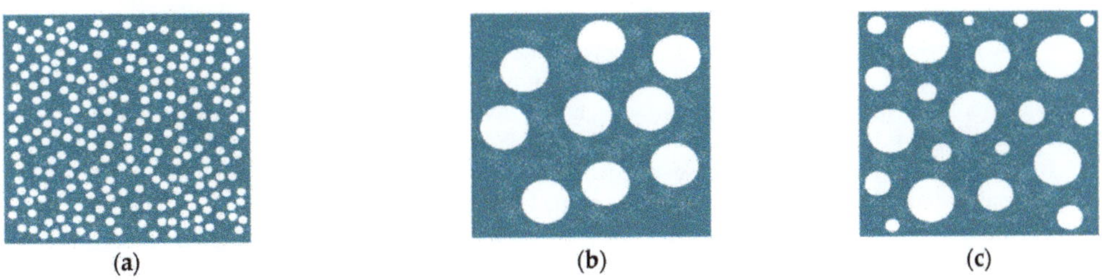

Figure 9. Schematic locations of pores with the diameters of 20 μm (**a**), 100 μm (**b**), and from 20 to 100 μm (**c**) in the computational domains at the porosity of 30%.

In the laboratory experiments, it was possible to visually estimate both characteristic pore sizes and porosity levels with certain errors (Figure 2). Therefore, large or small pores of identical sizes were varied in the calculations (Figure 9). In addition, a combination of them was used being typical for the studied samples. At the same porosity level, the number of small pores was evidently greater than that for large ones.

Lengths of the sides of the representative volume were 1000 μm for pores with a radius of 100 μm, but they were 500 μm for ones with a diameter of 20 μm, considering the convergence of the calculated elastic modulus values within 5%. The mechanical properties of the matrix in the representative volume were taken on the basis of the data reported for the PEEK/30GF composite [39]: the elastic modulus was 7000 MPa, the tensile strength was 80 MPa, and the Poisson's ratio was 0.39. The boundary conditions corresponded to the stretching of the computational domain along the X-axis, so the displacement along the Y-axis was specified at the top boundary. The bottom boundary was rigidly fixed and the displacements were zeroed along the X and Y axes. The lateral boundaries were free, so both normal and shear stresses were set to zero.

Figure 10 shows slightly nonlinear dependences of equivalent elastic modulus values versus the content of pores of various sizes. With the same porosity, the elastic modulus was lower in the presence of large pores than in the case of small ones. The discrepancy between the curves and their nonlinearity could be caused by the distribution of pores over the computational domain with increasing porosity levels. When pores of different sizes were present, with a fairly uniform distribution over the computational domain, the elastic modulus values were approximately in the middle between the two curves shown in Figure 10.

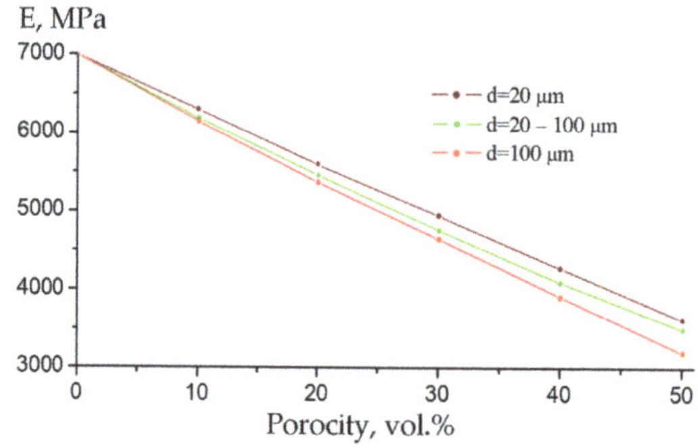

Figure 10. The elastic modulus versus porosity dependences for pores with different diameters *d*.

Figure 11a,b shows stress distributions over the representative volume with the dimensions of 1000 × 1000 μm in the presence of pores with diameters of 20 and 100 μm, respectively, at the porosity level of 30%. In addition, Figure 11c presents similar data for the representative volume 500 × 500 μm in size for pores of different diameters. Inside such pores, stresses were zero, but they were maximum at their boundaries in the matrix. The locations of the distribution of the maximum stresses near the pores corresponded to the solution of the Kirsch problem of stretching a plate with a round hole [40]. Theoretically, the magnitude of the stresses depended on the ratio of the pore radius to the distance between them in the elastic case: the smaller this ratio was, the lower the stress in the matrix at the same strains and, accordingly, the lower the elastic modulus was. Considering the nonlinear behavior of the PEEK/30GF composite, the real values were lower.

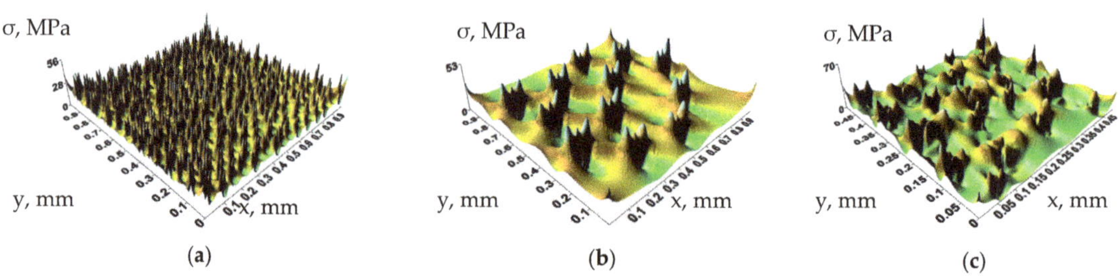

Figure 11. The stress distribution surfaces over the representative volume in the presence of pores with the diameters of 20 μm (**a**), 100 μm (**b**), and from 20 to 100 μm (**c**) at a porosity of 30%.

Based on the above, the low experimental values of the elastic modulus of the PEEK/30GF composite relative to the data reported by the feedstock manufacturer were reliably explained by the formation of pores in the AM samples.

5.2. Computed Micro-Tomography (Micro-CT)

Since the authors noted porosity as the key factor in reducing the mechanical properties of the studied samples, and in light of the use of additional modes 10–15 (Table 5), some of them were examined by the micro-CT. As discussed above, increasing the T_{ext} extruder temperature was supposed to reduce the molten polymer viscosity, so selected modes for the analysis differed primarily in this parameter, in particular mode 12 (T_{ext} = 380 °C/V_{3D} = 40 mm/min/ν = 5 rpm), mode 14 (T_{ext} = 440 °C/V_{3D} = 10 mm/min/ν = 5 rpm), and mode 10 (T_{ext} = 460 °C/V_{3D} = 20 mm/min/ν = 4 rpm).

The micro-CT analysis was carried out near the fracture surfaces of the dog-bone specimens with a height of the tomography area of ~8 mm. Three-dimensional views of the examined areas are shown in Figure 12, both from the side of the print head (Figure 12a–c) and from the side of the supporting table (Figure 12d–f). According to these data, no macro-discontinuities were observed, being characteristic of the FDM method due to the sequential deposition of melted filament beads.

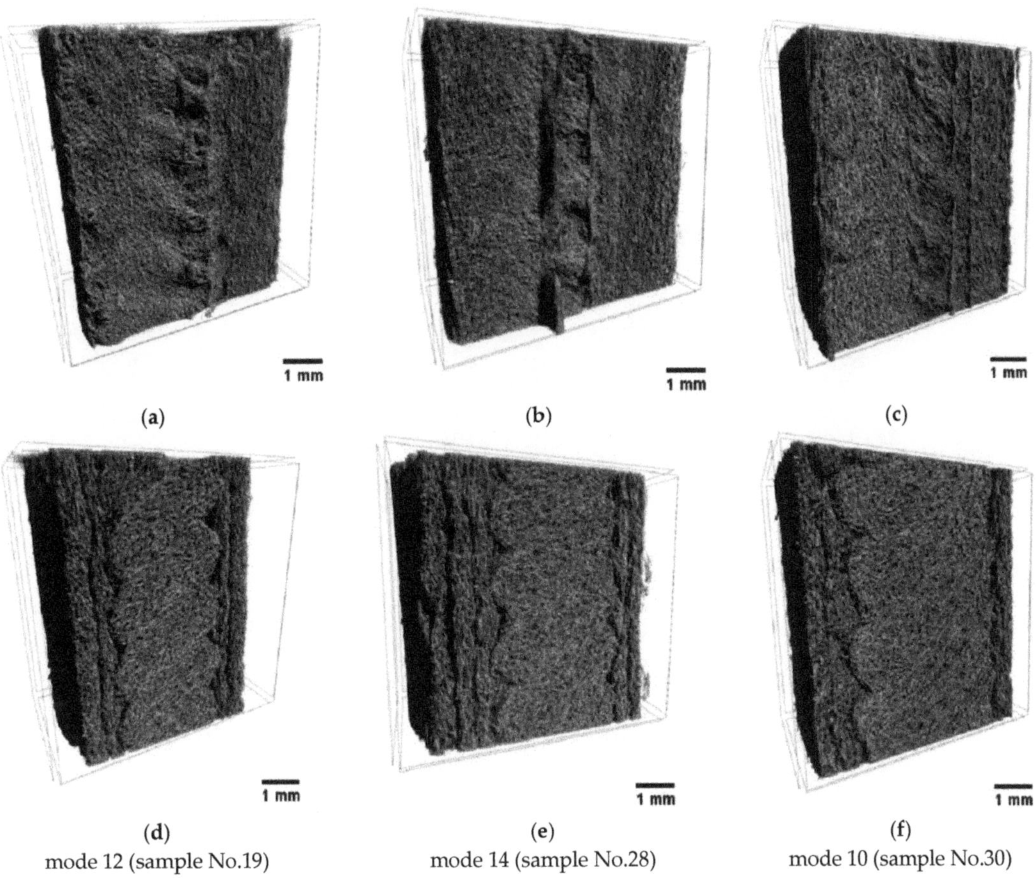

Figure 12. The three-dimensional micro-CT views of the samples, from both supporting table (**a–c**) and 3D-printing head (**d–f**) sides; mode 12 (**a**,**d**); mode 14 (**b**,**e**); mode 10 (**c**,**f**).

Figure 13 presents visualizations of the orthogonal projections of the PEEK/30GF composite to illustrate their internal (meso)structures. On the one hand, GFs were quite densely distributed throughout the bulk samples without a predominant orientation due to the high filling degree, which was noted above when analyzing the SEM micrographs of the fracture surfaces (Figure 2). At the same time, a denser structure of the sample was characteristic of mode 12 at the minimum T_{ext} extruder temperature (Figure 13d,g). In this sample, delamination was observed on the side of the supporting table (Figure 13a), which significantly weakened the section. According to the authors, this fact was the reason for reducing its mechanical properties. Note that the two other technological factors also affected the structures, so only their complex influence is to be discussed.

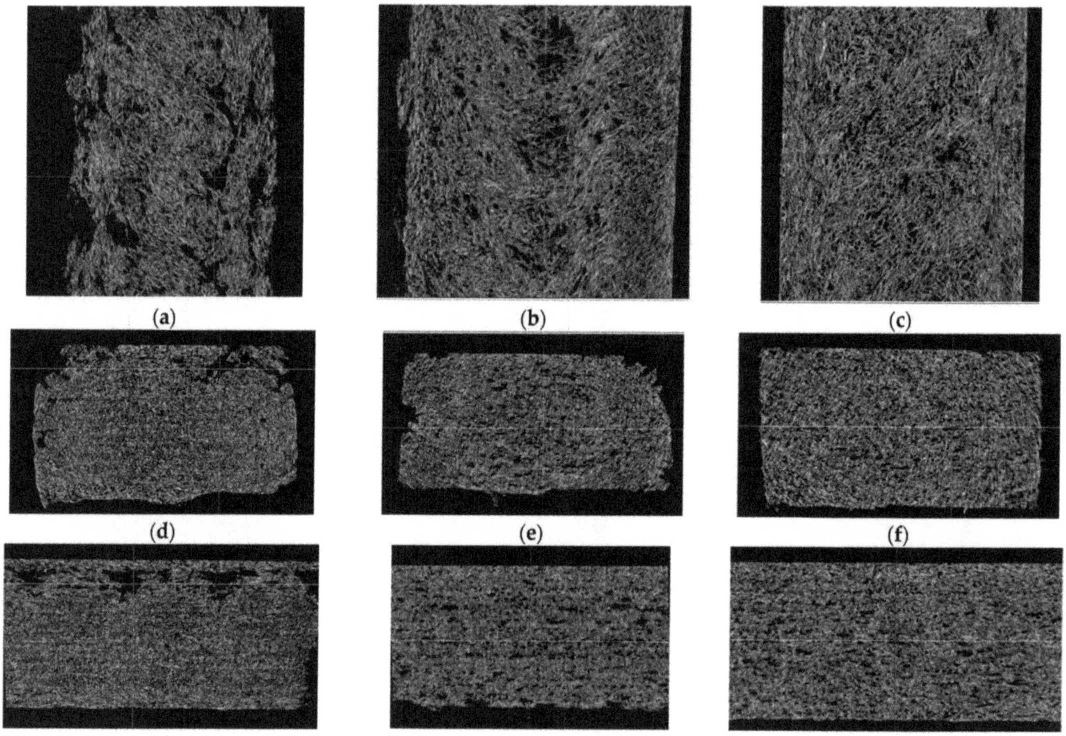

(a) **(b)** **(c)**
(d) **(e)** **(f)**
(g) mode 12 **(h)** mode 14 **(i)** mode 10

Figure 13. The orthogonal projections of the samples near the fracture surfaces at the image (slice) sizes of 7.5–8.0 mm (**a**–**c**), 7.5–4.5 mm (**d**–**f**), and 8.0–4.5 mm (**g**–**i**). Red denotes Z- axis section; Blue denotes X-axis section; Green denotes Y-axis section.

A less 'loose' (meso)structure with some 'variable-density' areas was observed for the intermediate T_{ext} extruder temperature of 440 °C (Figure 13b), while a more uniform and denser one throughout the analyzed volume was characteristic of the sample additively manufactured using mode 10 (Figure 13c). The pattern of the structures according to the other two orthogonal projections (Figure 13d–i) to a lesser extent reflected the difference in the above-discussed mechanical properties of these samples. Nevertheless, the sample's corners could contain large discontinuities, affecting the mechanical properties as well.

Thus, at the minimum T_{ext} extruder temperature of 380 °C, voids were found on the upper side of the sample (Figure 13d,g). As it increased, the samples possessed the more uniform macrostructures (Figure 13e,f,h,i). This fact indicated that the scatter in the mechanical properties revealed in the study could be caused by the macro-heterogeneity of the sample structures. In addition, it was consistent with the results of measuring the cross-sectional areas depending on the position of the section in height for the three samples (Figure 14). In these cases, the greatest dispersion was characteristic of mode 12, while the most negligible one was for mode 10. These results were contributed by both porosity and sample distortions, i.e., deviation from a given rectangular section.

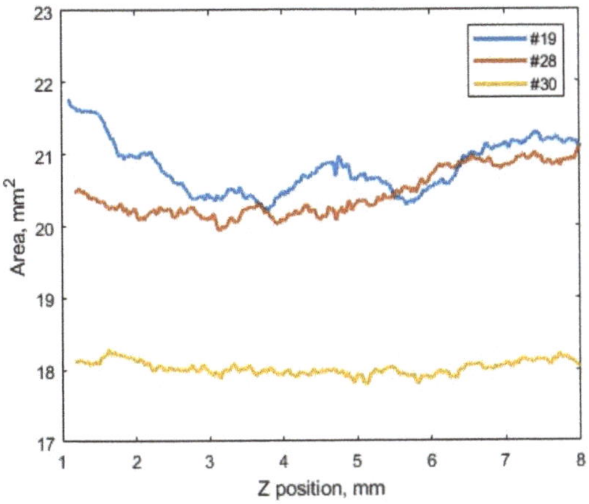

Figure 14. The comparative results of assessing the cross-sectional areas of the samples depending on the position of the height section (along the Z axis): mode 12 (sample No.19); mode 14 (sample No.28); mode 10 (sample No.30).

The above data characterized both separate scales of the structures and 'randomly' examined sections. Using the embedded software package, a quantitative analysis of the porosity levels was carried out throughout the entire investigated volumes. For the correctness of the calculation, volumes of interest (VOIs) were selected (cut out) and the porosity levels were assessed inside them (an example is shown in Figure 15).

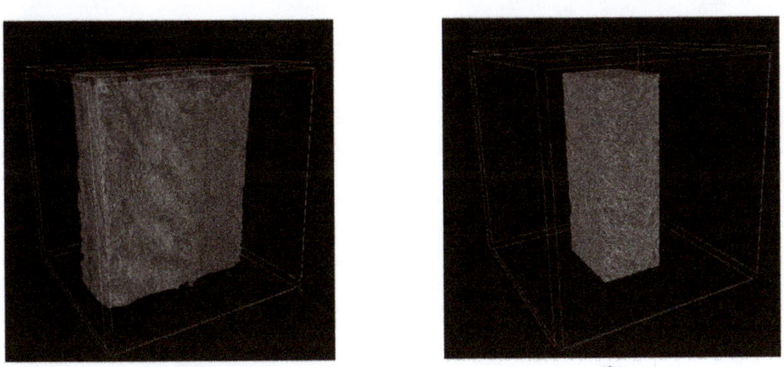

(a) (b)

Figure 15. Visualizations of a full tomogram (**a**) and a cut-out VOI (**b**).

For example, Figure 16 shows the sections used for assessing the porosity levels by applying the following analysis technique. Firstly, a threshold was selected according to which the tomograms were binarized (as a set of sections). This procedure solved the segmentation problem, where black areas corresponded to pores, and white ones reflected to the solid material. The ratios of the cross-sectional areas or volumes characterized the porosity levels presented in Table 9, decreasing with enhancing the T_{ext} extruder temperature.

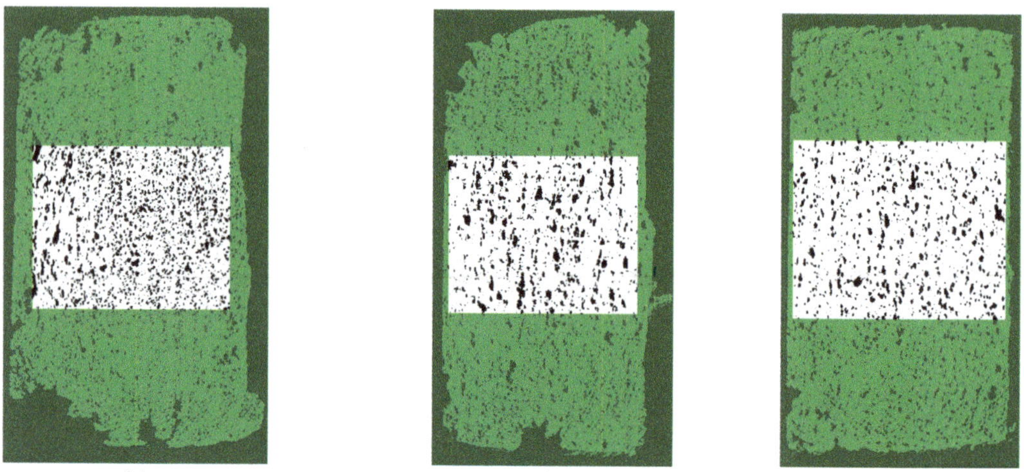

(a) Mode 12 (b) Mode 14 (c) Mode 10

Figure 16. Visualizations of the areas (sections) selected for each of the modes to calculate the porosity levels for the images with sizes of 4.5–7.5 mm (**a**–**c**).

Table 9. The porosity levels for the AM samples, according to the micro-CT analysis.

Mode	VOI, mm^3	Material Volume, mm^3	Porosity, %
Mode 12	59.76	50.25	16
Mode 14	57.15	50.26	12
Mode 10	57.15	51.71	9.5

The authors were aware that the approach applied for interpreting the obtained data, which was based on the dominant role of porosity, was idealized. However, it enabled the analysis of them and suggested a future research direction to improve the mechanical properties of high-viscosity thermoplastics, namely the porosity reduction. They also believed that it was important to once again discuss the possible influence of the orientation of GFs in the initial feedstock.

Figure 17 shows the micro-CT data of the original PEEK/30GF granulate, which enabled the conclusion that the vast majority of GFs were oriented along its axis. However, no preferential orientation of GFs was observed after 3D printing (Figure 13).

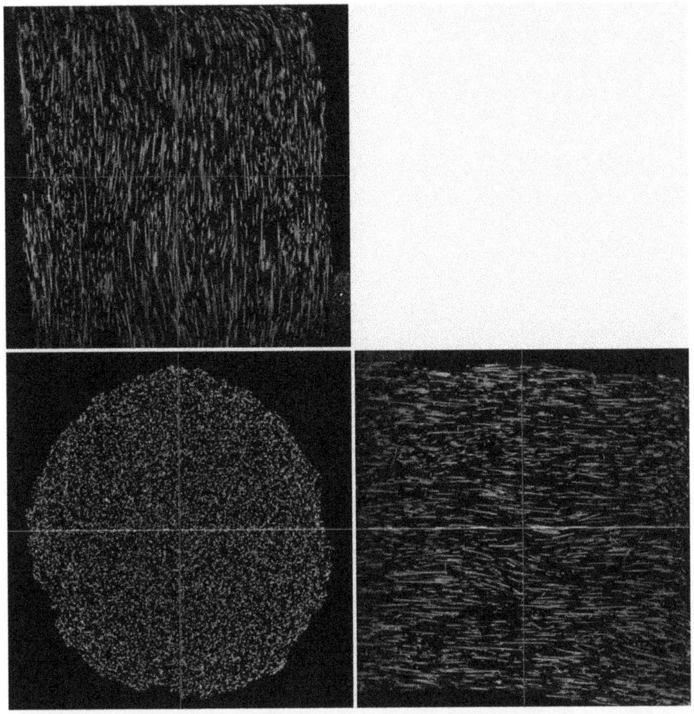

Figure 17. The orthogonal projections of an individual PEEK granule. Red denotes Z- axis section; Blue denotes X-axis section; Green denotes Y-axis section.

5.3. Effects of the 3D Printing Parameters

As a concluding comment, the authors return to the issue of the influence of the 3D printing parameters on the structures and the mechanical properties of the AM PEEK/30GF composites (a high-viscosity one, from the extrusion point of view).

It should be noticed, that the optimization of studied technological parameters did not allow to drastically increase the mechanical properties, including with the use of ANN simulation. However, it does not mean that they did not affect the structure and functional properties. The above-cited literature on FDM of PEEK and PEEK-based composites did focus on variation of the same 3D-printing parameters [30]. However, the use of low-viscosity polymers allowed for the reduction of the porosity and attained a wider range of variation of mechanical properties. Thus, the authors were able to compare the results of this study with those reported in the relevant papers [11,12,30].

Since it was not possible to avoid the high porosity of the 3D printed PEE/30GF composites by varying the input parameters with the use of an available 3D printer, the prospect might be the application of post-build processing, i.e., ultrasonic compaction with the use of US-welding machine. In this regard, the developed approach to ANN simulation will be efficient both to find out predicted local values of optimum parameters and to construct estimates of their range.

5.3.1. Extruder Temperature

The influence of this input parameter was most obvious. With its increasing up to 460 °C, the polymer viscosity decreased. More intense interlayer and intermolecular diffusion had to stimulate the mutual penetration of segments of macromolecules upon 3D printing. Due to the enhanced interlayer adhesion, the mechanical properties of the samples had to be improved. Rising the T_{ext} extruder temperature above a certain threshold level could cause thermal destruction of the polymer, which was incorporated into a priori knowledge for computer simulation using the ANNs.

5.3.2. Extrusion Rate

This factor could exert a dual effect on the structures and the mechanical properties of the additively manufactured samples. On the one hand, the PEEK/30GF composite had to possess a nonlinear viscosity change depending on the ν extrusion rate [41]. As a result, increasing the screw rotation speed could lead to the structuring of the molten polymer flow, while enhancing the internal energy of its segments could be accompanied by conformations of the supermolecular structure elements, improving the mechanical properties of the samples. On the other hand, the extruded molten material could be laid out more uniformly near the nozzle exit at low ν extrusion rates, providing more uniform structures and improving the mechanical properties.

5.3.3. Travel Speed

The V_{3D} travel speed factor to a certain extent was similar to the previous one since it determined the amount of the molten material squeezed out from the microextruder at a time. If it was assumed that denser structures were provided at low V_{3D} travel speeds, then a decrease in this parameter had to improve the mechanical properties of the AM samples. Nevertheless, its increasing reduced cooling rates of the additively manufactured samples, stimulating better spreading of the molten polymer, including when depositing a subsequent layer or a nearby bead. In such cases, enhancing the V_{3D} travel speed had to improve their mechanical properties.

6. Conclusions

The effect of the 3D printing parameters on the structures and the mechanical properties of the samples manufactured by the FDM method from the high-viscosity PEEK/30GF composite feedstock was investigated. It was shown that the formed macro- and microstructures limited the achievement of their high levels. In particular, the following conclusions were drawn.

1. The high viscosity of the molten polymer contributed to the great porosity levels of the AM samples which was proven in detail by the SEM micrographs. The presence of pores reduced the elastic moduli below 3 GPa, while the elongation at break values of ~5% were significantly higher than those for similar compression molded (hot-pressed) composites.
2. By computer simulation using two types of ANNs, the optimal combinations of the 3D printing parameters were determined for the PEEK/30GF composite: Extruder temperature: 450 ÷ 460 °C, Extrusion rate: 4.1 ÷ 4.59 rev/min, Travel speed; 13.92 ÷ 30.91 mm/s. When an ultra-small sample was utilized, the application of priory knowledge made it possible to achieve the appropriate simulation accuracy (MSE~0.06). In doing so, the RBFNN was able to construct a more realistic model, while for the FFNN the reasonable limitation of its size was implemented.
3. An interpretation of the differences in the results, predicted using the RBFNN and FFNN models, was proposed based on their operation principles. Generally, the determined optimal values were consistent with those obtained by the Taguchi method, and physically corresponded to the assumption formulated as the null hypothesis: higher quality of AM products from the PEEK/30GF composite was ensured by maximizing material feeding into the 3D-printing zone by reducing the V_{3D} travel speed of the moving extruder head with increasing both the T_{ext} extruder temperature and the ν extrusion rate.
4. The effect of porosity on the mechanical properties of the additively manufactured PEEK/30GF composites was assessed by implementing the FEM-based models of small, large, and mixed pores. The obtained results made it possible to explain the experimentally revealed relatively low level of strength.

5. The applied 3D printing parameters, primarily the T_{ext} extruder temperature, did not lead to a change in the chemical structure (in terms of oxidation) of the polymer matrix and can be used for additive manufacturing of products from the PEEK/30GF composite by the FDM method.
6. The micro-CT analysis of the AM samples enabled the conclusion that the optimal 3D printing parameters were the T_{ext} extruder temperature of 460 °C, the V_{3D} travel speed of 20 mm/min, and the ν extrusion rate of 4 rpm. These values correlated well with those obtained by computer simulation using the ANNs. In such cases, the homogeneous micro- and macro-structures were formed with minimal sample distortions and porosity levels within 10 vol.%.
7. The most likely reason for the great porosity levels was the expansion of the molten polymer when it was squeezed out from the microextruder nozzle of the 3D printer since the pressure in the chamber was caused by its high viscosity. Probably, the mechanical properties of such samples can be improved both by changing the 3D printing strategy to ensure the preferential orientation of GFs in the building direction and by reducing porosity via post-build treatment or ultrasonic compaction. The following research by the authors will be devoted to the implementation of these methods.

Supplementary Materials: The following supporting information can be downloaded at: https://www.mdpi.com/article/10.3390/polym16182601/s1, Table S1: Parameters of 3D-printing (input parameters); Table S2: Experimentally measured properties (output parameters); Figure S1: The lower magnification SEM micrographs of the PEEK/30GF composites additively manufactured using the modes presented in Table 2; Figure S2: The IR spectra of the samples of the PEEK/30GF composite additively manufactured using modes 3, 7, and 9, according to Table 2.

Author Contributions: Conceptualization, S.V.P., D.Y.S. and Y.V.D.; methodology, S.V.P. and Y.V.D.; software, D.Y.S.; validation, D.G.B., Y.V.D. and S.A.B.; formal analysis, V.O.A. and D.G.B.; investigation, Y.V.D., P.V.K., V.O.A., S.A.B., A.V.B. and D.G.B.; resources, S.V.P., P.V.K. and A.V.B.; data curation, Y.V.D., D.Y.S., S.A.B. and D.G.B.; writing—original draft preparation, D.Y.S. and S.V.P.; writing—review and editing, S.V.P.; visualization, D.Y.S., D.G.B. and S.A.B.; supervision, S.V.P.; project administration, S.V.P.; funding acquisition, P.V.K. All authors have read and agreed to the published version of the manuscript.

Funding: This research was supported by the Ministry of Science and Higher Education of the Russian Federation, project No. FSWM-2022-0018.

Institutional Review Board Statement: Not applicable.

Informed Consent Statement: Not applicable.

Data Availability Statement: The original contributions presented in the study are included in the article/Supplementary Material, further inquiries can be directed to the corresponding author.

Conflicts of Interest: The authors declare no conflicts of interest.

References

1. Kurtz, S.M. *PEEK Biomaterials Handbook*; Elsevier: Amsterdam, The Netherlands, 2011; 297p.
2. Wu, T.; Zhang, X.; Chen, K.; Chen, Q.; Yu, Z.; Feng, C.; Qi, J.; Zhang, D. The Antibacterial and Wear-Resistant Nano-ZnO/PEEK Composites Were Constructed by a Simple Two-Step Method. *J. Mech. Behav. Biomed. Mater.* **2022**, *126*, 104986. [CrossRef] [PubMed]
3. Friedrich, K. Polymer Composites for Tribological Applications. *Adv. Ind. Eng. Polym. Res.* **2018**, *1*, 3–39. [CrossRef]
4. Zhu, J.; Xie, F.; Dwyer-Joyce, R.S. PEEK Composites as Self-Lubricating Bush Materials for Articulating Revolute Pin Joints. *Polymers* **2020**, *12*, 665. [CrossRef]
5. Davim, J.P.; Cardoso, R. Effect of the reinforcement (carbon or glass fibres) on friction and wear behaviour of the PEEK against steel surface at long dry sliding. *Wear* **2009**, *266*, 795–799. [CrossRef]
6. Ketaspire PEEK Design and Processing Guide. Available online: https://www.solvay.com/sites/g/files/srpend221/files/2018-07/ketaspire-peek-design-and-processing-guide-en.pdf (accessed on 12 August 2024).

7. Maslavi, A.; Unal, H.; Olabi, M.N. Determination of "tribological performance working fields" for pure PEEK and PEEK composites under dry sliding conditions. *Wear* **2024**, *554–555*, 205464. [CrossRef]
8. Bashandeh, K.; Lan, P.; Polycarpou, A.A. Tribology of Self-Lubricating High Performance ATSP, PI, and PEEK-Based Polymer Composites up to 300 °C. *Friction* **2022**, *11*, 141–153. [CrossRef]
9. Massocchi, D.; Riboni, G.; Lecis, N.; Chatterton, S.; Pennacchi, P. Tribological Characterization of Polyether Ether Ketone (PEEK) Polymers Produced by Additive Manufacturing for Hydrodynamic Bearing Application. *Lubricants* **2021**, *9*, 112. [CrossRef]
10. Dontsov, Y.V.; Panin, S.V.; Buslovich, D.G.; Berto, F. Taguchi Optimization of Parameters for Feedstock Fabrication and FDM Manufacturing of Wear-Resistant UHMWPE-Based Composites. *Materials* **2020**, *13*, 2718. [CrossRef]
11. Duty, C.; Ajinjeru, C.; Kishore, V.; Compton, B.; Hmeidat, N.; Chen, X.; Liu, P.; Hassen, A.A.; Lindahl, J.; Kunc, V. What Makes a Material Printable? A Viscoelastic Model for Extrusion-Based 3D Printing of Polymers. *J. Manuf. Proc.* **2018**, *35*, 526–537. [CrossRef]
12. Kuba, D.; Matsuzaki, R.; Ochi, S.; Ogihara, S. 3D Printing of Composite Materials Using Ultralow-Melt-Viscosity Polymer and Continuous Carbon Fiber. *Compos. Part C Open Access* **2022**, *8*, 100250. [CrossRef]
13. Khaliq, M.H.; Gomes, R.; Fernandes, C.; Nóbrega, J.; Carneiro, O.S.; Ferrás, L.L. On the Use of High Viscosity Polymers in the Fused Filament Fabrication Process. *Rapid Prototyp. J.* **2017**, *23*, 727–735. [CrossRef]
14. Yang, Y.; Gu, S.; Liu, J.; Tian, H.; Lv, Q. Research and Development of a 3D Jet Printer for High-Viscosity Molten Liquids. *Micromachines* **2018**, *9*, 554. [CrossRef] [PubMed]
15. Antony, J. *Design of Experiments for Engineers and Scientists*, 2nd ed.; Elsevier: Waltham, MA, USA, 2014; p. 208.
16. Ferreira, S.L.C.; Bruns, R.E.; Ferreira, H.S.; Matos, G.D.; David, J.M.; Brandão, G.C.; da Silva, E.G.P.; Portugal, L.A.; dos Reis, P.S.; Souza, A.S.; et al. Box-Behnken design: An alternative for the optimization of analytical methods. *Anal. Chim. Acta* **2007**, *597*, 179–186. [CrossRef] [PubMed]
17. Rasheed, S.; Hashmi, I.; Zhou, Q.; Kim, J.K.; Campos, L.C. Central composite rotatable design for optimization of trihalomethane extraction and detection through gas chromatography: A case study. *Int. J. Environ. Sci. Technol.* **2023**, *20*, 1185–1198. [CrossRef]
18. de Aguiar, P.F.; Bourguignon, B.; Khots, M.S.; Massart, D.L.; Phan-Than-Luu, R. D-optimal designs. *Chemom. Intell. Lab. Syst.* **1995**, *30*, 199–210. [CrossRef]
19. Cerqueira, U.M.F.M.; Bezerra, M.A.; Ferreira, S.L.C.; Araújo, R.d.J.; da Silva, B.N.; Novaes, C.G. Doehlert design in the optimization of procedures aiming food analysis—A review. *Food Chem.* **2021**, *364*, 130429. [CrossRef]
20. Jahanbakhsh, M.; Farid, A.A.; Lotfi, M. Optimal flank wear in turning of Inconel 625 super-alloy using ceramic tool. *Proc. Inst. Mech. Eng. Part B J. Eng. Manuf.* **2016**, *232*, 208–216. [CrossRef]
21. Xie, J.; Qiao, Y.; Wang, Z.; Qi, Y.; Xu, Q.; Shemtov-Yona, K.; Chen, P.; Rittel, D. Application of the Taguchi Method to Areal Roughness-Based Surface Topography Control by Waterjet Treatments. *Appl. Surf. Sci. Adv.* **2024**, *19*, 100548. [CrossRef]
22. Kacker, R.N.; Lagergren, E.S.; Filliben, J.J. Taguchis's Orthogonal Arrays are Classical Design of Experiments. *J. Res. Natl. Inst. Stand. Technol.* **1991**, *96*, 577. [CrossRef]
23. Mallakpour, S.; Hussain, C.M. *Medical Additive Manufacturing: Concepts & Fundamentals*; Elsevier: Amsterdam, The Netherlands, 2024; p. 792. [CrossRef]
24. Zhang, Y.; Ling, C. A strategy to apply machine learning to small datasets in materials science. *NPJ Comput. Mater.* **2018**, *4*, 25. [CrossRef]
25. Stepanov, D.Y.; Tian, D.; Alexenko, V.O.; Panin, S.V.; Buslovich, D.G. Application of Neural Network Models with Ultra-Small Samples to Optimize the Ultrasonic Consolidation Parameters for 'PEI Adherend/Prepreg (CF-PEI Fabric)/PEI Adherend' Lap Joints. *Polymers* **2024**, *16*, 451. [CrossRef] [PubMed]
26. *ISO 527-2:2012*; Plastics—Determination of Tensile Propertie-Part 2: Test Conditions for Moulding and Extrusion Plastics. ISO: Geneva, Switzerland, 2021.
27. Osipov, S.; Prischepa, I. Algorithm of Estimation of the Degree of Porosity Homogeneity of Foamed Concretes by Local Volumes by X-ray Computed Tomography Method. *Materials* **2023**, *16*, 3244. [CrossRef] [PubMed]
28. The 'Bruker-microCT' Software Package. Available online: https://www.bruker.com/en/products-and-solutions/preclinical-imaging/micro-ct/3d-suite-software.html (accessed on 10 June 2024).
29. Ketaspire KT-880 GF30 BG20. Available online: https://resmart.com/ketaspire-kt-880-gf30-bg20.html (accessed on 10 June 2024).
30. Deng, X.; Zeng, Z.; Peng, B.; Yan, S.; Ke, W. Mechanical Properties Optimization of Poly-Ether-Ether-Ketone via Fused Deposition Modeling. *Materials* **2018**, *11*, 216. [CrossRef] [PubMed]
31. Demuth, H.B.; Beale, M.H.; De Jess, O.; Hagan, M.T. *Neural Network Design*, 2nd ed.; Martin Hagan: Stillwater, OK, USA, 2014.
32. Haykin, S. *Neural Networks: A Comprehensive Foundation*, 2nd ed.; Prentice Hall: Upper Saddle River, NJ, USA, 1998.
33. Shermergor, T.D. *Inhomogeneous Media Elasticity Theory*; Nauka Publisher: Moscow, Russia, 1977; 400p. (In Russian)
34. Zarubin, V.S.; Savelyeva, I.Y.; Sergeeva, E.S. Two-Way Estimates of Porous Solid Body Elasticity Moduli. *Eng. J. Sci. Innov.* **2017**, *12*. (In Russian) [CrossRef]
35. Pobedrya, B.E. *Mechanics of Composite Materials*; Moscow University Press: Moscow, Russia, 1984; 336p. (In Russian)
36. Segerlind, L. *Applied Finite Element Analysis*; John Willey & Sons: New York, NY, USA, 1976.
37. Zienkiewicz, O.C.; Taylor, R.L.; Zhu, J.Z. *The Finite Element Method: Its Basis and Fundamentals*, 7th ed.; Butterworth-Heinemann: London, UK, 2013.

38. Bochkareva, S.A.; Grishaeva, N.Y.; Lyukshin, B.A.; Lyukshin, P.A.; Matolygina, N.Y.; Panin, S.V.; Reutov, Y.A. A Unified Approach to Determining the Effective Physicomechanical Characteristics of Filled Polymer Composites Based on Variational Principles. *Mech. Compos. Mater.* **2019**, *54*, 775–788. [CrossRef]
39. The Mitsubishi Chemical Advanced Materials Group of Companies. Available online: https://www.mcam.com/en/products/shapes/advanced/ketron/ketron-hpv-peek (accessed on 10 June 2024).
40. Timoshenko, S.P.; Goodier, J.N. *Theory of Elasticity*; McGraw-Hill: New York, NY, USA, 1969; 567p.
41. Bratsykhin, E.A. *Technology of Plastics*, 2nd ed.; Chemistry: Leningrad, Russia, 1974; 352p. (In Russian)

polymers

MDPI

Article

Compression Behavior of 3D Printed Composite Isogrid Structures

Marina Andreozzi *, Carlo Bruni, Archimede Forcellese, Serena Gentili and Alessio Vita

Dipartimento di Ingegneria Industriale e Scienze Matematiche (DIISM), Università Politecnica delle Marche, Via Brecce Bianche 12, 60131 Ancona, Italy; c.bruni@staff.univpm.it (C.B.); a.forcellese@univpm.it (A.F.); s.gentili@pm.univpm.it (S.G.); alessio.vita@staff.univpm.it (A.V.)

* Correspondence: m.andreozzi@pm.univpm.it

Abstract: Composite materials, particularly carbon fiber-reinforced polymers (CFRPs), have become a cornerstone in industries requiring high-performance materials due to their exceptional mechanical properties, such as high strength-to-weight ratios, and their inherent lightweight nature. These attributes make CFRPs highly desirable in aerospace, automotive, and other advanced engineering applications. However, the compressive behavior of CFRP structures remains a challenge, primarily due to the material sensitivity to structural instability, leading to matrix cracking and premature failure under compressive loads. Isogrid structures, characterized by their unique geometric patterns, have shown promise in enhancing the compressive behavior of CFRP panels by providing additional support that mitigates these issues. Traditionally, these structures are manufactured using automated techniques like automated fiber placement (AFP) and automated tape laying (ATL), which, despite their efficacy, are often cost-prohibitive for small-scale or custom applications. Recent advancements in 3D-printing technology, particularly those involving continuous fiber reinforcement, present a cost-effective and flexible alternative for producing complex CFRP structures. This study investigates the compressive behavior of 3D-printed isogrid structures, fabricated using continuous carbon fiber reinforcement via an Anisoprint Composer A3 printer equipped with towpreg coextrusion technology. A total of eight isogrid panels with varying infill percentages were produced and subjected to buckling tests to assess their performance. The experimental results indicate a direct correlation between infill density and buckling resistance, with higher infill densities leading to increased buckling loads. Additionally, the failure modes were observed to shift from local to global buckling as the infill density increased, suggesting a more uniform distribution of compressive stresses. Post-test analyses using optical microscopy and scanning electron microscopy (SEM) revealed the presence of voids within the 3D-printed structures, which were found to negatively impact the mechanical performance of the isogrid panels. The findings of this study demonstrate that 3D-printed isogrid CFRP structures can achieve significant buckling resistance, making them a viable option for high-performance applications. However, the presence of voids remains a critical issue, highlighting the need for process optimizations in 3D-printing techniques to enhance the overall performance and reliability of these structures.

Keywords: 3D printing; composites; isogrid

1. Introduction

Composite materials are well known both in the scientific literature and in several industries for their remarkable mechanical performances conjugated to their lightness. Their applications vary from aerospace to automotive industries, where material performances are predominant with respect to costs. Carbon fiber-reinforced polymers, based on continuous filaments, are the preferred type of composites, as the highest specific characteristics are required [1]. Indeed, this kind of material presents very high stiffness, tensile strength, and fracture resistance. However, concerning compression behavior, CFRPs demonstrate some limitations. This is related to the tendencies of CFRP structures to buckle and their reinforcement fibers to micro-buckle, both compromising structural integrity [2]. Moreover, the polymer matrix, in which the fibers are embedded, is prone to cracking and delamination under compressive loads, thus reducing the material resistance [3]. To mitigate this problem, isogrid structures can be used to reinforce thin panels and to provide more compression resistance without adding significant weight [4–7]. For this reason, isogrid-reinforced panels are used in demanding applications such as fuselage for aircraft [8]. Isogrids are lattice structures typically produced by automated fiber placement or automated tape laying (ATL) [9]. These two manufacturing methods, which are based on the use of robotic arms, are suitable for manufacturing components of very large dimensions, such as those used in aircraft and the energy sector [10].

The advent of 3D-printing technology, particularly for thermoplastic matrices with continuous reinforcement, has significantly enhanced the ability to fabricate intricate structures with high precision, notably for smaller components [11,12], especially for components with limited dimensions. CFRPs can be 3D printed using a process similar to the one used for printing plastic products, specifically the fused filament fabrication (FFF) method [13]. However, handling the continuous reinforcements in the extrusion process is challenging. Indeed, extruding continuous filaments presents several drawbacks. The first drawback is fraying, which can occur during filament movement. It happens when the fibers start to unravel or separate, thus compromising the composite integrity and aesthetic appearance [14]. Another problem typical of continuous reinforcement extrusion is related to filament sewing, a phase necessary to produce components without shape constraints. Cutting a filament could lead to fiber damage, binder removal, and limitations in adhering to the predefined path for fiber deposition. Another issue worth noticing is filament impregnation. Indeed, using high-viscosity matrices like thermoplastics can result in incorrect impregnations within the heated nozzle. Even though the nozzle reaches a temperature that significantly reduces the viscosity of the thermoplastic matrix, the materials' rheology may prevent the thermoplastic from penetrating the fiber bundle. This results in voids within the 3D-printed composite, leading to a decrease in the mechanical performance [15].

Some examples of 3D-printed isogrid structures can be found in the scientific literature. Li et al. investigated the effect of geometric parameters on the compression strength of 3D-printed PVC isogrid, employing a numerical simulation to predict failure load [16]. Guajardo-Travino et al. investigated the effect of raster gaps and infill void in the compressive behavior of 3D-printed plastic isogrids [17]. However, all these studies were concerned with the use of unreinforced plastics for 3D printing isogrid structures. Some of the authors of these papers conducted research on isogrid structures 3D printed with short and discontinuous fibers. They studied the effect of reinforcing rib dimension and moisture absorption on the compressive behavior of isogrids produced in short carbon fiber-reinforced polyamide by an FFF process [18–20]. The studies found that varying the dimensions of the ribs could lead isogrids to fail under local or global buckling when subjected to compressive loads. In addition, they noted an improvement in compression resistance due to the structure drying after 3D printing.

Despite the widespread application of CFRPs, understanding the compressive behavior of 3D-printed CFRPs, particularly in isogrid configurations, remains limited. A notable gap exists in the scientific literature regarding the compression behavior of 3D-printed isogrid structures that utilize continuous reinforcement. Currently, two commercial desktop 3D printers, namely, the Markforged Mark Two and the Anisoprint Composer, are available for 3D printing with continuous CFRPs. The Anisoprint Composer in particular employs composite fiber co-extrusion (CFC) technology, which co-extrudes a composite material comprising a thermoplastic matrix and continuous fiber reinforcement [21]. To facilitate the handling of the continuous reinforcement, it is impregnated with a thermoset binder before the printing process [22]. As the fibrous reinforcement passes through the extrusion nozzle, it is impregnated with the thermoplastic matrix, resulting in a dual-matrix system.

Anisoprint dual-matrix technology offers numerous advantages, including the ability to design and manufacture lightweight yet extremely strong components with improved tensile, flexural, and torsional strength. This technology also allows for the customization of fiber reinforcement within the matrix, optimizing the design to meet specific structural and functional requirements. The control over the distribution of fibers and matrix enables modulation of the stiffness and mechanical strength in specific directions, significantly enhancing the performance of the final components compared to traditional manufacturing methods. Also, this dual-matrix configuration provides unique mechanical properties, making it essential to explore and understand its behavior, especially under compressive loads in complex geometries like isogrid structures [23,24].

This paper aims to analyze the impact of different infill percentages on the compressive behavior of 3D-printed isogrid structures fabricated from continuous carbon fiber-reinforced polymers using Anisoprint dual-matrix technology. If adding more material to a structure by increasing the infill percentage leads to an increase in its overall strength, this does not always guarantee an improvement in specific strength, which is the strength relative to the material weight. Indeed, previous studies [16,20] have shown that increasing material usage can reduce specific strength, particularly if the added material induces a different failure mode. To analyze the effect of the infill percentage on the strength and specific strength of isogrids, eight distinct structures were produced, each with varying infill percentages. Compressive tests were conducted on standard specimens, while buckling tests were performed specifically on the isogrid panels. In addition, the isogrid structures were weighed prior to testing to calculate their specific compressive resistance. These results were also compared with those of short carbon fiber-reinforced structures to better understand the role continuous fibers play in influencing buckling behavior. Then, detailed analyses were carried out on the tested structures to determine whether global buckling of the overall structure or local buckling of the reinforcements occurred. Finally, the fractured surfaces of the tested samples were examined using scanning electron microscopy (SEM) and high-resolution three-dimensional tomography to investigate the internal structures and defects within the 3D-printed composites.

2. Materials and Methods

2.1. Materials and 3D-Printing Process

In the present study, isogrid structures were manufactured using fused filament fabrication technology based on CFC, which enables the printing of thermoplastic polymer matrix composites reinforced with continuous carbon fibers. The printing system employed was the Anisoprint Composer A3, developed by Anisoprint Inc., Esch-sur-Alzette, Luxembourg, as illustrated schematically in Figure 1. The machine's print head consists of two extruders: one for extruding plastic material only (FFF) and the other for extruding composite material (CFC). Dual-matrix Anisoprint technology enables the simultaneous co-extrusion of a thermoplastic matrix and a bundle of thermoset resin-impregnated reinforcing fibers, allowing the fabrication of composite structures with optimized mechanical and physical properties. In the process under consideration, a bundle of epoxy resin-bonded carbon fibers and a thermoplastic filament are fed into a fusion chamber through two separate inlet tubes. Inside the chamber, the thermoplastic material is melted at a controlled temperature, allowing both the thermoset-impregnated fibers and the thermoplastic matrix to be extruded simultaneously through a single nozzle. This approach allows for the creation of a continuous dual-matrix composite structure, combining the advantages of the thermoplastic matrix, such as toughness and ease of melting, with the high mechanical properties of reinforced fibers impregnated with thermosetting resin.

CFC PA polyamide is a specialized thermoplastic developed by Polymaker for use with Anisoprint Technology in composite 3D printing. Its molecular structure of repeating amide bonds imparts notable mechanical strength and thermal stability. When melted, CFC PA's viscosity decreases significantly, enhancing the penetration between reinforcing fiber bundles during printing, which results in composites with improved mechanical properties. It presents a glass transition temperature (Tg) of 53.7 °C. With a low density of 1.03 g/cm^3 and rapid cooling and solidification rates, it enables precise layer placement without warping, maintaining dimensional accuracy. Its low hygroscopicity minimizes moisture absorption, reducing defects like bubbling or poor layer adhesion in open-air printing. Post-annealing, CFC PA exhibits a high tensile strength of 58 MPa and a Young's Modulus of 1442 MPa, making it suitable for structural components requiring stiffness and flexibility. It also has a heat deflection temperature of 104 °C at 0.45 MPa, allowing it to perform well under elevated temperatures. Compared to standard polyamides like PA6 or PA66, CFC PA offers better processing ease and performance advantages, making it ideal for high-performance applications in the aerospace, automotive, and engineering sectors. While not biodegradable, its durability contributes to sustainability by reducing the need for frequent part replacements.

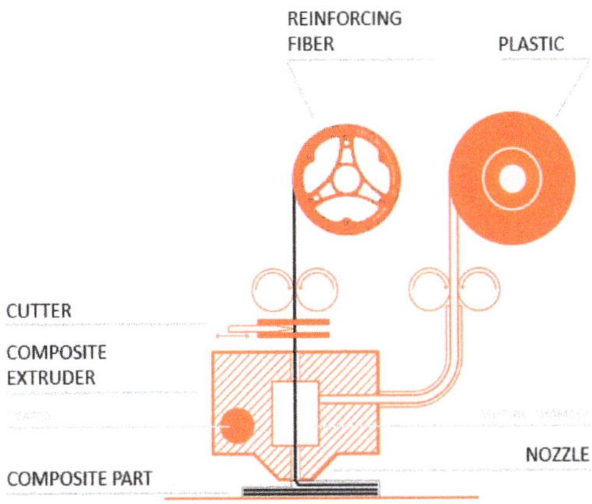

Figure 1. Scheme of the printing process of the Anisoprint A3 machine.

As reinforcement, continuous carbon fiber (CCF) produced by Anisoprint was utilized in this study. CCF is a carbon–epoxy composite composed of a bundle of 1500 carbon fibers, each with an average diameter of 7 μm, bonded together with an epoxy-based thermoset resin. According to the technical datasheet, CCF possesses an elastic modulus of 150 GPa, an ultimate tensile strength of 2200 MPa, and a fiber volume fraction of 60%. The printing process, which combines both the composite fiber co-extrusion polyamide and the CCF, results in a dual-matrix composite material. This composite presents a fiber volume fraction of 40%. This dual-matrix structure is formed due to the presence of both the thermoplastic matrix from the polyamide and the thermosetting resin from the carbon fiber, offering unique mechanical properties. The process parameters used to print the structures were derived by a previous work by the authors [25]. They are (i) macrolayer height of 0.32 mm, (ii) extrusion width of 0.65 mm, (iii) extruder temperature of 250 °C, and (iv) build plate temperature of 60 °C.

Short carbon fiber (SCF)-reinforced polyamide was also used to 3D print reference isogrid structures. The PA filament is filled with 15% carbon fibers, and it was used in an FDM process. According to the technical datasheet, its tensile strength and tensile modulus are equal to 103 MPa and 8.38 GPa, respectively.

2.2. Compression Test

Compression tests were conducted on 3D-printed continuous fiber bimatrix composite specimens to evaluate the compressive mechanical behavior of the material. The test was conducted according to ASTM D6641, and the specimens were made with a length of 140 mm, thickness of 3 mm, and width of 13 mm. The gage length of the specimen was 13 mm. The specimens were printed with the fibers arranged in the direction of application of the compressive load. To ensure the repeatability of the test, five specimens were tested.

A universal testing machine equipped with a 25 KN load cell was used to perform the tests. According to the standard, the appropriate test equipment was employed with a nominal rate of 1.3 mm/min.

2.3. Isogrid Structures

Isogrid structures were created with CAD software Autodesk Inventor Professional 2024, originating from a parallelepiped with dimensions of 106 × 80 × 8 mm (Figure 2). This geometry was imported into the proprietary Anisoprint slicing software called Aura v. 2.4.8, where the Isogrid infill pattern was set, enabling the desired lattice structure to be achieved. In addition, the following printing parameters were set: printing speed of 6 mm/s, microlayer height of 0.32 mm, extrusion width of 0.65 mm, and build plate temperature of 60 °C. These parameters are the ones suggested by Anisoprint for achieving the best printing results. The dual-matrix composite filament extrusion was performed at a temperature of 250 °C. The isogrids were 3D printed at the following infill percentages: 10%, 20%, 30%, 40%, 50%, 60%, 70%, and 80%. To increase the accuracy of the tests, three isogrids were printed for each infill percentage.

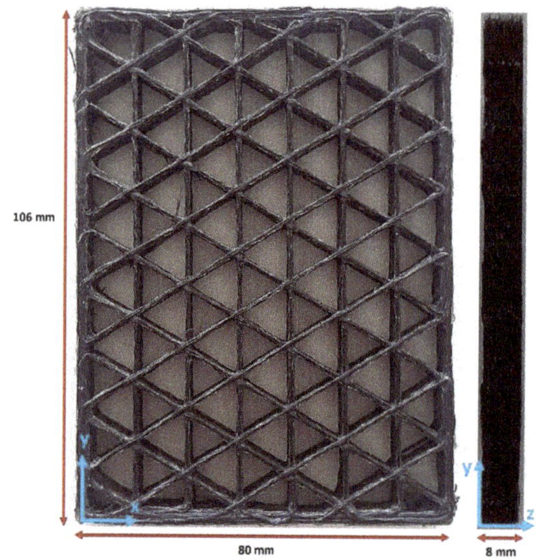

Figure 2. Example of the 30% isogrid.

Figure 3 illustrates the isogrid geometries at various infill densities. In the figure, the grey lines represent the reinforced perimeter of the structures, providing external support and stability, while the red lines indicate the reinforced infill, which contributes to the internal strength and buckling resistance of the isogrid configurations. These color-coded lines help visualize the distribution and arrangement of reinforcement within the isogrid structures, highlighting the differences in infill density and their potential impact on mechanical performance.

The reference isogrid structures in SCF were printed at 10% and 80% infill densities.

Before testing, the isogrid structures were weighted to determine their mass. The average values obtained from these measurements are presented in Table 1, both for continuous and short fiber-reinforced structures. These weight measurements are crucial for calculating the specific compressive resistance of the structures, allowing for a better understanding of the relationship between mass and mechanical performance.

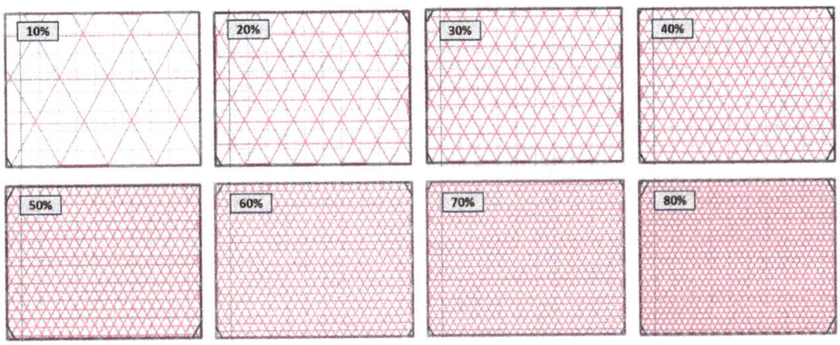

Figure 3. Infill densities of the isogrid structures.

Table 1. Average weight of isogrid structures.

% Infill	**10**		**20**	**30**	**40**	**50**	**60**	**70**	**80**	
	CCF	**SCF**	**CCF**	**CCF**	**CCF**	**CCF**	**CCF**	**CCF**	**CCF**	**SCF**
Weight [g]	12.72	10.82	19.73	26.79	33.54	39.04	46.12	52.19	58.23	51.2
Standard deviation	±0.5	±0.9	±1.2	±0.6	±0.8	±1.3	±0.7	±0.4	±0.6	±0.8

2.4. Tomography Analysis

The 3D-printed lattice structures were analyzed using the ZEISS METROTOM 1500 industrial tomograph, a device that allows for high-resolution three-dimensional imaging. This technology utilizes X-rays to penetrate the object, providing detailed information about the internal structure of the isogrid, including porosity, defects, and complex or hidden geometries that are not visible through surface inspection alone. For the test, the isogrid structures were positioned with their shorter side on a metal support and placed on the rotating plate of the machine, located between the X-ray source and the X-ray detector within the machine chamber. The characteristic parameters of the testing machine are detailed in Table 2. This setup enabled precise imaging and analysis of the internal features of the 3D-printed structures, contributing to a deeper understanding of their material integrity and performance.

Table 2. Tomograph settings used for the isogrid structure analysis.

Voltage	180 KW
Current strength	380 μA
Voxel	64.08 μm
Volume	303 × 1479 × 1824 Voxel
Filter	Cu 0.5 mm
Integration time	666 ms
X-ray beam size	68 μm

2.5. Buckling Test

To examine the behavior of 3D-printed isogrid structures made from carbon fiber-reinforced polyamide, buckling tests were conducted using an MTS 810 servo-hydraulic testing machine, with the mobile platen moving at a speed of 0.5 mm/min (as shown in Figure 4). The isogrid structures were positioned such that their shortest sides were in contact with the machine plates, allowing the compressive load to be applied parallel to the fiber orientation, which is critical for accurately assessing the material buckling behavior. During the tests, a load cell and an inductive displacement transducer (LVDT) were used to record the values of load the maximum load (Pmax) and displacement (Δh) throughout the testing process. Additionally, the maximum specific buckling load (Pmax/w) was calculated by dividing the maximum buckling load by the weight of the lattice structure. This provides a normalized measure of buckling resistance relative to the structure mass.

Figure 4. Buckling test of the 30% isogrid.

Images capturing the failure mechanisms of the structures at different infill percentages were taken using a specialized camera, allowing for a detailed visual analysis of how varying infill densities affect the structural integrity under compressive loading.

2.6. Scanning Electron Microscope and Optical Analyses

The examination of the isogrid morphology after the buckling tests was conducted using a Leica DMI8 optical microscope, provided by Leica Microsystems GmbH in Wetzlar, Germany. To gain more detailed insights into the fractured surfaces of the tested specimens, the field emission scanning electron microscope (FESEM) ZEISS SUPRA™ 40, equipped with a high-resolution GEMINI® lens, was employed. This advanced microscope enabled the capture of finely detailed images, revealing the microstructural characteristics and failure mechanisms of the material. Prior to conducting the SEM analysis, a metallization process was applied to the specimens to enhance their electrical conductivity, ensuring the acquisition of high-quality electron microscope images and facilitating accurate analysis of the material's internal structure.

3. Results

3.1. Compression Test

Figure 5 illustrates the typical force–displacement curves that describe the compressive behavior of dual-matrix continuous carbon fiber specimens, which were 3D printed using the co-extrusion technique. The curve exhibits an irregular pattern due to the progressive failure of the composite material, which occurs with the formation of bend bands during the test [26]. As observed, once the elastic region is surpassed, the curve not only becomes more irregular but also displays a noticeable first peak around 4 KN. This initial peak corresponds to the onset of partial fiber delamination within the composite. Following this, a second peak appears at the point of fracture, marking the final failure of the specimen. The average maximum force recorded during the tests was 4.62 ± 0.43 kN, while the ultimate compressive strength of the specimens was 81.77 ± 1.50 MPa. These values highlight the material performance under compressive loading, with the irregularities in the curve providing insights into the complex failure mechanisms at play during the deformation and eventual failure of the 3D-printed composite.

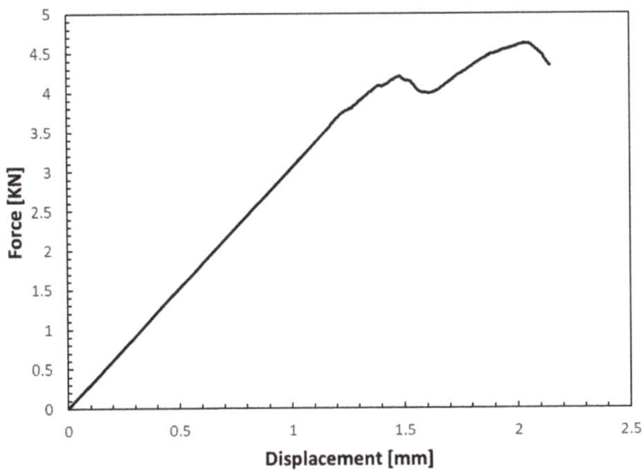

Figure 5. Typical compressive force–displacement curve.

3.2. Tomography Analysis

Figure 6 sequentially presents the tomography of five layers in the xz plane of the isogrid structure with 40% infill. The image highlights the emergence of macro-patterns at the intersections where the filaments are deposited. Specifically, at the nodes where filaments intersect, there is an overlap of dual filaments per layer, unlike along the ribs, where only single filaments are present. This overlap causes the filament to elevate slightly above the underlying layer as the deposition transitions from the rib to the node, creating areas of discontinuity within the structure. These discontinuity zones, which can potentially weaken the overall structural integrity, are located at different positions along the y-axis, as illustrated in the image. This observation is critical for understanding the internal structural dynamics and potential failure points within 3D-printed isogrid configurations.

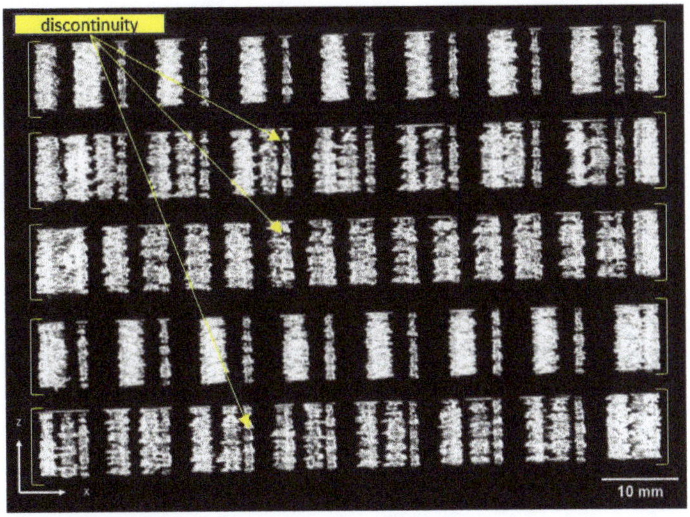

Figure 6. Tomography analysis of the 40% isogrid structure.

3.3. Buckling Tests

Figure 7 shows the average load–displacement curves obtained by the buckling test of isogrid structures at different infills. The analysis shows that the buckling load Pmax (maximum load supported by the structures) always increases as the percentage of infill increases. It can be observed that the curves corresponding to low infill percentages (10%, 20%, and 30%) show a similar behavior. In fact, after reaching the peak load, the curves show a step that becomes less evident as the infill density increases. The initial peak corresponds to the beginning of local buckling in the vertical ribs, while the subsequent peak is due to the onset of global buckling, caused by a greater displacement. In contrast, isogrids between 40% and 80% infill present a single peak corresponding to the onset of global buckling.

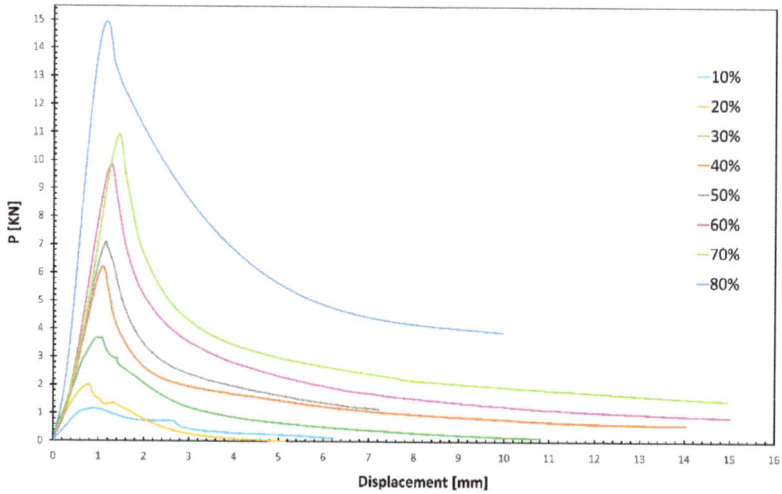

Figure 7. Average load–displacement curves.

Table 3 presents the average values of the buckling load and specific buckling load achieved for different infill densities, along with their respective standard deviations. Moreover, the results of the buckling test on the SCF-reinforced isogrid are reported as a reference. The data indicate that, similar to the trend observed with the maximum buckling load, the maximum specific buckling load also increases as the infill percentage increases. This relationship underscores the importance of infill density in enhancing the structural integrity and buckling resistance of the 3D-printed isogrid structures. The standard deviations provide an indication of the variability in the measurements, highlighting the consistency of the results across different samples. At 80% infill, the CCF structure shows an increase of 66% and 61.5% in maximum load and maximum specific load, respectively.

Table 3. Average maximum load and maximum specific load as the % infill changes.

% Infill	10		20	30	40	50	60	70	80	
	CCF	SCF	CCF	CCF	CCF	CCF	CCF	CCF	CCF	SCF
Maximum load (Pmax) [KN]	1.16 ± 0.12	0.85 ± 0.12	1.99 ± 0.23	3.12 ± 0.35	5.49 ± 0.32	7.45 ± 0.15	9.80 ± 0.24	11.93 ± 0.17	14.94 ± 0.20	7.08 ± 0.30
Maximum specific buckling load (Pmax/w) [KN/g]	0.091 ± 0.009	0.078 ± 0.016	0.101 ± 0.012	0.117 ± 0.013	0.164 ± 0.010	0.191 ± 0.004	0.212 ± 0.005	0.229 ± 0.003	0.257 ± 0.004	0.138 ± 0.005

The comparison with SCF indicates that the use of CFC significantly contributes to the buckling resistance, as the increases in the maximum specific buckling load for the 10% and for the 80% infill are equal to 16 and 85%, respectively.

The graphs in Figure 8 represent the trend, on a bi-logarithmic scale, of the buckling load (a) and of the specific buckling load (b) as the infill percentage varies. The rate of growth of the buckling load as a function of the infill % is lower at low percentages (up to 30%) and then increases at high percentages; in fact, two different slopes of the curve are evident. The change in slope becomes more pronounced when considering the specific buckling load, with an increase in the angular coefficient of the straight line, describing the behavior of the load as the infill density increases by 66% when going from 30% to 40%.

(a)

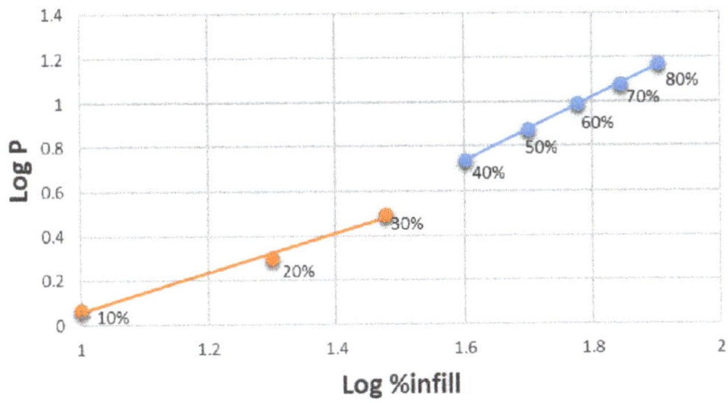

(b)

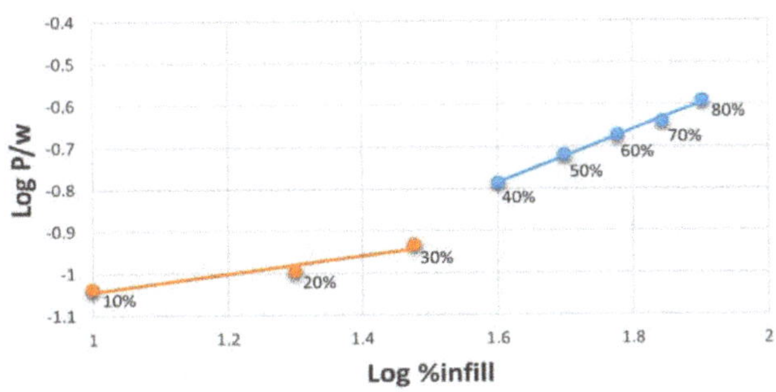

Figure 8. Buckling load (**a**) and specific buckling load (**b**) as a function of % infill, reported on a logarithmic scale: emphasis on the change in slope between lower 30% infills and higher.

The distinct behavior of structures with infill densities of up to 30% compared to those with higher densities can be attributed to the different buckling mechanisms these structures undergo. Specifically, structures with 10%, 20%, and 30% infill densities tend to fail through local buckling upon reaching the critical load. In these cases, failure is confined to specific zones within the structure, typically occurring on the xy plane, as illustrated in Figure 9. Conversely, at higher infill densities, the failure mechanism involves the entire structure, leading to global buckling. This global failure is characterized by deformation and buckling that extends across the entire cross-section of the isogrid.

Visual inspection of the isogrids after failure further supports these observations. Structures with up to 30% infill show localized failure zones, whereas those with higher infill densities exhibit more widespread failure, involving the entire structure. Additionally, in the higher-density structures, fiber pull-out is observed not only on the xy plane but also in the z direction, indicating a more complex and extensive failure mechanism, as shown in Figure 10. These findings highlight the critical role of infill density in determining the buckling behavior and overall structural integrity of 3D-printed isogrid configurations.

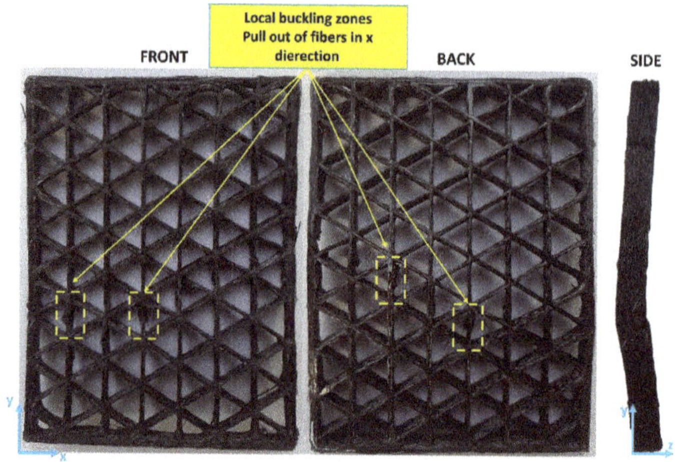

Figure 9. The tested 30% infill isogrid structure.

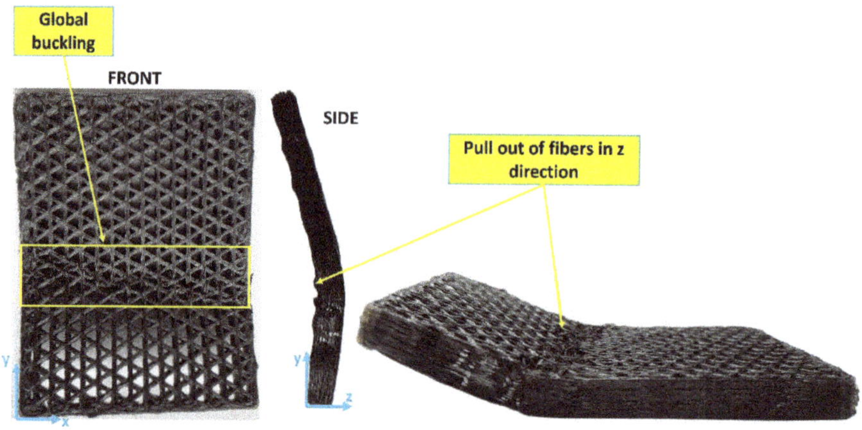

Figure 10. The tested 60% infill isogrid structure.

3.4. Optical and Scanning Electron Microscope Analyses

The morphology of the isogrids after the buckling tests was observed through an optical microscope. Figure 11 shows the 16× and 12.5× magnifications of the isogrids at 30% and 60% infill. Comparison of the two images highlights distinct failure mechanisms, which are related to the different modes through which the two structures reach buckling, either local or global. Indeed, the optical micrograph of the 30% isogrid shows a failure of the fibers on the xy plane (also observable in the 10% and 20% structures), while the 60% isogrid exhibits a different failure mechanism, with most of the fibers failing on the xz and yz planes.

The image of the fracture surfaces of the 3D-printed lattice structure was also observed by exploiting SEM (Figure 12). The presence of macrovoids between the deposited fiber bundles is evident (recognizable at 500× magnification), which leads to a non-homogeneous structure. Single carbon fibers can also be noted. On the other hand, the 1000× magnification highlights the presence of microvoids between the single filaments. The presence of macro- and microvoids can be attributed to deposition issues and low and inhomogeneous impregnation of the fiber during extrusion. These defects also affect the mechanical properties of the material itself. Moreover, they compromise the material mechanical properties and make it difficult to achieve the expected mechanical performances. Furthermore, the voids significantly impact the material's ability to absorb humidity, further affecting its overall integrity and functionality.

(a) 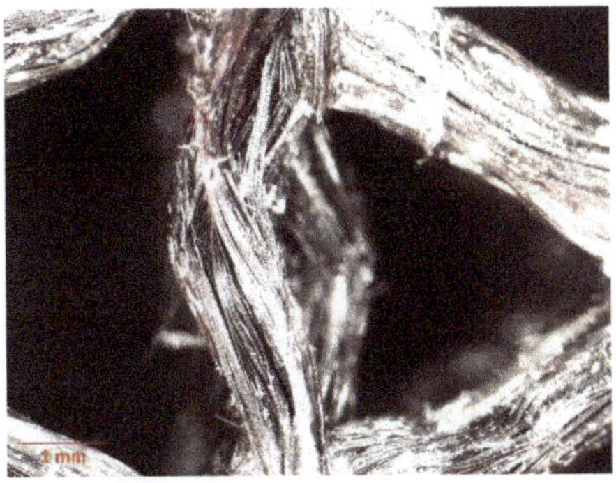

(b) 

Figure 11. (**a**) Infill 30%, 16×, local buckling; (**b**) infill 60%, 12.5× global buckling.

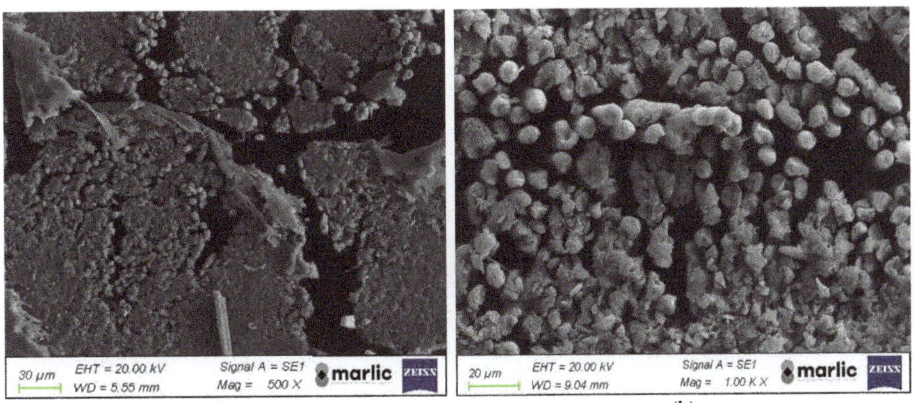

(a) (b)

Figure 12. (**a**) The 500× and (**b**) 1000× SEM magnification of the 3D-printed composite.

4. Conclusions

In this study, composite isogrid structures made from continuous carbon fiber-reinforced polyamide were fabricated using a 3D-printing process. The research focused on analyzing the effect of varying infill densities on the buckling behavior of these isogrid structures. The key findings from this investigation are summarized as follows:

- The compressive behavior of the 3D-printed specimens closely resembles that of traditional composite materials, demonstrating comparable mechanical responses under load.
- During the buckling tests, the load applied to the isogrid structures increases with displacement, reaching a peak value that corresponds to the onset of buckling. This behavior is consistent with theoretical predictions for buckling phenomena.
- The failure modes of the isogrid structures are influenced by the infill density. Structures with lower infill density predominantly fail through local buckling, characterized by deformation confined to specific areas. Conversely, structures with higher infill density tend to fail through global buckling, where the entire structure deforms uniformly.

- The maximum buckling load increases as the lattice infill density increases, regardless of the type of buckling failure. This indicates that denser structures are better able to withstand compressive forces before failing.
- The rate at which the buckling load increases is modest at lower infill densities but becomes significantly steeper at higher densities. This suggests that there is a threshold density beyond which the structural integrity is markedly enhanced.
- The specific buckling load, which normalizes buckling resistance against the structure's weight, shows a trend similar to that of the absolute buckling load. This implies that as the structure's weight increases, so does its resistance to buckling, making heavier but denser structures more robust.
- Continuous carbon fibers significantly increase both the maximum buckling load and the maximum specific buckling load compared to short carbon fibers.
- Microscopy analyses provide insights into the failure mechanisms, clearly distinguishing between local and global buckling modes in the lattice structures. These observations are crucial for understanding how to design against specific types of failure.
- Scanning electron microscopy images reveal the presence of both macro- and microvoids within the composite structure. These voids are critical defects that negatively impact mechanical performance, indicating areas for potential process improvements.

This research aims at demonstrating that specific strength can still be improved with increased material usage, even when different buckling modes occur. These findings are critical for the design of stronger and lighter 3D-printed composite structures, particularly in industries like aerospace and automotive engineering, where both weight reduction and strength are essential.

Isogrid lattice structures manufactured from composites via 3D printing exhibit high buckling loads, particularly when local buckling is minimized. However, the presence of voids remains a significant challenge, underscoring the need for further process optimization to reduce void formation and enhance structural performance. Future research will focus on reducing the porosity and the void quantity in the structures by optimizing 3D-printing parameters and by performing post-processing operations in order to produce more efficient components, and on exploring the impact of moisture absorption on the buckling resistance of these isogrid structures.

Author Contributions: Conceptualization, M.A., C.B., S.G. and A.V.; Methodology, A.V.; Validation, A.V.; Formal analysis, A.V.; Investigation, A.V.; Resources, C.B.; Writing—original draft, M.A., C.B. and A.V.; Supervision, A.F. All authors have read and agreed to the published version of the manuscript.

Funding: This research received no external funding.

Institutional Review Board Statement: Not applicable.

Data Availability Statement: The original contributions presented in the study are included in the article, further inquiries can be directed to the corresponding author.

Acknowledgments: The authors acknowledge the Marlic laboratory for their assistance with the SEM analysis.

Conflicts of Interest: The authors declare no conflict of interest.

References

1. Chung, D.D.L. *Carbon Fiber Composites*; Butterworth-Heinemann: Oxford, UK, 1978; Volume 9, ISBN 0750691697.
2. Elumalai, E.S.; Krishnaveni, G.; Sarath Kumar, R.; Xavier, D.D.; Kavitha, G.; Seralathan, S.; Hariram, V.; Premkumar, T.M. Buckling Analysis of Stiffened Composite Curved Panels. *Mater. Today Proc.* **2020**, *33*, 3604–3611. [CrossRef]
3. Barbero, E.; Tomblin, J. A Phenomenological Design Equation for FRP Columns with Interaction between Local and Global Buckling. *Thin-Walled Struct.* **1994**, *18*, 117–131. [CrossRef]
4. Kim, T.D. Fabrication and Testing of Thin Composite Isogrid Stiffened Panel. *Compos. Struct.* **2000**, *49*, 21–25. [CrossRef]

5. Sorrentino, L.; Marchetti, M.; Bellini, C.; Delfini, A.; Sette, F. Del Manufacture of High Performance Isogrid Structure by Robotic Filament Winding. *Compos. Struct.* **2017**, *164*, 43–50. [CrossRef]
6. Wodesenbet, E.; Kidane, S.; Pang, S. Optimization for Buckling Loads of Grid Stiffened Composite Panels. *Compos. Struct.* **2003**, *60*, 159–169. [CrossRef]
7. Andrianov, I.V.; Kalamkarov, A.L.; Weichert, D. Buckling of Fibers in Fiber-Reinforced Composites. *Compos. B Eng.* **2012**, *43*, 2058–2062. [CrossRef]
8. Antony, A. Effect of RIB Orientation in Isogrid Structures: Aerospace Applications. *Int. J. Sci. Technol. Eng.* **2017**, *3*, 319–327.
9. Totaro, G. Optimal Design Concepts for Flat Isogrid and Anisogrid Lattice Panels Longitudinally Compressed. *Compos. Struct.* **2015**, *129*, 101–110. [CrossRef]
10. Zhang, L.; Wang, X.; Pei, J.; Zhou, Y. Review of Automated Fibre Placement and Its Prospects for Advanced Composites. *J. Mater. Sci.* **2020**, *55*, 7121–7155. [CrossRef]
11. Gibson, I.; Rosen, D.; Stucker, B. *Additive Manufacturing Technologies: 3D Printing, Rapid Prototyping, and Direct Digital Manufacturing*, 2nd ed.; Springer: Cham, Switzerland, 2015; pp. 1–498. [CrossRef]
12. Gibson, I.; Rosen, D.W.; Stucker, B. *Additive Manufacturing Technologies: Rapid Prototyping to Direct Digital Manufacturing*; Springer: New York City, NY, USA, 2010; ISBN 9781441911193.
13. Kim, J.; Kang, B.S. Enhancing Structural Performance of Short Fiber Reinforced Objects through Customized Tool-Path. *Appl. Sci.* **2020**, *10*, 8168. [CrossRef]
14. Pandelidi, C.; Bateman, S.; Piegert, S.; Hoehner, R.; Kelbassa, I.; Brandt, M. The Technology of Continuous Fibre-Reinforced Polymers: A Review on Extrusion Additive Manufacturing Methods. *Int. J. Adv. Manuf. Technol.* **2021**, *113*, 3057–3077. [CrossRef]
15. Lutz, A.; Harmia, T. Impregnation Techniques for Fiber Bundles or Tows. *Polypropyl. AZ Ref.* **1999**, *2*, 301–306. [CrossRef]
16. Li, M.; Lai, C.; Zheng, Q.; Han, B.; Wu, H.; Fan, H. Design and Mechanical Properties of Hierarchical Isogrid Structures Validated by 3D Printing Technique. *Mater. Des.* **2019**, *168*, 107664. [CrossRef]
17. Guajardo-Trevino, A.M.; Ahuett-Garza, H.; Orta-Castanon, P.; Daniel Urbina-Coronado, P.; Saldana, C.; Kurfess, T.R. Effects of Deposition-Strategy-Induced Raster Gaps and Infill Voids on the Compressive Strength of 3D Printed Isogrid Structures. *Manuf. Lett.* **2022**, *31*, 15–19. [CrossRef]
18. Forcellese, A.; Simoncini, M.; Vita, A.; Di Pompeo, V. 3D Printing and Testing of Composite Isogrid Structures. *Int. J. Adv. Manuf. Technol.* **2020**, *109*, 1881–1893. [CrossRef]
19. Forcellese, A.; di Pompeo, V.; Simoncini, M.; Vita, A. Manufacturing of Isogrid Composite Structures by 3D Printing. In Proceedings of the Procedia Manufacturing; Elsevier: Amsterdam, The Netherlands, 2020; Volume 47, pp. 1096–1100.
20. Di Pompeo, V.; Forcellese, A.; Mancia, T.; Simoncini, M.; Vita, A. Effect of Geometric Parameters and Moisture Content on the Mechanical Performances of 3D-Printed Isogrid Structures in Short Carbon Fiber-Reinforced Polyamide. *J. Mater. Eng. Perform.* **2021**, *30*, 5100–5107. [CrossRef]
21. Adumitroaie, A.; Antonov, F.; Khaziev, A.; Azarov, A.; Golubev, M.; Vasiliev, V.V. Novel Continuous Fiber Bi-Matrix Composite 3-D Printing Technology. *Materials* **2019**, *12*, 3011. [CrossRef]
22. Azarov, A.V.; Antonov, F.K.; Vasil'ev, V.V.; Golubev, M.V.; Krasovskii, D.S.; Razin, A.F.; Salov, V.A.; Stupnikov, V.V.; Khaziev, A.R. Development of a Two-Matrix Composite Material Fabricated by 3D Printing. *Polym. Sci.-Ser. D* **2017**, *10*, 87–90. [CrossRef]
23. Andreozzi, M.; Bianchi, I.; Gentili, S.; Mancia, T.; Simoncini, M. Effect of Fabric Areal Weight on the Mechanical Properties of Composite Laminates in Carbon-Fiber-Reinforced Polymers. *J. Compos. Sci.* **2023**, *7*, 351. [CrossRef]
24. Azarov, A.V.; Antonov, F.K.; Golubev, M.V.; Khaziev, A.R.; Ushanov, S.A. Composite 3D Printing for the Small Size Unmanned Aerial Vehicle Structure. *Compos. B Eng.* **2019**, *169*, 157–163. [CrossRef]
25. Andreozzi, M.; Gentili, S.; Forcellese, P.; Bellezze, T.; Corinaldesi, V.; Luzi, F.; Vita, A. Effect of Moisture Content on the Mechanical Performance of 3D Printed Continuous Reinforced Two-Matrix Composite. *Int. J. Adv. Manuf. Technol.* **2024**, *133*, 5117–5126. [CrossRef]
26. Lee, S.H.; Waas, A.M. Compressive Response and Failure of Fiber Reinforced Unidirectional Composites. *Int. J. Fract.* **1999**, *100*, 275–306. [CrossRef]

polymers

Article

Study on Acoustic Emission Characteristics and Damage Mechanism of Wind Turbine Blade Main Spar with Different Defects

Yanan Zhang [1,*], Shaojie Xue [1], Chuanyong Chen [2], Tianchang Ma [3] and Bo Zhou [3]

1 School of Mechanical Engineering, Shenyang Ligong University, Shenyang 110159, China
2 School of Energy Engineering, Zhejiang University, Hangzhou 310058, China; chen_cy@zju.edu.cn
3 School of Mechanical Engineering, Shenyang University of Technology, Shenyang 110870, China; ma.tc@sut.edu.cn (T.M.)
* Correspondence: yanan_zhang@zju.edu.cn

Abstract: This paper aimed to understand the AE signal characteristics and damage mechanism of wind turbine blade main spar materials with different defects during the damage evolution process. According to the typical delamination and wrinkle defects in wind turbine blades, the GFRP composite with defects is artificially prefabricated. Through acoustic emission experiments, the mechanical properties and acoustic emission characteristic trends of wind turbine blade main spar composites with different defects under tensile loading conditions were analyzed, and the damage evolution mechanism of different defects was explained according to the microscopic results. The results show that the existence of artificial defects will not only affect the mechanical properties of composite materials but also affect the damage evolution process of the materials. The size and location of delamination defects and the different aspect ratio of the wrinkle defects have a certain influence on the damage mechanism of the material. K-means cluster analysis of AE parameters identified the damage models of GFRP composites. The types of damage modes of delamination defects and wrinkle defects are the same, and the range of characteristic frequency is roughly the same. This study has important reference significance for structural damage monitoring and damage evolution research of wind turbine blade composites.

Keywords: wind turbine blade; main spar; glass/epoxy composites; acoustic emission; damage mechanism; damage modes

1. Introduction

The most important part of producing wind energy is the wind turbine blade, whose production costs make up roughly 23% of the unit cost [1]. The majority of wind farms are situated in isolated locations, making maintenance and monitoring of the turbines challenging. Inaccurate assessments made too late could result in catastrophic damage to wind turbine blades under demanding operating conditions, catastrophic accidents, and enormous financial losses. The profile structure of large wind turbine blades is depicted in Figure 1, and more than 90% of the components of wind turbine blades are composed of composite materials [2].

About 80% of the weight of the wind turbine blades is supported by the main spar, with the remaining 20% being supported by the leading and trailing edges [3]. The primary spar's structure is laminate, and to increase its strength and stiffness, a high unidirectional strength glass fiber fabric is utilized as the reinforcing phase [4]. About 90% of the total brandishing stiffness is contributed by the main spar, and web plates provide internal support for the two shells to guarantee adequate structural stability [5]. In blade fracture accidents, the main spar fracture is the most common fracture form. The main spar's local stress concentration can be caused by the combination of inadequate vacuum perfusion and the main spar's lay-up properties, which can readily result in blade manufacturing errors. Defects will decrease the material's fatigue resistance, which will exacerbate damage initiation and expansion under complex random loads and hasten the main spar's fracture process. Currently, blade quality is mostly in danger due to manufacturing faults in primary spars, which also play a significant role in early failures and blade fracture accidents [6]. Delamination, wrinkles, air bubbles, and a lack of adhesive are examples of defects that can occur during the main spar laminate preparation process because of the reinforcement, matrix, and interface between them [7]. For example, the fiber cloth is not set flat during the cloth layer laying process, which leads to the formation of wrinkle faults. Because the fiber and resin have different thermal expansion coefficients and rates of moisture absorption, there is a chance that they will expand to different degrees during the curing process, which will ultimately result in the formation of lamination defects. In addition, poor air inflow or exhaust during vacuum perfusion leads to poor resin infiltration of the molded leaves [8].

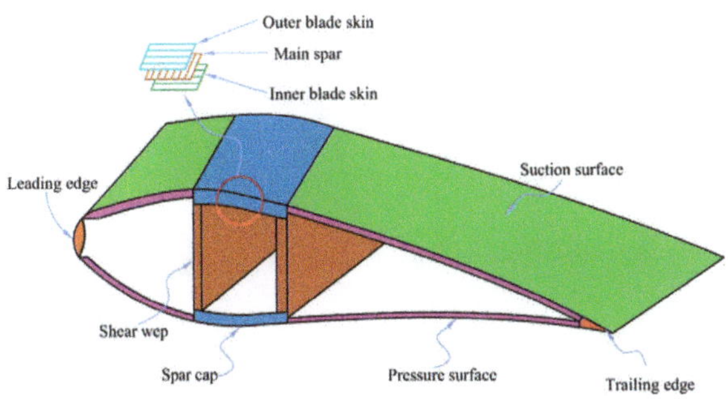

Figure 1. Profile structure of wind turbine blade.

The combined influence of unstable components like delamination is the primary cause of blade damage. The sort of blade defect that most affects the structural performance of the blade is the delamination defect. Lamination defects are among the most prevalent types of defects in blades because the composite material used to make blades is made up of multiple layers, making it simple for defects to occur between them [9]. Wrinkle defects are a common main process defect found in many manufacturers' goods because they are reasonably easy to arise in the blade layering process and can have a significant impact on the reliability of composite material products [10]. In particular, if there are wrinkle defects in the root, trailing edge, main spar, and other structural areas, in the operation of wind turbines, the interlayer cracking of fiber-reinforced composite materials in the wrinkle area will cause lamination damage of blades in serious cases [11]. At present, according to the characteristics of quality accidents in blades, wind power blades basically have three major risk sources: raw materials, manufacturing process defects, and design factors. Regardless of the maintenance faults or scrapping caused by quality reasons, process defects occupy a dominant position [12,13]. Typical manufacturing process defects of the wind turbine blade main spar structure are shown in Figure 2.

We aim to investigate the influence of manufacturing defects on the healthy life of wind power blades. At present, due to the limitations of manufacturing levels and process technology, the manufacturing defects of wind power blades cannot be completely avoided.

Therefore, it is very important for the structural health monitoring of wind power blades to analyze the signal characteristics, damage modes, and damage mechanisms of different defects with the help of sensors. Over the recent years, a remarkable number of scientific papers demonstrate the capability of AE in nondestructive testing (NDT), structure health monitoring (SHM), condition monitoring (CM), and fault diagnosis for RE generation, transmission, transformation, and storage systems [14]. Due to the high sensitivity of acoustic emission detection, it can effectively identify and monitor the weak matrix cracks and fiber damage components and characteristics in structures. The generation of acoustic emission signals changes with the stress, time, and temperature of the tested component, so dynamic information of structural damage and damage evolution with the above variables can be obtained, and long-term continuous state monitoring can be conducted. At present, acoustic emission monitoring technology has been widely used in damage monitoring, damage source location, and damage identification [15].

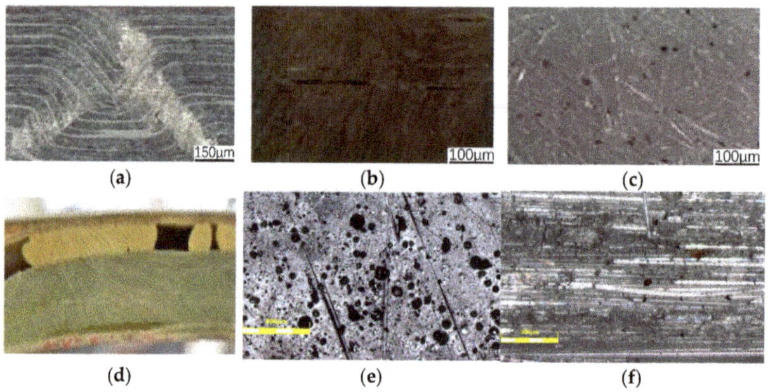

Figure 2. Typical manufacturing process defects of wind turbine blade main spar structure: (**a**) wrinkle defects; (**b**) delamination defects; (**c**) air bubble defects; (**d**) lack glue defects; (**e**) pore defects; (**f**) inclusion defects [16].

In the study of damage evolution, Nikhakht et al. [17] conducted a series of experiments on glass fiber reinforced plastic material samples with different layerings through acoustic emission data analysis and microscopic imaging methods. The results show that matrix cracking is the first failure mode, and its frequency ranges from 50 kHz to 200 kHz. The damage evolution depends on the fiber orientation. In addition, the load displacement curve, acoustic emission data, and microscopic imaging results have a good correlation in the initial, evolution, and expansion stages of damage. Loutas [18] et al. conducted tensile tests on braided carbon/carbon composites, focusing on the influence of the manufacturing process on the fiber/matrix interface, and conducted online acoustic emission monitoring. The results show that the impact is related to the damage mechanism under different load levels. Bourchak et al. [19] conducted tests on carbon fiber reinforced plastic composite laminates under static and fatigue loads, and used acoustic emission for monitoring. The results show that the AE energy is correlated with the damage observed by ultrasonic scanning and microscope in terms of damage type, location, and effective parameters of damage accumulation. Michalcova et al. [20] used optical equipment and acoustic emission technology to monitor the crack growth of a double-cantilever spar with plain carbon fiber reinforced composite materials in the environmental laboratory, and studied the relationship between acoustic emission energy and damage degree. The results show that the emission energy decreases when the temperature increases and the crack grows.

In addition to damage evolution research, acoustic emission technology is also widely used in wind power blade damage monitoring. Xu et al. [21] used acoustic emission analysis technology to monitor the fatigue test of 59.5 m long blades, extracted the original AE signal features through wavelet packet decomposition, identified the damage mode, and verified the robustness of the method. Tang et al. [22] conducted damage monitoring for the fatigue test of 45.7 m long blades based on AE technology, installed AE sensors near the web and trailing edge, respectively, and introduced initial defects. With the increase in fatigue test times, the growth of fatigue damage caused by primary defects and the location of damage were successfully detected. Zhou Bo et al. [23] used the blind deconvolution separation method to extract the features of acoustic emission signals collected in the fatigue test of 3.95 m long reduced ratio blades, and obtained the characteristics of weak cracks in blade skin and their changing trend with fatigue load. The damage evolution of composite materials is a complex mechanical process with highly nonlinear, multi-excitation source characteristic coupling and a large number of random factors. Different defects have significantly different AE signal characteristics, components, and damage modes in the process of damage evolution, so the physical meaning expressed by their damage modes is different [24]. Therefore, as a necessary prerequisite for the realization of wind power blade health monitoring and condition assessment, it is necessary to clarify the acoustic emission signal characteristics, damage modes, and damage mechanisms of different defects in the damage evolution process, so as to improve the interpretation of defect damage behavior.

In summary, the study of acoustic emission signal characteristics and damage mechanisms of different defective main spar composite materials during the damage process can provide support for the realization of wind turbine blade condition monitoring, contribute to the realization of wind turbine blade health condition monitoring, reduce the incidence of major safety accidents, and reduce the huge economic losses caused by late, untimely repair. In this paper, the artificial prefabrication of GFRP composites containing defects was carried out to solve the most harmful lamination and wrinkle defects in wind power blades. The damage evolution of GFRP laminates with lamination and wrinkle defects was studied using the acoustic emission monitoring technique. The influence of the size and position of different defects on the mechanical properties of materials was analyzed, and the acoustic emission signal characteristics and damage modes of different defects during the damage process were obtained by means of acoustic emission sensor signals, and the damage mechanism of different defects was analyzed by combining the acoustic emission signal characteristics and microscopic results during the damage process. This work has important reference significance for structural damage monitoring and damage evolution mechanism research of wind power blade composites.

2. Experiment Methods

2.1. Specimen Preparation

Because the thickness of the main spar of the wind turbine blade is much smaller than the plane size of the blade, the change in surface bending deformation can be ignored. Moreover, the normal deflection of the main spar surface is constant along its thickness direction, and the stress perpendicular to the plane direction of the main spar can be ignored. Therefore, the main spar structure of the blade conforms to the applicable conditions of the laminate theory, and the use of GFRP laminate specimens in the study conforms to the practice. According to GB/T25383-2010 wind turbine blade quality standards, wind turbine blades should be tested for process defects during processing and completion, such as wrinkle defects, delamination defects, air bubble defects, lack glue, etc. It is necessary to increase the strength check especially for delamination defects and wrinkle defects, which shows that delamination defects and wrinkle defects are important factors affecting the production quality of wind turbine blades [25].

Vacuum-Assisted Resin Infusion (VARI) was used to make laminates with defective blades. The technique was Resin Transfer Molding (RTM), which is a high-performance and low-cost vacuum-assisted perfusion method developed on the basis of RTM [26]. At present, VARI has become one of the main manufacturing processes for wind power blades, and the manufacturing method is shown in Figure 3. Under the same process conditions, the strength, stiffness, hardness, and other physical properties of the parts obtained by the vacuum perfusion method will be increased by 30–50% compared with the traditional hand paste process, and the porosity can be reduced to less than 0.2%, and the cost can be saved by 50%. In addition, compared with the traditional resin molding method, the vacuum perfusion method can reduce the porosity of composite parts and save 50–70% of the mold cost [27].

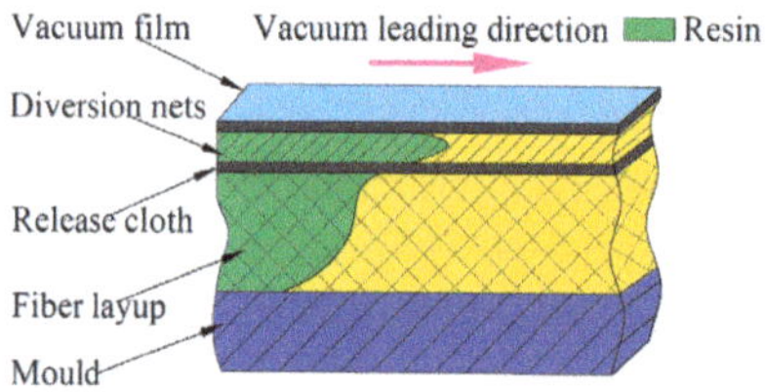

Figure 3. Diagram of resin introduction layer structure.

Lay a pre-prepared fiber layer on the bottom mold, cover the mold plate after laying, and ensure that the bottom and upper mold can be fully closed. A vacuum bag is used to wrap and seal, and then a vacuum pump is used to pump air to negative pressure. Finally, the resin liquid is injected into the whole mold through the grease injection port, and the resin is fully permeated through the layer structure. The concrete implementation steps of the specimen are shown in Figure 4. Glass fiber unidirectional cloth (ECW600-1270, 600 g/m^2) is selected as the lay-up material. The epoxy resin used in the preparation process of GFRP composites is Araldite LY 1564 SP, with a density of 1.1~1.2g/cm^3 and a concentration of 100% transparent liquid. The curing agent model is Aradur 3486, the density is 0.94~0.95 g/cm^3, and the concentration is 100% blue liquid. The mass ratio of the epoxy resin and curing agent in the resin matrix is 100:34.

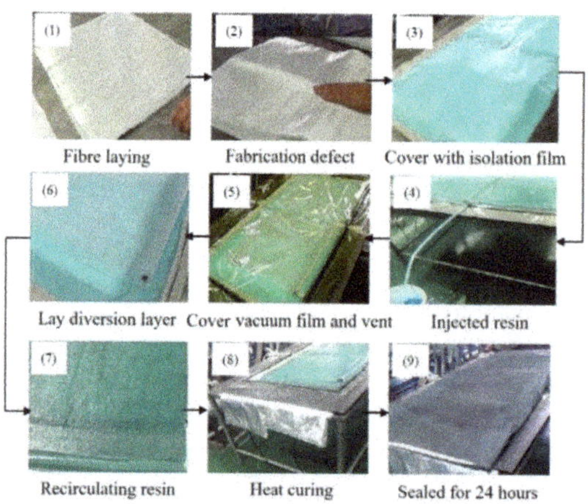

Figure 4. GFRP composite manufacturing process.

Since epoxy resin is a thermosetting resin, the resin liquid and curing agent should be fully stirred and mixed before heating in the preparation process. It is worth noting that in the curing heating process, the temperature cannot be set too high, as too high temperature heating will lead to changes in material properties. At the same time, too low temperature control will lead to poor fluidity of the resin liquid and difficulty penetrating the fiber layer structure. In addition, too low a temperature is not conducive to the removal of bubbles in the resin, and the remaining bubbles will lead to an increase in porosity. After repeated experiments, the curing heating temperature should be controlled at 70~80 °C.

The prefabrication method of artificial lamination defects is made by placing polytetrafluoroethylene film in the fiber prefabrication block. In the process of high temperature curing, the polytetrafluoroethylene film is deposited on the surface of the resin base, resulting in a better fusion between the resin base infiltrated on the surface and the fiber layer, thus forming a weak interface. In the cooling process, the weak interface is separated due to the different thermal expansion coefficient between the polytetrafluoroethylene film (PTFE) and the composite material, so as to simulate the delamination defect. The specific operation process of the artificial layered defect production method in this paper is as follows:

(1). First lay the bottom glass plate mold and keep it clean, and lay the layer from the bottom to the top.
(2). When the delamination operation is carried out to the layered defect design layer, place the PTFE film with a thickness of 0.5 mm.
(3). After the layer is finished, lay the stripping cloth, diversion net, and vacuum bag.
(4). Connect the covered paving structure to the vacuum resin filling system, and check the air tightness of the filling device and mold.

After the inspection is qualified, the grease is injected and finally the curing operation is carried out. The schematic material samples of the four combinations are shown in Figure 5. Specimen A1 has no delamination defect and specimen A2 has PTFE film with a size of 25 mm × 25 mm sandwiched between the third and fourth layers. The PTFE films are 25 mm × 15 mm and 25 mm × 25 mm in size, respectively, sandwiched between the 12th and 13th layers of specimens A3 and A4.

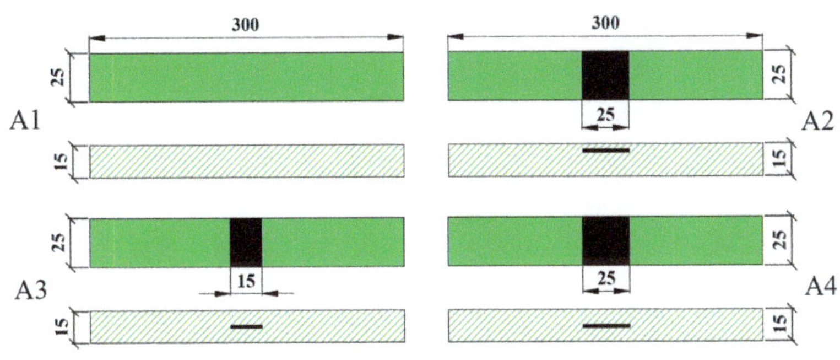

Figure 5. Size diagram of delamination defect specimens (units: mm).

When the wrinkle defect is prefabricated manually, the copper wire of uniform material is generally placed between the fiber cloth and the mold. In the process of vacuum perfusion, the pre-placed copper wire can change the angle of the fiber cloth layer and prevent the infiltration of resin on the wrinkle area of the fiber cloth. After cooling and curing, the pre-placed copper wire is removed by electric winch traction, and the prepared resin is injected into the empty slot of the copper wire for re-curing, so that the area is rich in resin, thus simulating the wrinkle defect. In this paper, copper wire with a diameter of φ

= 2 mm is used to make wrinkles when preforming wrinkle defects. The specific operation process is as follows:

(1). First clean the sign of the mold plate, and then evenly apply the release agent to the surface of the mold plate.
(2). Place the copper wire with a diameter of φ = 2 mm on the mold plate and place the glass fiber cloth on the copper wire, and perform the lay-up operation from the bottom to the top. The copper wire placed at different layer positions produces different aspect ratios, which are, respectively, labeled Z1 (2:0.2), Z2 (5:1), and Z3 (1.5:0.3). The specific size parameters are shown in Figure 6. About 1 m of copper wire is reserved during operation to facilitate fixing on the electric winch.
(3). After the layer is finished, lay the stripping cloth, diversion net, and vacuum bag.
(4). Use an electric winch to remove the copper wire and fill the copper wire empty tank with pre-stirred resin liquid for curing operation. This area is a rich resin area.
(5). Cut the prepared GFRP composite material into a laminate with a predetermined size, polish the gap flat, and finally paste reinforced aluminum sheets at both ends of the specimen.

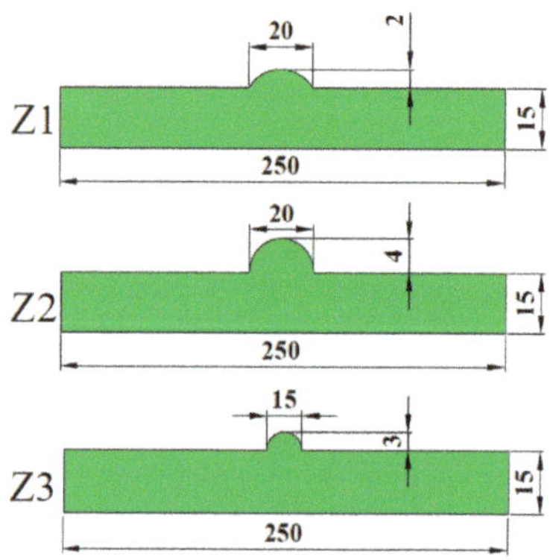

Figure 6. Size diagram of wrinkle defect specimens (units: mm).

The microscopic morphology of the resin aggregation layer with artificial prefabricated wrinkle defects is shown in Figure 7a. It can be seen that the boundary between the resin aggregation area of artificial wrinkle defects and the composite material forms an "Arched" transition zone. The change in the fiber layering angle in this region makes the composite susceptible to stress concentration interference during loading. As shown in Figure 7b after the natural wrinkle defect is polished, the fiber structure near the boundary of the resin gathering area changes, which is consistent with the interface structure at the artificial prefabricated wrinkle defect. In Figure 7c, it can be seen that the fibers near the layered interface at the macro level do not fully fuse with the resin matrix due to the presence of the PTFE film. Natural stratification easily weakens the adhesion of the material interface, resulting in the destruction of the interface structure, and its microscopic morphology is shown in Figure 7d. By comparing the defect morphology of artificial lamination and natural lamination, it can be seen that the artificial lamination defect composite method adopted in this paper can simulate the natural lamination defect better.

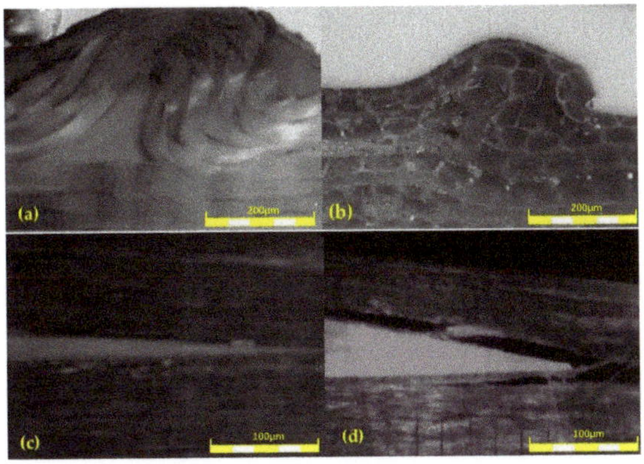

Figure 7. Morphological features of defects: (**a**) artificial wrinkle defects; (**b**) natural wrinkle defects; (**c**) artificial delamination defects; (**d**) natural delamination defects.

2.2. Acoustic Emission Test Method

The tensile tests of composites were performed according to the ASTM standard D3039 partly (Standard Test Method for tensile properties of Ploymer Matrix Composite Materials) [28]. The tensile test was carried out on a WAW-300B servo universal test machine. The maximum test force is 300 kN, the relative error of the test force indication value is $\leq \pm 1\%$ of the value, and the test force measurement range is 2~100% of the maximum test force. The control range of constant displacement is 0.01 mm/s, and the test system is shown in Figure 8. An acoustic emission sensor was installed at the defect of the specimen and fixed with tape. In order to provide a good acoustic coupling between the specimen surface and the sensor, silicone grease was used as the coupling agent and a pencil fracture test was performed. In order to ensure the effective elimination of electrical and mechanical noise, the appropriate threshold is fixed at 40 dB through repeated testing. The initial load is applied to eliminate the noise interference in the loading process of the reinforced aluminum sheet.

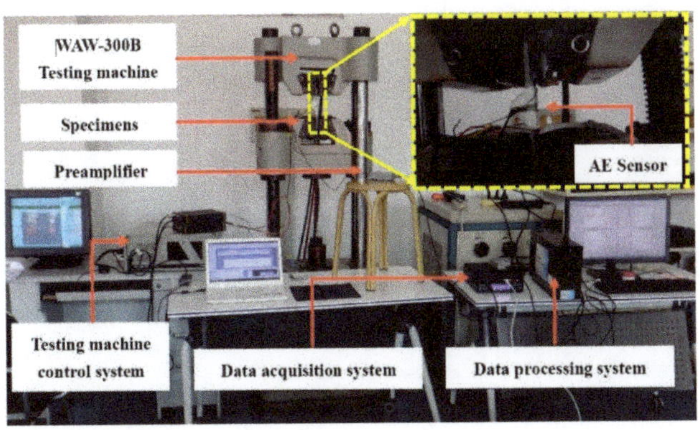

Figure 8. Acoustic emission testing system.

It is worth noting that the quality of acoustic emission data obtained by the experiment mainly depends on the timing parameters of the time-domain waveform. The PDT (peak definition time), HDT (hit definition time), and HLT (hit lock time) are parameters used to select acoustic emission event characteristics. The main acoustic emission parameters and definition methods obtained in the experiment are shown in Figure 9. In the experiment, the center frequency of the acoustic emission sensor is 150 kHz and the gain of the preamplifier is 40 dB. The sampling frequency is set to 3 MHz, the peak definition time is 30 μs, the peak definition time is 150 μs, the hit definition time is 150 μs, and the hit lock time is 300 μs [29].

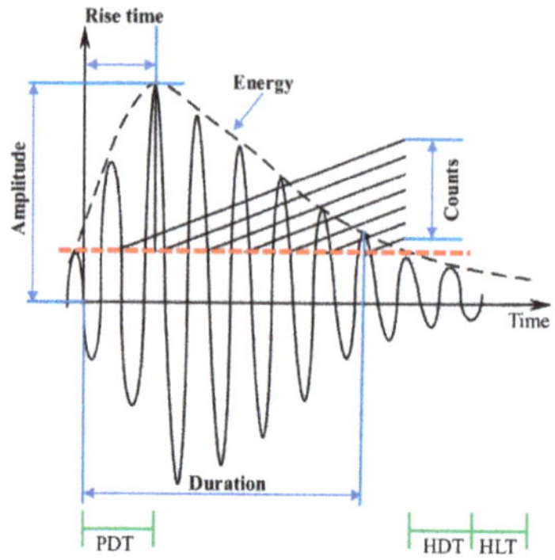

Figure 9. Definition of acoustic emission parameters.

The sensor selects four symmetrical sensor channels, and the position of the sound source is the real lead break position, that is, the red line position, as shown in Figure 10. Through the arrival time of the acoustic emission signal received by the sensor and the distance of the sensor ruler, the calibration of the acoustic emission wave speed is completed. Considering the complex structure of the composite material itself, the acoustic velocity calibration of the acoustic emission signal needs to be measured several times, and the average value is taken. In order to solve the attenuation problem of acoustic emission signals, the sensor signals at different positions are tested during the lead-breaking test. The amplitude of the lead-breaking signal collected by the sensor should not be less than two times of the threshold amplitude. Due to the close distance between the acoustic emission sensor and the defect position in this paper, the attenuation effect of the acoustic emission signal is relatively small. The sound velocity calculation method is shown in Equation (1).

$$\overline{V} = \left(\frac{L_1}{t_1} + \frac{L_2}{t_2} + \frac{L_3}{t_3 - t_1} + \frac{L_4}{t_4 - t_2}\right)/4 \tag{1}$$

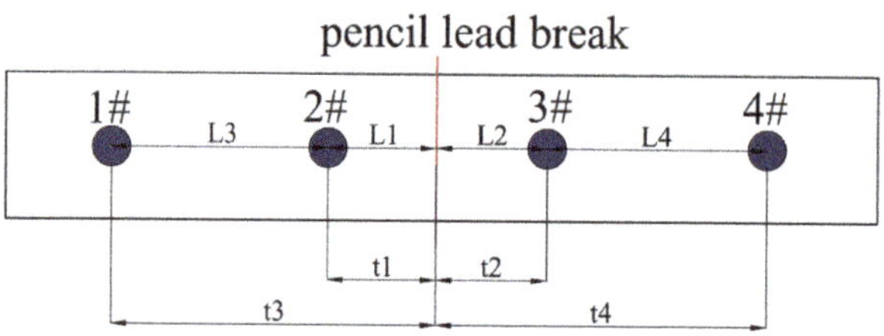

Figure 10. Sound velocity calibration for acoustic emission lead break test.

In the lead-breaking experiment, a 0.5 mm pencil was used, the lead length was about 2.5 mm, and the angle between the lead and the specimen surface was about 30°. The signal frequency of the experimental interrupt lead test is distributed in 100–400 kHz. When the characteristic frequency of the broken lead signal is 150 kHz, the amplitude of the signal is the largest, and the other low amplitude signals may be the interference of environmental noise.

In the test process, the number of sensors is first determined by the attenuation of the pencil lead-breaking signal in the pre-test. The specific operation method is as follows:

Two AE sensors, 40 mm apart, were arranged on the top and bottom of the laminate specimen, and a pencil lead-breaking test was carried out at the defect center. The AE wave velocity and attenuation amplitude of the laminates were 4423.7 m/s and 0.22~0.38 dB/cm, respectively, and the maximum time difference of AE signal attenuation in the pencil lead-breaking test was determined ($\Delta tmax$ = 0.0084 ms).

An acoustic emission lead-breaking waveform is shown in Figure 11. As can be seen from the figure, the time difference between sensor 1 and sensor 2 to receive the acoustic emission signal of broken lead is 0.0018 ms (<$\Delta tmax$ = 0.0084 ms). The test results show that both sensors can effectively receive acoustic emission signals, and there is no significant difference in signal attenuation. Therefore, the use of an acoustic emission sensor in the experiment can also effectively capture the damage characteristics, which is consistent with the experimental conclusion of Favretto [30].

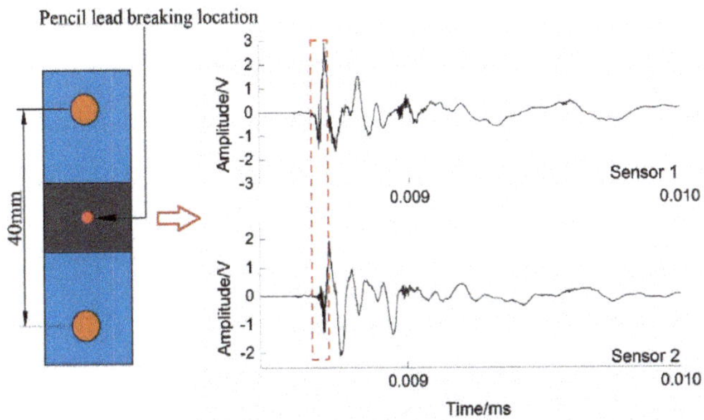

Figure 11. The schematic of pencil lead break test.

3. Experimental Results

3.1. K-Means Clustering Analysis of AE Signals

The acoustic emission waveform is stored by a computer and acquisition system, and the corresponding characteristic parameters such as energy, amplitude, and count are calculated according to the acoustic emission waveform. Since these characteristic parameters are derived from acoustic emission acquisition waveforms, the damage characteristics of different defects can be obtained by analyzing them to a certain extent. The acoustic emission characteristic parameters obtained in the experiment mainly include: amplitude, energy, count, and peak frequency, etc. All acoustic emission characteristic parameters are regularized [0, 1], and the characteristic parameter matrix is established. Due to the complexity of composite structure, the AE source has diversity and uncertainty, and different AE sources can produce completely different AE signals. However, in real situations, these feature parameters may exhibit high similarity. This is because they are all different ways of quantifying physical phenomena produced by the same acoustic emission source, and these physical phenomena are inherently related to each other. A fully linked geometric distance metric was used to cluster the feature parameters, and the results show that the AE number, energy, peak frequency, and amplitude height have some similarity in describing the damage trend, as shown in Figure 10. Therefore, the feature parameters with high similarity are selected for K-means clustering analysis.

The K-means clustering algorithm classifies the data features in the sample space, and uses the optimization iteration function to correct the clustering results in the initial supervised learning sample. The K-means clustering algorithm is a clustering algorithm based on the optimal criterion function, and the criterion function is as follows [31]:

$$J = \sum_{i=1}^{k} \sum_{x \epsilon C_i} \|x - m_i\|^2 \quad (2)$$

where x is the acoustic emission parameter data sample, k is the number of clusters, C_i is the sample set of the Class i cluster, m_i is the mean vector of cluster i, and satisfies the following:

$$m_i = \frac{1}{n} \sum_{x \in x_i} x_i \quad (3)$$

where n_i is the number of samples in C_i. Criterion function J is used to describe the total error square generated by the k cluster sample set $C_1, C_2, \cdots, C_k$ represented by different clustering centers $m_1, m_2, \cdots, m_k$. The smaller the J value, the higher the similarity of data in the same cluster sample set. The K-means clustering algorithm uses the iterative function to continuously adjust the clustering center, so that the error sum of squares criterion function **J** of each cluster sample set can obtain the minimum value. Therefore, the number of clusters k and the initial cluster center should be determined before using this method, as shown in Figure 12.

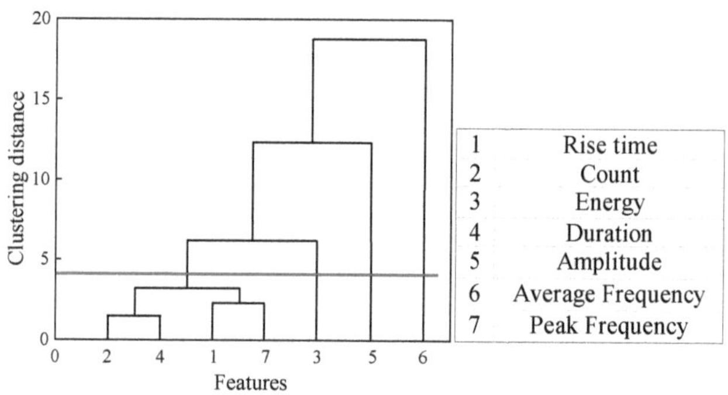

Figure 12. Correlation tree diagram of acoustic emission signal parameters.

Due to the high sensitivity of acoustic emission sensors to weak signals, it is also susceptible to noise interference. During the experiment, the influence of environmental noise should be avoided as much as possible, but the noise source of the experimental equipment itself cannot be avoided. The AE data obtained at a high sampling frequency are very large, so principal component analysis (PCA) is adopted to preprocess the AE features, reduce the information dimension, and screen effective AE parameters.

The comatrix is established after the normalization of the acoustic emission characteristic parameters selected in the experiment. C is set to represent the comatrix of acoustic emission characteristic data. The eigenvalue λ_i can be calculated by the equation $det(C - \lambda_i) = 0$. The columns of the eigenvector matrix A satisfy the equation $C = ADA^T$, where the matrix D is the eigenvalue $\lambda_1, \lambda_2, \ldots, \lambda_m$ forming a diagonal matrix, and $\lambda_1 \geq \lambda_2 \geq \ldots \geq \lambda_m$. PCA can transform multidimensional AE data into a series of linearly uncorrelated new features called principal components. Let A be a matrix of eigenvectors, a_{ij} is the element that makes up A, and then each principal component can be represented as follows:

$$Pd_i = \sum a_{ij} \overline{p_j} \quad (4)$$

where $\overline{p_j}$ is the initial set of normalized AE signal parameters. The AE signals after PCA dimensionality reduction were analyzed by K-means clustering. The number of clusters was evaluated using the Davis–Bouldin and Silhouette indices, and the higher the *DB* index and the lower the *SI* index, the best number of clusters for the sample data [32,33].

$$DB(k) = \frac{1}{k}\sum_{i=1}^{k} max(\frac{DW_i + DW_j}{DC_{ij}}) \tag{5}$$

$$SI = \frac{1}{n}\sum_{i=1}^{n} \frac{b(x_i) - a(x_i)}{max\{a(x_i), b(x_i)\}} \tag{6}$$

where DW_i represents the average distance of the center of mass between adjacent clusters. DW_j represents the average distance of the center of mass between the cluster center and other clusters. $(DW_i + DW_j)/DC_{ij}$ represents the similarity between different clusters, which is the clustering distance. $a(x_i)$ represents the average distance between x_i and other clusters, and $b(x_i)$ represents the average distance between x_i and neighboring clusters.

Reference [34] compared the clustering sensitivity of different acoustic emission parameters. The results show that the acoustic emission amplitude, average frequency, and energy have high variance values, which means that these three acoustic emission parameters will obtain better clustering results. The clustering number results of composite specimens are shown in Figure 13. It can be seen from the evaluation results of the *DB* and *SI* index that the optimal cluster number is 4.

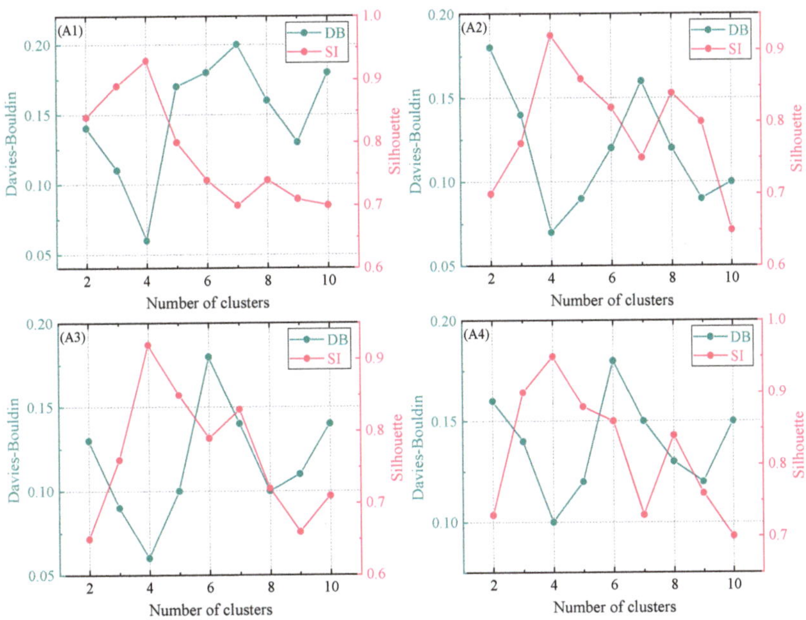

Figure 13. Evaluation of the number of clusters: specimens A1, A2, A3, and A4.

The clustering results and AE signal quantity of samples with delamination defects and wrinkle defects are shown in Tables 1 and 2. The matrix cracking stage is represented by Cluster 1's characteristic center frequency of 63.2–65.8 kHz, the fiber/matrix stripping stage by Cluster 2's characteristic center frequency of 136.9–142.87 kHz, the delamination of the interface stage by Cluster 3's characteristic center frequency of 254.46–284.3 kHz, and the fiber breakage stage by Cluster 4's characteristic center frequency of 365.2–397.75 kHz, according to the clustering results of delamination defect specimens. The typical center frequencies of Cluster1 matrix cracking and Cluster2 fiber/matrix stripping were 69.38–73.22 kHz and 137.24–148.53 kHz, respectively, in the clustering results of wrinkle defect specimens. The typical center frequency of Cluster4 fiber breakage is 392.83–402.27 kHz, while the characteristic center frequency of Cluster3 interface delamination is 281.61–287.20 kHz, which corresponds to the interface delamination stage.

In addition, the low frequency component whose peak frequency is less than 30 kHz is due to mechanical vibration interference during loading. The amplitude of acoustic emission signals can represent the intensity of sound source features, and the amplitude distribution is related to the deformation mechanism inside the material. According to Tables 1 and 2, during the damage evolution process, glass fiber composites containing lamination defects and wrinkle defects will generate a large number of high-amplitude AE signals. In the initial stage of damage evolution, the matrix cracking and fiber/matrix stripping damage characteristics of layered defect specimens are more active than those of non-defective specimens. The reason is that the existence of delamination defects will affect the strength of the interlayer structure, leading to the rapid deterioration of matrix cracking and fiber/matrix spalling during the tensile process. By comparing the defect specimen with the no-defect specimen, it can be found that the damage characteristics of the layered defect and the wrinkle defect are basically the same. However, the central frequency of the damage mode characteristics is different, which indicates that the composite with defects has a certain influence on the damage characteristic frequency and produces fluctuations.

Table 1. Cluster analysis results of delamination defect specimens.

	Specimen A1		Specimen A2	
	Amplitude/dB	Center frequency/kHz	Amplitude/dB	Center frequency/kHz
Cluster1	40.1-91.8	71.31	40.1-91.5	63.20
Cluster2	40.3-86.5	131.84	40.3-83.2	142.87
Cluster3	41.5-72.2	297.60	41.3-71.5	254.46
Cluster4	42.2-66.3	407.25	41.6-67.8	365.20
	Specimen A3		Specimen A4	
	Amplitude/dB	Center frequency/kHz	Amplitude/dB	Center frequency/kHz
Cluster1	40.1-92.3	65.80	40.1-91.2	65.41
Cluster2	40.3-86.5	136.90	40.3-86.5	138.62
Cluster3	41.6-68.4	284.30	41.7-70.2	267.94
Cluster4	42.5-62.3	397.75	42.2-65.3	385.83

Table 2. Cluster analysis results of wrinkle defect specimens.

	Specimen Z1		**Specimen Z2**		**Specimen Z3**	
	Amplitude/dB	**Center Frequency/kHz**	**Amplitude/dB**	**Center Frequency/kHz**	**Amplitude/dB**	**Center Frequency/kHz**
Cluster1	40.1–89.2	71.65	40.1–91.2	73.22	40.1–92.3	69.38
Cluster2	40.5–90.3	144.42	40.3–86.5	148.53	40.3–86.5	137.24
Cluster3	41.0–75.8	284.25	41.7–70.2	281.61	41.6–68.4	287.20
Cluster4	42.2–57.3	398.65	42.2–65.3	392.83	42.5–62.3	402.27

As shown in Figure 14, acoustic emission waveforms and spectrum characteristics of different damage modes were extracted according to the clustering results. The damage characteristic spectrum of real wind turbine blades and the acoustic emission waveform characteristics of glass/epoxy composites were compared. By comparing the damage pattern recognition results of the reference and this paper, the characteristic frequency range of the four damage patterns is basically consistent with the reference results. The time-domain waveform duration of high-frequency signals is shorter, and the time-domain waveform duration of low-frequency signals is longer [35–37]. It is worth noting that the acoustic emission signals of matrix cracks in real wind turbine blades in reference [36] are compared. In this paper, the acoustic emission signal amplitude representing the low-frequency matrix crack fluctuates greatly, as shown in Table 1. The reason for this phenomenon is that in the damage process, the matrix usually starts with small cracks and gradually expands into larger cracks, which leads to obvious amplitude fluctuation. In addition, due to the influence of defects, the matrix structure is more likely to produce high-amplitude acoustic emission characteristics during the damage process.

The relationship between tensile load and peak frequency and time is shown in Figure 15. The grouping boundaries of Cluster 1 and Cluster 2 AE signals in the low frequency stage have overlap in the peak frequency clustering results. This behavior is caused by the low frequency stage's peak amplitude being very modest and the signal there being highly contaminated by noise. Therefore, the data partitioning accuracy in the process of cluster analysis will be affected [38]. It can be seen from Figure 15 that in the damage mode of the low frequency stage, the entire stretching process happens virtually concurrently with matrix cracking and fiber stripping. Due to the constant evolution and quick spread of faults, many fiber breakage characteristic frequency signals emerge prior to reaching the fault point. The findings indicate that four damage modes will manifest simultaneously and a large number of high-frequency signals will manifest when the composite is close to failure.

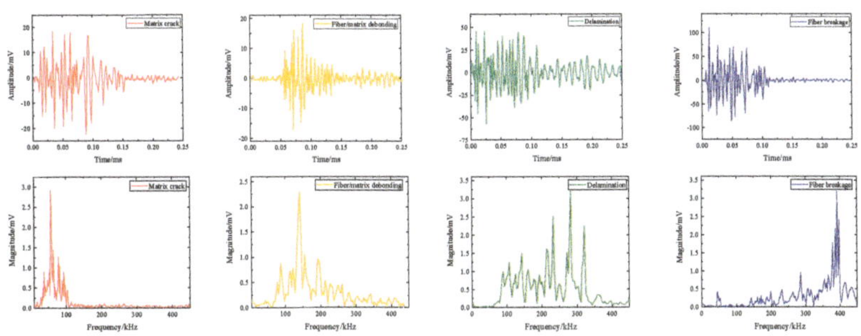

Figure 14. Time domain and frequency domain signals of different damage modes.

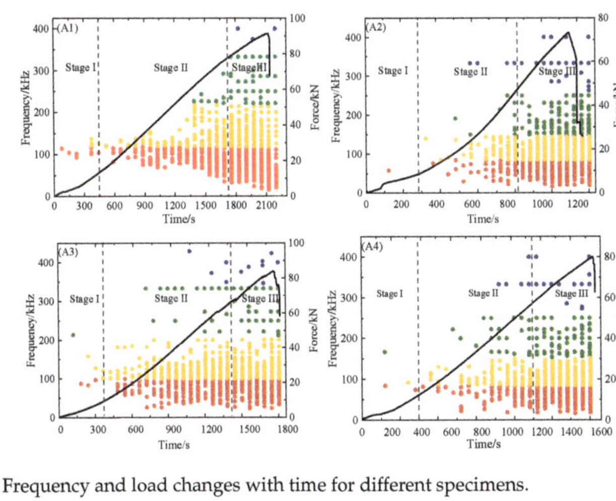

Figure 15. Frequency and load changes with time for different specimens.

According to the results of clustering in Table 1, AE signals of fiber bundle breakage are characterized by peak frequency and low amplitude, and samples with delamination defects generate AE signals of peak frequency earlier than standard specimens. In the second stage of damage evolution, fiber/matrix stripping and propagation lead to the continuous emergence of high-frequency AE signals. For sample A1 shown in Figure 15 (A1), AE signals with high-frequency fiber breakage began to appear when the load was increased to about 80 kN. Similarly, for specimens A2, A3, and A4 with delamination defects in Figure 15 (A2–A4) with AE characteristics of high-frequency fiber breakage, the corresponding loads were 45 kN, 75 kN, and 55 kN, respectively. Compared with specimens A1, A3, and A4, the high-frequency fiber breakage feature of specimen A2 appears at about 550 s, which is earlier than the peak frequency and irreversible damage time of other specimens. This is because the closer the lamination defect is to the surface of the specimen, the lower the structural bearing capacity of the specimen.

According to the above analysis, defect depth and size are related to AE characteristic frequency. As shown in Figure 15, the depth and size of defects in the damage evolution process of GFRP composites have a certain influence on the frequency characteristics of fiber breakage. In layered defects, when the defect depth is the same, the characteristic frequency of the fiber breakage of the specimen with a large defect size is smaller than that of the specimen with a small defect size. When the defect size is the same, the fracture characteristic frequency of the defect depth in the surface layer is smaller than that in the middle layer. Similarly, the effect of wrinkle defects with different aspect ratios on damage patterns is roughly the same as that of layered defects. Therefore, the frequency characteristics of the damage mode of composite materials are related to the physical properties of the material itself, but different defect sizes and depths will affect the frequency range of the damage mode characteristics.

As shown in Figure 16, the peak frequency of acoustic emission signals in damage stage I is 93 kHz, and the frequency components in this stage are mainly low-frequency characteristics, and there are almost no high-frequency fiber breakage components larger than 350 kHz. In stage II, the acoustic emission peak frequency is mainly 93 kHz, 150 kHz, and 210 kHz, and there is a frequency characteristic of 290 kHz. In this stage, the main damage is matrix cracking, fiber breakage, and interface delamination. In stage III, in addition to the low and medium frequency features, a high-frequency feature with a peak frequency of 378 kHz appears, indicating that the damage component of fiber breakage is very active in this stage. The spectrum characteristics of the above transmitted signals are basically consistent with the clustering results in Figure 15 (A1).

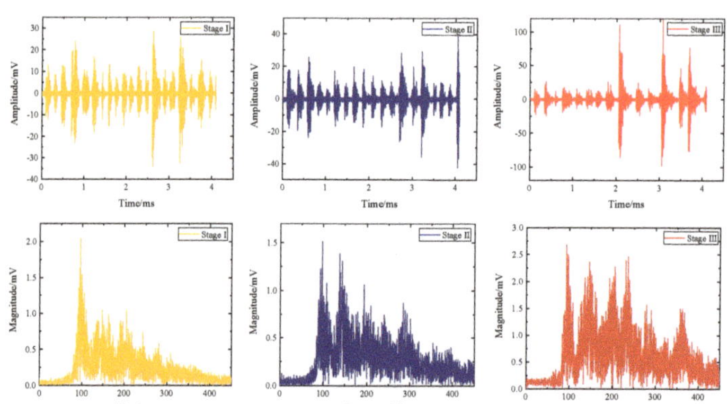

Figure 16. Acoustic emission signal spectrum characteristics of specimen A1 at different damage stages.

3.2. Acoustic Emission Characteristics of Delamination Defects

According to the tensile test of the composite material specimens with layered defects, the average failure loads of layered specimens A1, A2, A3, and A4 are determined to be 92.8 kN, 76.41 kN, 85.24 kN, and 79.73 kN, respectively. The displacement change trend of the composite material sample in the tensile test is shown in Figure 17. In the experiment, the ultimate breaking load of specimens A1, A2, A3, and A4 is significantly lower than that of specimen A1 when the layered defects are present. This shows that lamination defects can reduce the ultimate strength of composite structures, and the closer the lamination defects are to the surface of composite materials, the greater the impact on the structural strength of composite materials. In addition, when the location of the lamination defect is the same, the larger the size, the greater the influence on the structural strength of the composite.

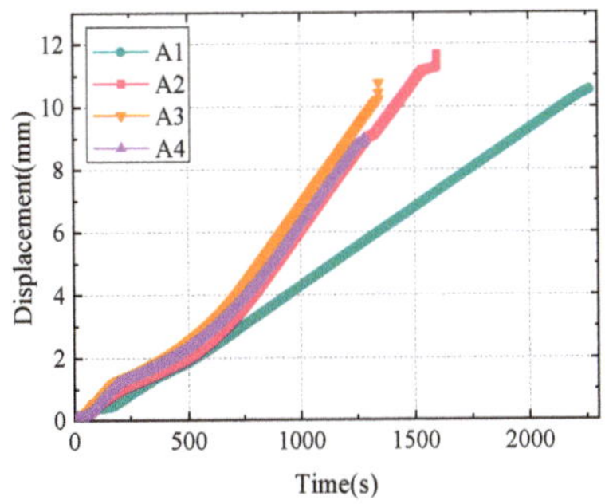

Figure 17. Displacement curve of specimens in tensile test.

AE amplitude, accumulation count, energy, and count parameters were selected as AE feature analysis in the tensile test results. Figure 18 displayed the correlation between the specimen's accumulated count variations over time and the AE accumulation energy. A low energy amplitude change is nevertheless visible in the first phase, despite the load being smaller and the rising rate being slower. This behavior demonstrates that the AE signal gathered at this point is essentially a noise signal, making it vulnerable to interference from the noise signal during the initial loading stage. From specimen A1 in Figure 18 (A1), it can be found that AE energy amplitude fluctuates greatly with the increase in time, which is related to the significant increase in sample damage degree. When t = 1600 s, the cumulative count and energy of the AE signal increase rapidly. This phenomenon shows that a large number of high-energy AE signals appear with the beginning and accumulation of damage in the composite material, which indicates that the early energy accumulation is partially released at this stage, and the irreversible damage evolution begins.

Compared with specimen A1 without delamination defects, the AE characteristics of delamination specimens A2, A3, and A4 shown in Figure 18 (A2–A4) are significantly different. As can be seen from Figure 18 (A2), due to the existence of stratification defects, the AE energy peak and accumulation count of specimen A2 showed an obvious upward trend when t = 400 s. The reason is that the delamination defect has a great impact on the structural strength of the material, which directly leads to the irreversible damage caused by the delamination defect earlier in the damage process. By comparing specimens A2 and A4, it can be seen from Figure 18 (A4). When the depth of defect increases, the peak value of AE energy and the active trend of the accumulation count during the damage process decrease to a certain extent. The results show that when the size of the lamination defect is the same, the closer the position of the defect is to the surface of the specimen, the greater the influence on the structural strength of the main spar composite.

It can be seen from Figure 18 (A3) that when t = 1050 s, the AE energy peak and accumulation count of specimen A3 show an obvious upward trend, and the activity of the AE characteristics of specimen A3 in the stable damage stage is significantly weaker than that of specimens A2 and A4. By comparing the emission energy peak and accumulation count of specimens A2, A3, and A4, it can be seen that the size of layered defects has the greatest influence on the damage evolution of materials, followed by the depth of defects. The existence of delamination defects in the damage process of the material will lead to the damage of the fiber layer, and the defect part will be damaged in advance under the action of a high strength load. Furthermore, the entire fiber sliding off at the bridging contact, the stripping of the fiber–matrix interface, and the damaged defects influence the surrounding matrix tissue. High-energy AE signals are produced as the fibers break and pull out, reducing the structural strength within the layer. This can be explained by the significant quantity of energy released when the specimen's internal structural damage rapidly deteriorates.

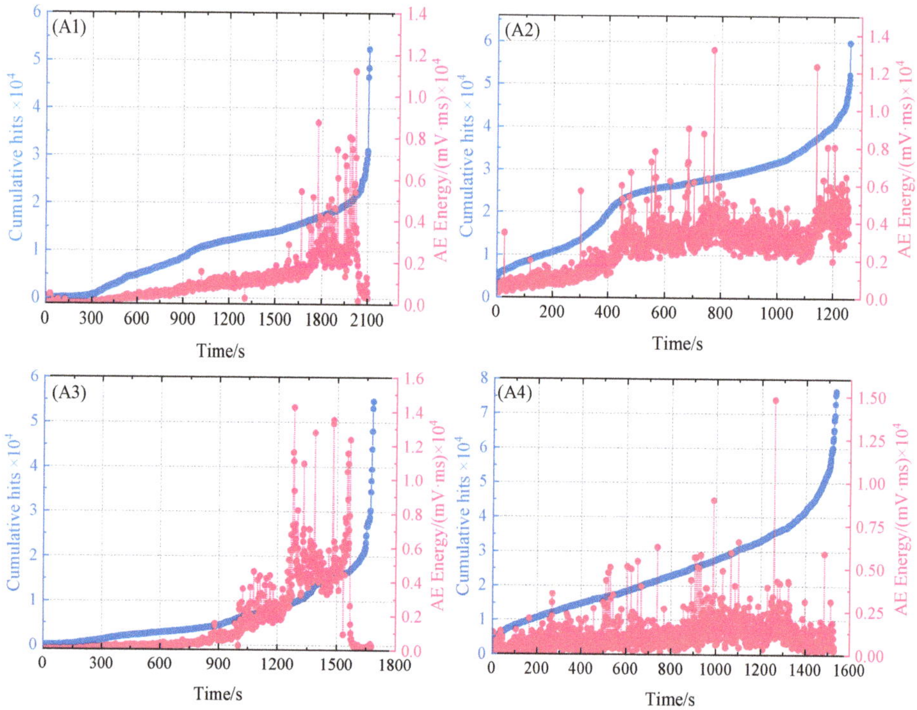

Figure 18. AE accumulation energy and accumulation count distribution over time: specimens A1, A2, A3, and A4.

According to the load curve of layered defective specimens and the AE impact count history diagram in Figure 19, it can be seen that the AE impact count history diagram in the damage evolution process can be divided into three stages: (1) In the first stage, although a small amount of acoustic emission impact number was generated in the initial stage of loading, its fluctuation range was small. According to the analysis of acoustic emission frequency variation in the loading process in Figure 15, it can be seen that the matrix structure of the composite material will produce random distributed micro-cracks and sporadic fiber breakages at the initial loading stage, resulting in the corresponding acoustic emission signals. However, such damage has little effect on the overall stiffness of the specimen, so the acoustic emission impact count presents a gentle linear increase. (2) In the second stage, with the increase in load, the internal structural damage of the composite material is aggravated, and the acoustic emission count increases rapidly at this time. Due to the influence of the stress state of the fiber and the interface, the fiber bundle tends to stretch and rotate in the direction of stretching. At this time, the matrix is subjected not only to tensile stress, but also to extrusion stress. Under the combined action of these stresses, the matrix presents plastic shear failure. This destruction reduces the bearing capacity of the material structure to some extent, resulting in an active AE accumulation count. (3) In the third stage, due to the tensile stress and shear stress of the fiber, the fiber breakage and fiber stripping damage are very severe. Therefore, the acoustic emission signal increases rapidly and reaches the maximum amplitude. Most of the acoustic emission signals in the whole loading process come from the third stage, which is consistent with the mechanical analysis of the specimen.

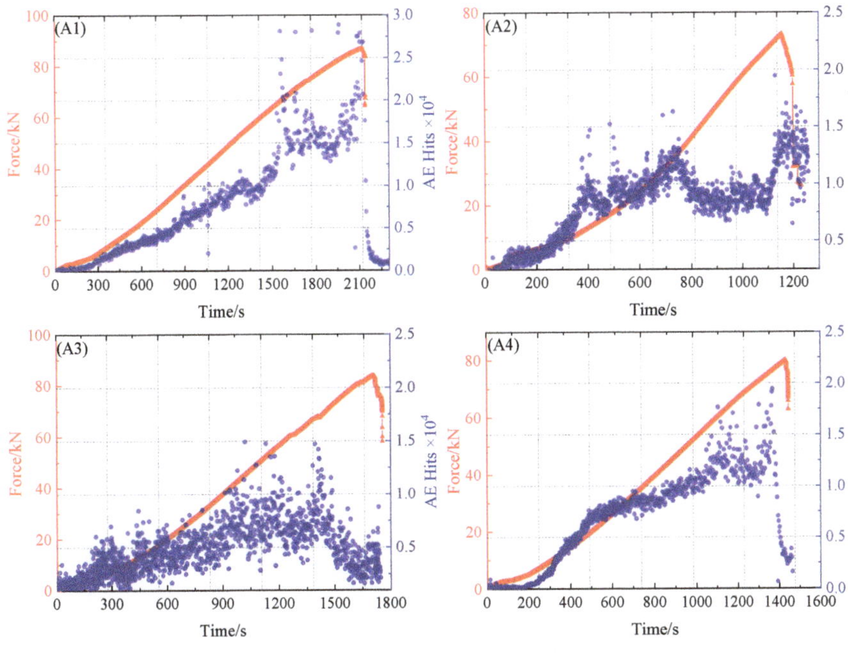

Figure 19. Tensile force and acoustic emission impact history of delamination defect specimens: specimens A1, A2, A3, and A4.

3.3. Acoustic Emission Characteristics of Wrinkle Defects

The aspect ratio parameters of wrinkle defect specimens Z1, Z2, and Z3 are shown in Figure 6. The experimental conditions, parameter settings, and number of sensors in the experimental process are the same as those in the layered defect test method. AE amplitude and energy are selected as AE feature analysis in the tensile test results. The tensile load–AE relative energy curve of wrinkle defect specimens is shown in Figure 18. The fracture failure loads of wrinkle specimens Z1, Z2, and Z3 are 87.6, 76.2, and 82.1 kN, respectively. By comparing Figure 20 (Z1, Z2), when the wrinkle height is the same, wrinkle defects with different aspect ratios have different effects on the mechanical properties of composite material specimens. When the wrinkle aspect ratio is smaller, the mechanical properties of the specimens will decrease more. Compared with Figure 20 (Z2 and Z3), when the wrinkle aspect ratio is the same, the height change in wrinkle defects has different effects on the mechanical properties of composite material specimens. The greater the wrinkle height value, the more the mechanical properties of the specimens decrease.

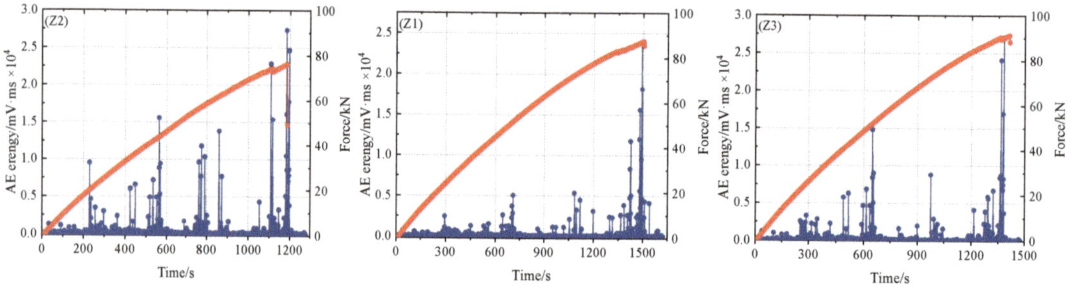

Figure 20. Relationship between load and energy of wrinkle defect.

As can be seen from Figure 18, the overall trend of acoustic emission energy increases with the increase in load. At the early stage of loading, the internal damage of the composite material has no obvious change, and the acoustic emission energy is relatively low. However, with the increase in load, the AE energy shows a gradually increasing trend, which indicates that the damage accumulation and evolution inside the composite material frequently release high-energy characteristics. When the fracture failure load is reached, the instability failure occurs inside the composite structure, and the variation trend of acoustic emission energy reaches a peak.

As shown in Figure 21, the acoustic emission amplitude and accumulation count of different wrinkle defect specimens during the whole loading process can be divided into two stages: damage accumulation and damage destruction. In the initial loading stage, there are some high-amplitude signals, but the cumulative count rises slowly, indicating that the AE signals collected by the sensor in the initial stage contain noise signals [39].

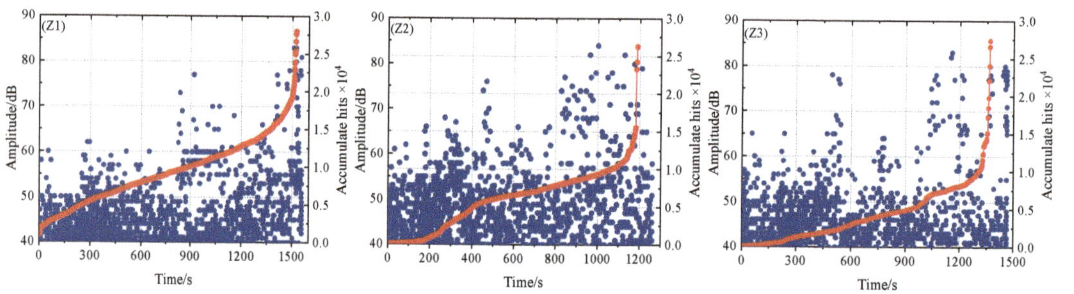

Figure 21. Time history of AE amplitude and accumulated hits of wrinkle defect.

In addition, different from single-medium metal materials, the damage evolution of composite materials is a complex random damage process. In the process of damage evolution, random fiber breakage damage characteristics are usually accompanied by higher energy. Therefore, discrete peak values are associated with random fiber breakage damage within the material. Although random fiber breakage occurs during the damage process, a small amount of fiber breakage does not affect the structural strength of the material.

With the increase in load, some acoustic emission signals with an amplitude higher than 60dB appear, which may be related to the evolution of matrix cracking damage in the rich resin region of wrinkle defects. In the damage accumulation stage, the acoustic emission signals are mainly distributed in the amplitude range of 50~65 dB, and the acoustic emission accumulation count increases steadily. During the damage and failure stage, acoustic emission signals with an amplitude ranging from 50 dB to 65 dB increased significantly, indicating that the matrix damage in the specimen structure was aggravated, the accumulation and count of acoustic emission accelerated, and high-amplitude acoustic emission signals greater than 70 dB continuously appeared until the specimen was broken. In addition, by comparing the variation trend of AE characteristics of different wrinkle defects in Figure 21, the AE accumulation and counting of wrinkle specimens Z2 and Z3 in the damage and failure stage are significantly faster than that of wrinkle specimen Z1, which is related to the aspect ratio of wrinkle defects. The smaller the aspect ratio of the wrinkle, the more the mechanical properties of the material decline and the more obvious the structural deformation during the loading process, resulting in more active acoustic emission characteristics in the damage and failure stage.

3.4. Damage Mechanism Analysis of Specimens with Different Defects

The count of acoustic emission signals and the statistics of energy parameters in the experimental drawing process are shown in Table 3. The AE amplitude distribution of GFRP composites with different defects during the stretching process is generally between 40 and 90 dB. The AE count increases significantly during matrix cracking and fiber stripping, which is related to the active degree of damage characteristics. Acoustic emission energy characteristics can be seen in the phase of amplitude less than 60 dB, and the energy release is very small. However, with the intensification of damage evolution, the AE energy increases gradually, especially in the high amplitude range, and the emission of AE energy is more active. As shown in Table 3, the AE characteristics of different defect types overlap. Compared with delamination specimen A2 and wrinkle specimen Z1, the AE impact percentage of both is 0.5% in the range of 85~90 dB.

Table 3. Acoustic emission events and energy percentages of different defective specimens.

Specimens	Percentage of Acoustic Emission Impacts (%)				Percentage of Acoustic Emission Energy (%)			
	40~55 dB	55~70 dB	70~85 dB	85~90 dB	40~55 dB	55~70 dB	70~85 dB	85~90 dB
A2	65.1	27.3	7.1	0.5	0.3	3.5	25.8	70.4
A3	61.9	29.3	8.1	0.7	0.8	3.8	26.6	68.8
A4	63.5	27.7	8.2	0.6	0.6	4.8	25.4	69.2
Z1	66.1	25.7	7.7	0.5	0.7	4.2	23.2	71.9
Z2	67.9	24.1	7.6	0.4	0.5	3.8	28.2	67.5
Z3	68.9	23.9	6.8	0.4	0.9	5.2	25.7	68.2

In the whole stretching process, with the increase in load, the proportion of the high-amplitude signal in the acoustic emission signal increases. The acoustic emission characteristics vary significantly with the load in the general trend, especially when the composite material specimen is in the instability and failure stage, and the "crackling" emitted by the specimen can be clearly heard, accompanied by an acoustic emission amplitude higher than 70 dB. In the whole stretching process, the AE counts with an amplitude higher than 70dB account for 7.2~8.8% of the total AE events, and the AE energy accounts for 93.9~96.2%. The reason is that in the uniaxial tensile process of the GFRP composite with a uniaxial lamination structure, the force direction of the structure is consistent with the direction of the fiber. However, the matrix inside the material first succumbed to the force and produced plastic deformation. The existence of glass fiber inhibits the crack expansion of the matrix to a certain extent, and the strain energy inside the material accumulates and releases through the interface between the fiber and the matrix, resulting in interface debonding and delamination, which leads to the weakening of the interface properties of the material. The matrix crack, along with the shear failure, causes a large area of fiber stripping and releases a lot of energy. Due to the particularity of a one-way GFRP-laminated structure, fiber is the main load bearer of the structure, and the matrix mainly plays the role of fixing the fiber and transferring load. Therefore, the fracture failure mechanism of materials is mainly caused by a large number of fiber breakages and longitudinal splitting, rather than interface debonding [40].

Obviously, when the unidirectional GFRP composite is subjected to tensile load, the interface properties of the fiber and matrix are weakened, which are manifested as matrix cracking, fiber stripping, and interface delamination. The experimental analysis shows that during the evolution of the tensile damage of GFRP composites, the acoustic emission characteristics change in different stages, and the intensity, energy, and count of acoustic emission signals are different in each stage. In the initial stage of fiber failure, the amplitude, count, energy, and duration of fiber failure are low, while the AE signal of fiber breakage shows high intensity in the instability stage. The whole test process of the AE signal can be divided into three stages: (1) At the stable stage, the friction and damage evolution degree between the fiber matrix inside the specimen is low, and the acoustic emission activity inside the specimen is weak. (2) During the slow rising period, the internal damage of the specimen began to intensify, evolve, and accumulate gradually, and then rose slowly, mainly because the internal damage of the specimen accumulated to a certain extent and released energy, resulting in small-scale macroscopic damage. (3) During the accelerated rise period, when the stress concentration reaches a certain extent, irreversible damage evolution occurs inside the specimen, and the internal damage increases sharply. AE accumulation and count are characterized by accelerated release, indicating that the damage enters the instability failure expansion.

The metallographic microscope of the fracture position of the layered defect specimen is shown in Figure 22. The damage mode and failure form of the sample can be clearly observed through a scanning electron microscope. Figure 22(A1) shows the fracture micromorphology of specimen A1, from which fiber stripping and fracture can be clearly seen. Because A1 is a no-defect specimen, the fiber inside the structure has high bonding strength with the matrix, and the stress is evenly distributed along the fiber direction under the action of load. Therefore, there is no obvious bending deformation of glass fiber after sample A1 fracture. Figure 22(A2) is the microtopography of the fracture of the A2 specimen, from which it can be observed that a large area of fiber breakage failure is observed, and some fibers are not sufficiently bonded to the matrix, resulting in the fracture of nearby fibers. Furthermore, a large number of fiber breakages distribute a portion of the load to the adjacent fibers, causing an unequal distribution of stress, as indicated by the random distribution of fiber breakages. Figure 22(A3) shows the fracture micromorphology of the A3 specimen. The fibers separate from the matrix as a result of the interface between the local fibers and the matrix breaking. Due to the existence of delamination defects, the strength of the interlayer structure is reduced, and numerous fibers are pushed out when the matrix's adherence to the fiber disintegrates. Figure 22(A4) shows the fracture microstructure of A4. Due to the large size of the delamination defect of this specimen, the bearing capacity of the fiber and the binding effect between the fiber and the matrix are the primary determinants of the internal stress distribution of the material under load. The stretching of the fiber bundle itself is primarily responsible for the distortion of the fiber when it starts to experience stress. However, the fiber bundle and matrix will slip and dislocate when the external force is raised further, resulting in severe fiber deformation. At this point, the material structure's bearing capacity significantly decreases due to large-area fiber breakage and interface delamination, ultimately leading to failure.

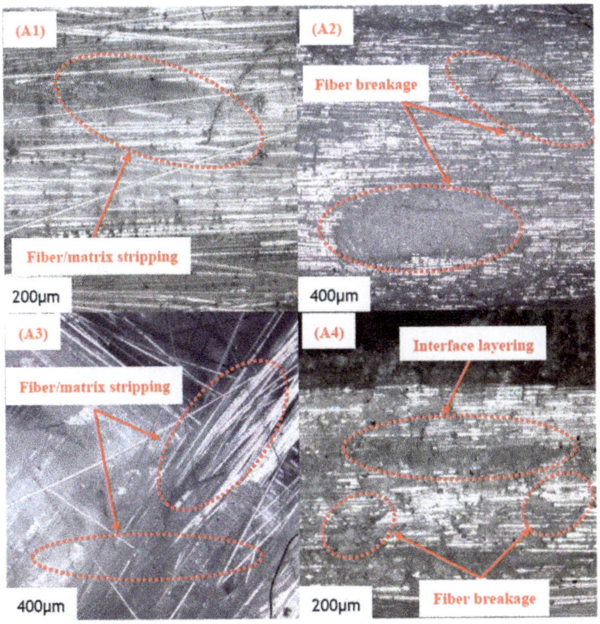

Figure 22. Micrograph of delamination defect specimens after fracture.

The metallographic microscope of the fracture position of the wrinkle defect specimen in the experimental results is shown in Figure 23. Different from the layered defects, the fibers are bent at a certain angle due to the existence of wrinkle defects, and the fibers are scattered and accompanied by distortion after the specimen is broken. Fiber breakage occurs at the interlayer of the coated fiber, and interface debonding occurs as a result of the debonding and sliding between the fiber and the matrix, as shown in Figure 23(Z1). Considering the mechanical characteristics of wrinkle defects, the rich resin region needs to bear a combination of axial tensile, compressive, and bending loads under the action of loads [41]. Therefore, it can be seen from Figure 23(Z2) that matrix cracks are more likely to accumulate in the rich resin region of wrinkle defects, and the stress concentration of fiber bedding bending is more likely to lead to interlayer cracking damage.

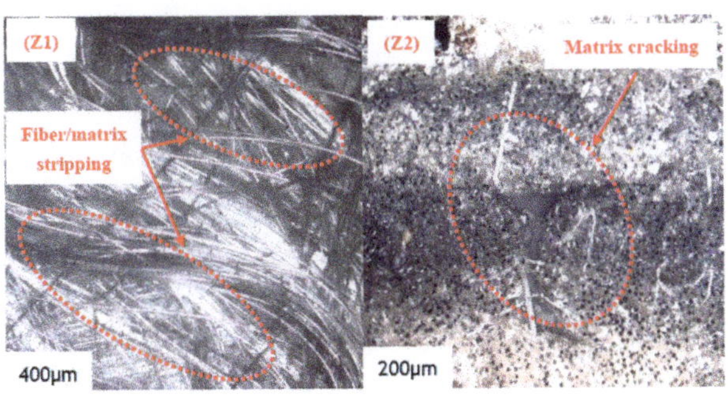

Figure 23. Micrograph of wrinkle defect specimens after fracture.

4. Conclusions

In view of the typical layered defects and wrinkle defects in the main spar structure of wind turbine blades, this paper uses the vacuum perfusion process (VARI) to fabricate GFRP composite materials with artificial prefabrication defects, and describes the manufacturing process and artificial prefabrication defect method of the main spar composite materials of wind turbine blades in detail. It provides an experimental reference for subsequent research on the damage evolution identification of wind power blade girder composites.

Acoustic emission experiments are used to analyze the mechanical properties and acoustic emission characteristics of wind turbine blade girder composites with various flaws under tensile loading. In addition to affecting the mechanical characteristics of composite materials, the presence of engineered defects will also impact the damage progression of material. For delamination defects and wrinkle defects, the different size of defects and the position of the layer will affect the damage evolution trend of the material.

The characteristic frequencies of the damage modes of GFRP composites were identified by K-means clustering analysis of acoustic emission parameters. The types of damage modes of wrinkle defects and delamination defects are the same, and the range of characteristic frequency is roughly the same. It is worth noting that the existence of defects will lead to certain fluctuations in the range of cluster center frequency and characteristic frequency, which should be paid attention to in the health monitoring of damage characteristics. The fracture failure mechanism of the material is mainly caused by a large number of fiber breakages and longitudinal splitting, rather than interface debonding.

Author Contributions: Data curation, Y.Z.; Formal analysis, C.C.; Funding acquisition, B.Z.; Methodology, S.X.; Software, S.X.; Writing—original draft, Y.Z.; Writing—review and editing, T.M. and B.Z. All authors have read and agreed to the published version of the manuscript.

Funding: This work was supported by the National Natural Science Foundation of China (Grant No. 51575361).

Institutional Review Board Statement: Not applicable.

Data Availability Statement: The original contributions presented in the study are included in the article, further inquiries can be directed to the corresponding author.

Conflicts of Interest: The authors declare no conflicts of interest.

References

1. Kumar, Y.; Ringenberg, J.; Depuru, S.S.; Devabhaktuni, V.K.; Lee, J.W.; Nikolaidis, E.; Andersen, B.; Afjeh, A. Wind energy: Trends and enabling technologies. *Renew. Sustain. Energy Rev.* **2016**, *53*, 209–224. [CrossRef]
2. Martin, R.W.; Sabato, A.; Schoenberg, A.; Giles, R.H.; Niezrecki, C. Comparison of nondestructive testing techniques for the inspection of wind turbine blades spar caps. *Wind. Energy* **2018**, *21*, 980–996. [CrossRef]
3. Liu, W.; Tang, B.; Han, J.; Lu, X. The structure healthy condition monitoring and fault diagnosis methods in wind turbines: A review. *Renew. Sustain. Energy Rev.* **2015**, *44*, 466–472. [CrossRef]
4. Chehouri, R.; Younes, A.; Ilinca, A.; Perron, J. Review of performance optimization techniques applied to wind turbines. *Appl. Energy* **2015**, *142*, 361–388. [CrossRef]
5. González-González, A.; Cortadi, A.J.; Galar, D.; Ciani, L. Condition monitoring of wind turbine pitch controller: A maintenance approach. *Measurement* **2018**, *123*, 80–93. [CrossRef]
6. Broberg, P.H.; Lindgaard, E.; Thompson, A.J.; Belnoue, J.P.H.; Hallett, S.R.; Bak, B.L. Parametric study on the effect of material properties, tool geometry, and tolerances on preform quality in wind turbine blade manufacturing. *Compos. Struct.* **2024**, *344*, 118324. [CrossRef]
7. Movaghghar, A.; Vov, G. A method of estimating wind turbine blade fatigue life and damage using continuum damage mechanics. *Int. J. Damage Mech.* **2012**, *21*, 810–821. [CrossRef]
8. Zhao, L.; Gong, Y.; Zhang, J. A survey on delamination growth behavior in fiber reinforced composite laminates. *Acta Aeronaut. Astronaut. Sin.* **2019**, *40*, 10–7527.
9. Hameed, Z.; Hong, Y.S.; Cho, Y.M. Condition monitoring and fault detection of wind turbines and related algorithms: A review. *Renew. Sustain. Energy Rev.* **2009**, *13*, 1–29. [CrossRef]
10. Yu, F.; Zhou, B.; Xin, W. Fatigue life prediction for the main spar with wrinkle defects of a wind turbine blade. *Fatigue Fract. Eng. Mater. Struct.* **2021**, *44*, 2764–2780. [CrossRef]

11. Marin, J.C.; Barroso, A.; Paris, F. Study of fatigue damage in wind turbine blades. *Eng. Fail. Anal.* **2009**, *16*, 656–668. [CrossRef]
12. Du, Y.; Zhou, S.; Jing, X.; Peng, Y.; Wu, H.; Kwok, N. Damage detection techniques for wind turbine blades: A review. *Mech. Syst. Signal Process.* **2020**, *21*, 106445. [CrossRef]
13. Overgaard, L.; Lund, E.; Thomsen, O.T. Structural collapse of a wind turbine blade. Part A: Static test and equivalent single layered models. *Compos. Part A Appl. S.* **2010**, *41*, 257–270. [CrossRef]
14. He, Y.; Li, M.; Meng, Z.; Chen, S.; Huang, S.; Hu, Y.; Zou, X. An overview of acoustic emission inspection and monitoring technology in the key components of renewable energy systems. *Mech. Syst. Signal Process.* **2021**, *148*, 107146. [CrossRef]
15. Ding, S.; Yang, C.; Zhang, S. Acoustic-signal-based damage detection of wind turbine blades—A review. *Sensors* **2023**, *23*, 4987. [CrossRef]
16. Zhang, Y. Study on Damage Evolution in Wind Turbine Blades Based on Acoustic Emission Signal Processing. *Shenyang Univ. Technol.* **2020**, *6*, 36–37.
17. Nikbakht, M.; Yousefi, J.; Hossein, H.T. Delamination evaluation of composite laminates with different interface fiber orientations using acoustic emission features and micro visualization. *Compos. Part B-Eng.* **2017**, *113*, 185–196. [CrossRef]
18. Loutas, T.H.; Kostopoulos, V. Health monitoring of carbon/carbon, woven reinforced composites. Damage assessment by using advanced signal processing techniques. Part I: Acoustic emission monitoring and damage mechanisms evolution. *Compos. Sci. Technol.* **2009**, *69*, 265–272. [CrossRef]
19. Bourchak, M.; Farrow, I.R.; Bond, I.P. Acoustic emission energy as a fatigue damage parameter for CFRP composites. *Int. J. Fatigue* **2007**, *29*, 457–470. [CrossRef]
20. Michalcova, L.; Kadles, M. Crack growth monitoring of CFRP composites loaded in different environmental conditions using acoustic emission method. *Procedia Eng.* **2015**, *114*, 86–93. [CrossRef]
21. Xu, D.; Liu, P.; Chen, Z. Damage mode identification and singular signal detection of composite wind turbine blade using acoustic emission. *Compos. Struct.* **2021**, *255*, 2954–2966. [CrossRef]
22. Tang, J.; Soua, S.; Mares, C. An experimental study of acoustic emission methodology for in service condition monitoring of wind turbine blades. *Renew. Energ.* **2016**, *99*, 170–179. [CrossRef]
23. Zhou, B.; Zhang, Y.; Chen, C. Acoustic emission detection of fatigue cracks in wind turbine blades based on blind deconvolution separation. *Fatigue Fract. Eng. Mater. Struct.* **2017**, *40*, 959–970.
24. Zhang, Y.; Zhou, B.; Yu, F.; Chen, C. Cluster analysis of acoustic emission signals and infrared thermography for defect evolution analysis of glass/epoxy composites. *Infrared Phys. Technol.* **2021**, *112*, 581–595. [CrossRef]
25. *GB/T25383*; Wind Turbine Generator System—Rotor Blades. Chinese National Standard: Beijing, China, 2010.
26. van Oosterom, S.; Allen, T.; Battley, M.; Bickerton, S. An objective comparison of common vacuum assisted resin infusion processes. *Compos. Part A* **2019**, *125*, 105528. [CrossRef]
27. Liu, Q.; Zhao, L.; Peng, Z.; Chen, Z. Effect of vacuum leakage process deficiency on quality and performance of VARI molded composite. *Acta Mater. Compos. Sin.* **2014**, *31*, 1046–1052.
28. *ASTM/D3039*; Standard Test Method for Tensile Properties of Polymer Matrix Composite Materials. Información De Contacto: Norwood, MA, USA, 2008.
29. Monti, A.; El Mahi, A.; Jendli, Z. Mechanical behaviour and damage mechanisms analysis of a flax fiber reinforced composite by acoustic emission. *Compos. Part A Appl. S* **2016**, *90*, 100–110. [CrossRef]
30. Favretto Cristini, N.; Hégron, L.; Sornay, P. Identification of the fragmentation of brittle particles during compaction process by the acoustic emission technique. *Ultrasonics* **2016**, *67*, 178–189. [CrossRef]
31. Huang, X.; Wang, B.; Yang, C. Evaluating Damage Evolution of Three-dimension Needled C/SiC Composite Based on Acoustic Emission Signal Analysis. *China J. Inorg. Mater.* **2018**, *33*, 609–616.
32. Kumar, C.S.; Arumugam, V.; Santulli, C. Characterization of indentation damage resistance of hybrid composite laminates using acoustic emission monitoring. *Compos. Part B-Eng.* **2017**, *111*, 165–178. [CrossRef]
33. Morizet, N.; Godin, N.; Tang, J.; Maillet, E.; Fregonese, M.; Normand, B. Classification of acoustic emission signals using wavelets and random forests: Application to localized corrosion. *Mech. Syst. Signal Process.* **2016**, *70*, 1026–1037. [CrossRef]
34. Mohammadi, R.; Najfabadi, M.A.; Saeedifar, M.; Yousefi, J.; Minak, G. Correlation of acoustic emission with finite element predicted damages in open-hole tensile laminated composites. *Compos. Part B-Eng.* **2017**, *108*, 144–152. [CrossRef]
35. Roundi, W.; Abderrahim, E.M.; Abdellah, E.G.; Rebiere, J.L. Acoustic emission monitoring of damage progression in Glass/Epoxy composites during static and fatigue tensile tests. *Appl. Acoust.* **2018**, *132*, 124–134. [CrossRef]
36. Tang, J.; Soua, S.; Mares, C.; Gan, T. A pattern recognition approach to acoustic emission data originating from fatigue of wind turbine blades. *Sensors* **2017**, *17*, 2507. [CrossRef] [PubMed]
37. Jia, H.; Zhang, L.; Wang, J. Damage pattern recognition of wind turbine blade composite material based on acoustic emission technology. *Chin. J. Renew. Energy Resour.* **2022**, *40*, 67–72.
38. Arumugam, V.; Saravanakumar, K.; Santulli, C. Damage characterization of stiffened glass-epoxy laminates under tensile loading with acoustic emission monitoring. *Compos. Part B-Eng.* **2018**, *147*, 22–32. [CrossRef]

39. Zhao, G.; Zhang, L.; Tang, C. Clustering of AE signals collected during torsional tests of 3D braiding composite shafts using PCA and FCM. *Compos. Part B-Eng.* **2019**, *161*, 547–554. [CrossRef]
40. Zhou, W.; Zhao, W.; Zhang, Y.; Ding, J. Cluster analysis of acoustic emission signals and deformation measurement for delaminated glass fiber epoxy composites. *Compos. Struct.* **2018**, *195*, 349–358. [CrossRef]
41. Leong, M.; Overgaard, L.C.; Thomsen, O.T.; Lund, E.; Daniel, I.M. Investigation of failure mechanisms in GFRP sandwich structures with face sheet wrinkle defects used for wind turbine blades. *Compos. Struct.* **2012**, *94*, 768–778. [CrossRef]

polymers

MDPI

Article

Research on Interlayer Toughening and Damage Detection of Laser-Induced Graphene and Short Kevlar Fibers Aramid Fiber/Epoxy Resin Composites

Baolai Wang, Weidong Tian *, Chao Wang and Qi Wang

Yantai Research Institute, Harbin Engineering University, Yantai 264000, China; wangbaolai1979@sina.com (B.W.); 18066241035@163.com (C.W.); 13314486617@163.com (Q.W.)
* Correspondence: tian20010127@126.com; Tel.: +86-178628727789

Abstract: The poor interlaminar fracture toughness is a critical limiting factor for the structural applications of aramid fiber/epoxy resin composites. This study investigates the effects of laser-induced graphene (LIG) and short Kevlar fibers on the interfacial toughness and damage detection of aramid composite materials. Mode II tests and tensile tests were conducted to evaluate mechanical properties and damage detection using the piezoresistive characteristics of LIG. The results indicate that LIG combined with short Kevlar fibers significantly enhances the interfacial toughness of the composites, achieving a 381.60% increase in initial Mode II fracture toughness. Although LIG reduced the tensile strength by 14.02%, the addition of short Kevlar fibers mitigated this effect, preserving the overall mechanical performance. Scanning electron microscopy (SEM) analysis revealed enhanced toughening mechanisms, including increased surface roughness, altered crack propagation paths, and fiber bridging. Additionally, LIG enabled real-time damage monitoring, showing a significant increase in resistance upon delamination or crack propagation and a marked increase in resistance upon the tensile fracture. This research indicates that the synergistic effects of LIG and short Kevlar fibers not only enhance the interlaminar toughness of aramid composites but also provide a novel strategy for effective damage detection in fiber-reinforced materials.

Keywords: laser-induced graphene; short Kevlar fibers; aramid fiber/epoxy resin composites; fracture toughness; tensile strength; damage detection

1. Introduction

Aramid fibers possess excellent flame resistance, heat resistance, high toughness, impact resistance, and fatigue resistance. High-performance composites made from aramid fibers are widely used in various fields, including sporting goods, aerospace, marine applications, and ballistic protection, particularly in applications such as helmets and armored vehicles [1,2]. However, the smooth and inert surface of aramid fibers results in poor adhesion with the resin matrix, which adversely affects the interlaminar mechanical properties and the overall load-bearing capacity of the structure, thereby limiting the practical application of aramid fibers [3–6]. Therefore, it is crucial to conduct in situ damage detection and interlaminar toughening studies for aramid fiber-reinforced composites.

Over the past 30 years, extensive research has been conducted on interlaminar toughening by scholars. Sela et al. [7] enhanced the fracture toughness of structures by doping second phases such as ceramics and rubber to toughen the resin. Subsequently, resin toughening techniques have continuously developed, with thermoplastic resins, carbon nanotubes, and thermoplastic nanofibers utilized as reinforcing materials for toughening modifications [8–10]. However, these methods can adversely affect the mechanical properties of the resin matrix. Additionally, Z-fiber toughening techniques such as Z-pin, 3D weaving, and stitching technology have been proposed and applied to delay delamination in composites [11–13]. Nonetheless, these approaches face challenges such as high processing difficulty, expensive manufacturing costs, and potential impacts on the in-plane mechanical properties of the structure. In response to the numerous issues associated with these methods, many researchers have shifted their focus to interlaminar toughening, which includes particle toughening, film toughening, and short fiber toughening. Minh Hoang Nguyen [14] introduced toughening particles into the interlayer of carbon fiber composites, resulting in an approximately 100% increase in the ultimate shear strength. Salvatore Giacomo Marino et al. [15] inserted 30% and 60% thermoplastic acrylonitrile butadiene styrene (ABS) overlapping films between the layers of carbon/glass–epoxy composites. The pseudoplastic strain increased to 0.67% and 0.74%, respectively, while the carbon fiber volume fraction decreased by 12.7% and 16.5%, demonstrating improved interlaminar Mode II fracture toughness. Both methods can improve interfacial toughness; however, they present significant processing challenges.

Compared to other toughening methods, short fiber toughening only requires spreading short fibers between the panels, which has garnered significant attention from researchers due to its operational simplicity. Scholars have investigated the effects of inserting various types and contents of short fibers on interlaminar toughness and in-plane mechanical properties. Dong Quan et al. [16] found that incorporating ultraviolet-irradiated poly-etherether-ketone (PEEK) fibers into the interlayers of carbon fiber composites significantly improved Mode I fracture toughness from 185 J/m^2 to 726 J/m^2. Melike Kilicoglu et al. [17] demonstrated that inserting polyamide 6/polycaprolactone (PA6/PCL) mixed fibers between the layers of carbon fiber composites could markedly enhance Mode I fracture toughness, particularly achieving optimal results at a 60/40 ratio, where the initiation and propagation toughness increased by 69% and 59%, respectively. Park et al. [18] indicated that surface-modified short Kevlar fibers significantly increased the Mode II fracture toughness of carbon fiber composites with increasing crack length, from around 1.0 kJ/m^2 to 3.0–4.0 kJ/m^2. Zheng Hao et al. [19] concluded that short Kevlar fibers exhibited the best toughening effect, with the highest Mode I fracture toughness reaching 1.336 kJ/m^2, an increase of 94.8%, determining that the fiber length was 6 mm, while finding that short fiber insertion did not affect in-plane mechanical properties. A comprehensive analysis reveals that current short fiber toughening technology is primarily applied to carbon fiber-reinforced epoxy composites, with short Kevlar fibers demonstrating the most effective toughening performance. In contrast, aramid fiber composites face limitations due to inadequate surface activation, which hinders effective bonding with the resin matrix. Therefore, it is crucial to modify the surface of aramid fibers to enhance roughness and improve their bonding efficacy.

Currently, the modification methods for aramid fibers primarily include physical and chemical modifications. These methods involve the application of plasma or the introduction of specific chemical agents to improve interfacial properties. However, these approaches may damage aramid fibers and reduce their tensile strength [20,21]. In 2014, LIN [22] first employed a carbon dioxide infrared laser system to directly generate laser-induced graphene (LIG) on the surface of commercial polyimide films (PI). This method has garnered significant attention due to its simplicity and the fact that it does not require chemical modification. Jalal Nasser et al. [23] demonstrated that transferring LIG to the surface of carbon fibers and found that the LIG layer generated at 200 DPI pulse density increased Mode I and Mode II fracture toughness by 41% and 69%, respectively, while maintaining tensile performance, confirming the effectiveness of LIG in improving interface toughness. LIG also exhibits excellent piezoresistive properties and high electrical conductivity, enabling resistance-based sensing methods for the detection of strain or damage in insulating materials. Typically, an electric current is applied to measure voltage, allowing resistance changes to be calculated based on Ohm's law; a sudden increase in resistance is detected when the material sustains damage. LA Groo et al. [24] integrated LIG into glass fibers to monitor damage accumulation and propagation during tensile fatigue loading through resistance changes. This resistance-based sensing differs from using external sensors or embedded fiber sensors as it does not add to the structural volume or weight [25,26]. Subsequently, Chyan et al. [27] discovered that LIG could be directly generated on the surface of aramid fibers, facilitating its application in aramid fiber composites. Lori Anne Groo et al. [28] generated graphite structures directly on the surface of aramid fibers, demonstrating that these structures could enable in situ monitoring of tensile and bending strains by measuring resistance. Given the critical ballistic performance of aramid fibers, Kelsey Steinke et al. [29] utilized LIG for embedded impact sensing in aramid fiber-reinforced composites, employing resistance measurements to obtain information about the impact severity. Since the quality of LIG is influenced by factors such as laser power, scanning speed, and pulse frequency, Iman Naseri et al. [30] investigated the effects of these laser parameters on the quality of LIG on the surface of aramid fibers, indicating a significant impact of laser settings on LIG quality. However, while LIG has been shown to enhance interlaminar fracture toughness and serve as a resistance sensor, the effects of short Kevlar fibers combined with LIG on aramid fiber-reinforced composites remain unexplored.

This study investigates the effects of LIG and short Kevlar fibers on the mechanical properties of aramid fiber/epoxy resin composites while employing LIG for damage tracking during loading tests. Graphene was generated on the surface of aramid fibers using laser induction, and short Kevlar fibers were introduced into the interlayers. The end-notched flexure (ENF) test was utilized to evaluate the impact of these modifications on the interfacial fracture toughness of the composites, determining the capability of LIG to track delamination damage. Scanning electron microscopy (SEM) was employed to observe the fracture surfaces, revealing the toughening mechanisms involved. Additionally, tensile tests were conducted to analyze the influence of short Kevlar fibers and LIG on the in-plane mechanical properties of the composites while also assessing the ability of LIG to monitor damage induced by tensile loading.

2. Materials and Methods

2.1. Materials

In this study, a 1500D meta-aramid woven aramid fiber fabric produced by DuPont was selected as the panel material, with an areal density of 200 g/m^2. Due to its high strength and abrasion resistance, this material is widely used in various applications. The smooth surface of the aramid fibers means that the adhesive strength at the fiber-matrix interface primarily depends on the roughness of the fiber surface after treatment and the bridging effect of the inserted short fibers.

The epoxy resin system used in this study was prepared by mixing Z105 epoxy resin (from West System) with slow hardener 206 (from West System) in a ratio of 5:1, ensuring thorough impregnation of the aramid fiber fabric surface. To ensure adequate bonding between the short Kevlar fibers and the aramid fiber panel, it is imperative to use sufficient epoxy resin to guarantee that the short fibers are completely wetted within the voids between the panel fibers.

2.2. Short Aramid Fibers Preparation

The short fibers utilized in this study are Kevlar 49, produced by DuPont. According to the research conducted by Zheng Hao et al. [19], the maximum fiber bridging region and the best toughening effect occur when the length of the short fibers is 6 mm. Therefore, Kevlar 49 fibers were cut to a length of 6 mm, with a length tolerance maintained within ±0.5 mm. The chopped short Kevlar fiber bundles were then placed in a blender equipped with dull blades and mixed at high speed until the fibers transitioned from bundles to a uniformly dispersed fluff-like consistency. Finally, the short Kevlar fibers were thoroughly impregnated with the mixed epoxy resin to form a uniformly distributed short fiber film, based on the desired fiber density. For instance, in this study, short Kevlar fibers treated with epoxy resin were laid between the aramid fiber panels, resulting in a short Kevlar fiber film with a mass per unit area of 6 g/m^2 after curing. The preparation process of the short Kevlar fiber film is illustrated in Figure 1.

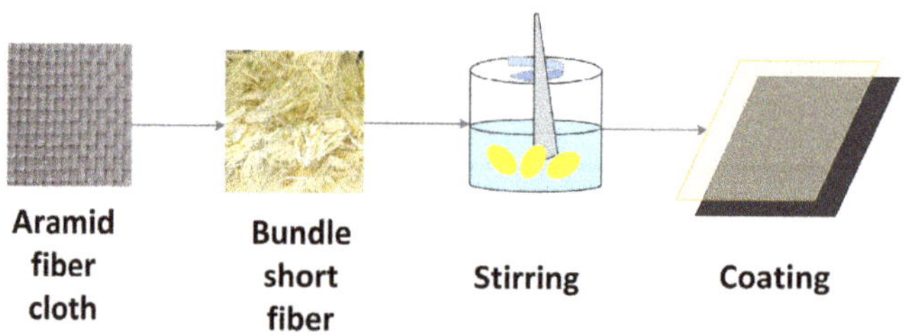

Figure 1. Preparation process of short Kevlar fiber film.

2.3. Generation of LIG on Aramid Surface and Characterization

According to the research by Kelsey Steinke et al. [29], graphene was generated on the surface of aramid fibers. Prior to laser treatment, aramid fabric sheets were cut to dimensions of 300 mm × 300 mm, with a cutting tolerance maintained within 1 mm. The edges were then hardened and fixed to prevent the shedding of fine fibers. The cut aramid fabric was first cleaned in a 95% ethanol solution using electromagnetic waves for 15 min, followed by drying in an oven at 90 °C for 60 min to remove surface impurities and residual moisture. A CO_2 infrared laser engraving machine (80W-4060, from Ketai Laser, Liaocheng, China) was employed to generate graphene on the surface of the aramid fibers. The laser processing was conducted in raster mode, with a laser power set to 12 W and a scanning speed of 300 mm/s. The distance between the laser output and the aramid fiber surface was maintained at 14 mm. To ensure stable laser output, the laser water-cooling was employed for water circulation temperature control, maintaining the operating temperature of the laser generator within the range of 20–23 °C. This process facilitated the uniform generation of graphene arrays on the surface of the aramid fibers. The preparation process of graphene on the aramid fiber surface is illustrated in Figure 2. After the laser treatment, a Renishaw 2000 high-precision Raman spectrometer, operating at a wavelength of 532 nm, was employed to analyze the surface of the aramid fibers before and after treatment.

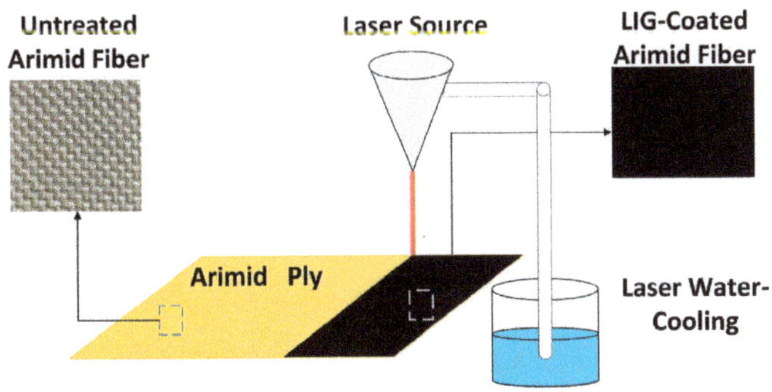

Figure 2. Generation of graphene on the surface of aramid fibers.

2.4. Model II Specimen Fabrication

The Mode II fracture toughness of the LIG-coated and short Kevlar fibers that toughen aramid fiber-reinforced composites were assessed following the ASTM D7905. A total of 24 LIG-coated aramid fiber prepregs were fully impregnated with a resin mixture containing short Kevlar fibers. After impregnation, 12 prepregs were arranged with the LIG-coated facing upwards, while the remaining 12 prepregs were oriented with the LIG coating facing downwards, achieving a symmetrical layout. To introduce a pre-crack within the laminate, a 70 mm long Teflon film was inserted between the 12th and 13th layers, resulting in an initial delamination length of 50 mm. After impregnation, the aramid fiber fabric is placed into a 300 mm × 300 mm mold and cured under a constant pressure of 0.3 MPa. The mold is heated from room temperature to 60 °C in 20 min, held at this temperature for 4 h, and then allowed to cool to room temperature naturally. After the epoxy resin fully cured, the 6 mm thick laminate was cut into 190 mm in length and 25 mm in width. A thin layer of white paint was sprayed on the edges of the specimens to facilitate visual observation of the crack initiation and propagation. For in situ resistance measurements, silver paint rings were affixed to each side of the laminate and connected to 33-gauge copper wires with silver paint and electrical tape. This methodology establishes a robust experimental basis for examining the toughening mechanisms in composite materials.

2.5. Mode II Testing

The completed specimens were tested using an Instron universal testing machine under quasi-static loading at a rate of 5 mm/min. A three-point bending setup with a 5 kN load cell recorded data, with a span of 70 mm and a crack tip distance of 50 mm from the left support. Loading was ceased once the crack propagated to the load center. For the assessment of failure behavior, the critical strain energy release rate G_C was predominantly used to characterize the interlaminar toughness of the composite materials, while the Mode II delamination fracture toughness was calculated using the following equation [31]:

$$G_{IIC} = \frac{9P\delta a^2}{2W(2L^3 + 3a^3)} \times 10^3 \quad (1)$$

where P is the critical load; δ is the corresponding critical displacement; a is the initial effective crack length; W is the specimen width; and $2L$ is the span length.

During the delamination testing, a 3 mA DC current was applied via a power supply (from Kuaiqu Electron, Shenzhen, China) to the copper wire connected to the outermost silver paint ring. A digital multimeter measured the voltage at the innermost silver paint ring. A schematic of the loading and in situ monitoring setup is presented in Figure 3. After testing, fracture surfaces were analyzed using a TESCAN CLARA (from Tescan,

Brno, The Czech Republic) scanning electron microscope (SEM). A representative fracture surface of the panel was selected for gold coating to improve conductivity. The analysis was conducted with an acceleration voltage of 20 kV, and magnification varied from 20× to 1000×.

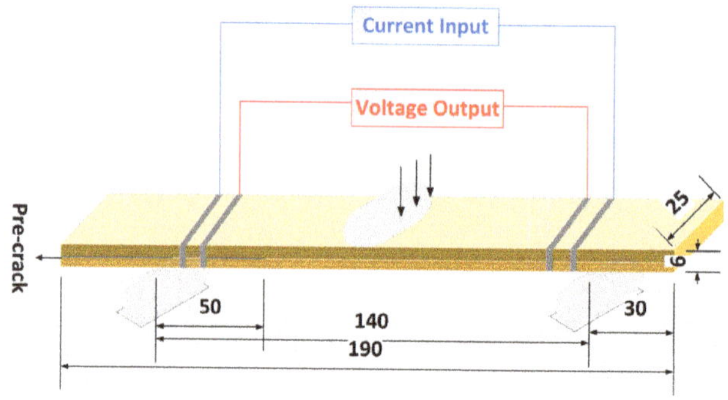

Figure 3. Schematic diagram of Mode II specimens.

2.6. Tensile Specimen Fabrication

Three layers of LIG-coated aramid fiber fabric were thoroughly impregnated using a mixed epoxy resin containing short Kevlar fibers oriented with the LIG side up, using the same molding process as the Mode II specimens. After the epoxy resin had fully cured, the 0.9 mm thick laminates were cut into specimens measuring 160 mm in length and 12.25 mm in width according to ASTM D3039. A layer of silver paint was applied to the edges of each specimen, and 33-gauge copper wire was connected to facilitate the measurement of electrical resistance changes during tensile testing.

2.7. Tensile Testing

The same loading frame and loading rate used in the Mode II delamination tests were employed until specimen failure occurred. During the tensile testing, resistance measurements were simultaneously taken using the same method as in the delamination tests. A schematic diagram of the loading and in situ monitoring of the tensile specimens is shown in Figure 4.

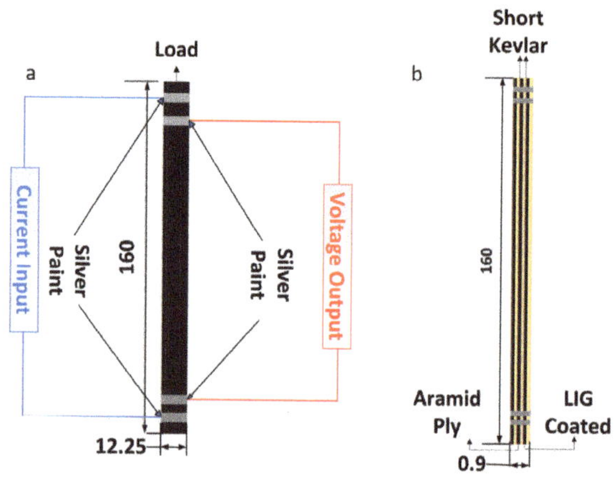

Figure 4. (**a**) Schematic diagram of the tensile specimens (front view); (**b**) Schematic diagram of tensile specimens (side view).

3. Results and Discussion

3.1. LIG Chemical Characterization

Raman spectroscopy plays a crucial role in the analysis of graphene due to its ability to accurately characterize the microstructure and chemical composition of the material [32,33]. In particular, Raman spectroscopy effectively identifies key features of graphene, including the D peak, G peak, and 2D peak. The D peak (approximately 1350 cm^{-1}) reflects structural defects in graphene, arising from the breathing modes of hybridized carbon atoms. The G peak (approximately 1582 cm^{-1}) represents in-plane vibrations of carbon atoms, indicating the strength and integrity of the graphite structure, and is considered the primary characteristic peak of graphene. The intensity ratio between the D peak and the G peak (I_D/I_G) serves as an important indicator of the defect density and crystallinity of graphene. The 2D peak (approximately 2700 cm^{-1}), also known as the G′ peak, represents the stacking arrangement of carbon atoms between layers and is independent of structural defects; its full width at half maximum is primarily used to assess the number of graphene layers. Consequently, by analyzing these three characteristic peaks in Raman spectra, we can determine whether graphene has been generated on the surface of aramid fibers and evaluate the quality of the produced graphene.

Figure 5 presents the Raman spectra of the aramid fiber surface before and after treatment. A comparison reveals that the Raman spectrum prior to treatment (Figure 5b) displays several characteristic peaks of the aramid fibers at 1182 cm^{-1}, 1274 cm^{-1}, 1327 cm^{-1}, 1515 cm^{-1}, 1599 cm^{-1}, and 1647 cm^{-1} [34]. In contrast, the Raman spectrum after laser induction (Figure 5a) clearly shows the D band, G band, and 2D band of graphene, located at 1325 cm^{-1}, 1563 cm^{-1}, and 2679 cm^{-1}, respectively. Moreover, I_D/I_G indicates that the graphene generated through laser induction exhibits a high degree of graphitization, minimal structural defects, and a high level of crystallinity. Therefore, under these laser parameters, high-quality graphene can be produced on the surface of aramid fibers.

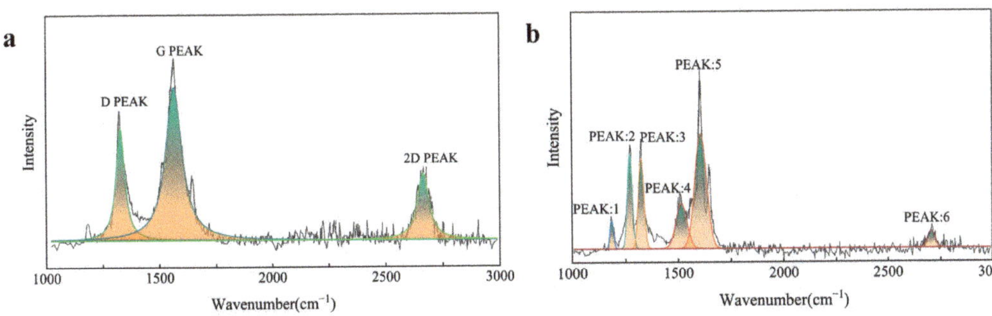

Figure 5. (**a**) Raman spectrum of aramid fibers LIG-coated; (**b**) Raman spectrum of untreated aramid fibers.

3.2. Interlaminar Fracture Toughness

Due to the susceptibility of composite laminates to interlaminar fracture failure, the application of LIG onto the surface of aramid fibers can significantly enhance the Mode II fracture toughness. Furthermore, its piezoresistive properties enable effective monitoring of delamination failure in the laminates.

This study conducted a detailed analysis of five sets of experimental data, selecting three sets with outstanding results for further evaluation. The average load and average fracture toughness were calculated for these data sets. Figure 6a presents the average load displacement curves for the four types of specimens. It is evident from the figure that all ENF specimens exhibit a similar trend; they demonstrate a linear relationship in the elastic phase before delamination occurs. As the displacement load increases, the curves begin to show nonlinearity, indicating the onset of pre-crack delamination, with the cracks propagating relatively slowly. Once the load reaches its peak, rapid crack propagation occurs, leading to a decrease in load until complete failure of the specimen.

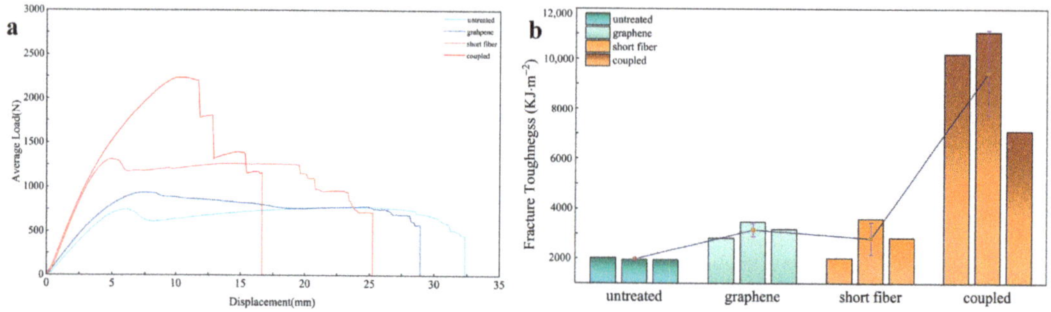

Figure 6. (**a**) Average load displacement curves for the four types of specimens; (**b**) Fracture toughness at the onset of delamination.

In engineering applications, there is often a greater focus on the toughness value at the delamination initiation state [35]. This study identifies the critical point in the linear phase of the load displacement curve as the initiation point for delamination [36]. This method is straightforward and offers high precision. The average initiation fracture toughness for the four types of specimens was calculated using Equation (1), with results presented in Table 1. The data indicate that the average initiation fracture toughness values for the untreated laminate, laminate only LIG-coated, laminate only short Kevlar fibers toughened, and the hybrid laminate are 1964.46 KJ $\times$ m^{-2}, 3152.53 KJ $\times$ m^{-2}, 2809.18 KJ $\times$ m^{-2}, and 9460.79 KJ $\times$ m^{-2}, respectively. Correspondingly, the fracture toughness values improved by 60.48%, 43.00%, and 381.60%.

Table 1. Ultimate load and fracture toughness of four types of specimens.

Type	Ultimate Load (N)	Fracture Toughness (KJ $\times$ m^{-2})
Untreated	752.60 $\pm$ 32.45	1964.46 $\pm$ 35.11
LIG	947.42 $\pm$ 33.92 (25.89%)	3152.53 $\pm$ 266.43 (60.48%)
Short Kevlar fibers	1333.47 $\pm$ 233.10 (752.60%)	2809.48 $\pm$ 647.82 (43.00%)
LIG–short Kevlar fibers	1757.91 $\pm$ 513.91 (133.58%)	9460.49 $\pm$ 1704.91 (381.79%)

The results indicate that both LIG and short Kevlar fibers can enhance the delamination fracture toughness of laminates. The LIG coating increases the roughness of the aramid fiber surface, thereby improving the adhesion strength between the panel and the resin matrix. It also absorbs energy during interfacial debonding and slip processes, allowing the interfacial layer to better resist crack propagation under shear forces. Moreover, the high modulus of LIG prevents further crack extension through its own elastic deformation, thereby enhancing the fracture toughness of the laminate. In the case of short Kevlar fibers, their inherent toughness allows them to absorb energy through elastic deformation during crack propagation, preventing direct crack penetration, thereby preventing direct crack penetration. Additionally, the fracture and pullout behaviors of the short Kevlar fibers contribute to energy absorption during crack growth, thus enhancing interfacial toughness [35]. However, compared to the laminate containing only LIG, the laminate with only short Kevlar fibers lacks surface modification on the aramid fibers. This smooth fiber surface fails to form effective bonding with the epoxy resin, resulting in a less pronounced bridging effect of the short fibers; thus, the improvement in fracture toughness for the laminate with only short Kevlar fibers is minimal. In contrast, the combined effect of the LIG-coated and short Kevlar fibers significantly enhances the structural fracture toughness.

To better evaluate the synergistic toughening effect of LIG and short Kevlar fibers, the Mode II interlaminar fracture toughness of carbon fiber/epoxy composites containing only LIG generated at a pulse frequency of 200 DPI [23] and those containing only 10 g/m^2 short Kevlar fibers [37] were compared (Table 2). The results showed that the fracture toughness of the carbon fiber/epoxy composites containing only LIG increased by 69%, while the composites with only short Kevlar fibers exhibited a 61.8% improvement. Both of these values were lower than the 381.79% increase in fracture toughness achieved by the synergistic effect of LIG and short Kevlar fibers. Notably, whether containing only LIG or only short Kevlar fibers, the carbon fiber/epoxy composites exhibited superior toughening effects compared to the aramid fiber/epoxy composites. This can be attributed to the stronger interfacial bonding between carbon fibers and epoxy resin, which facilitates efficient load transfer to the fibers under external forces, thereby enhancing the overall toughness of the composite material.

Table 2. Comparison of fracture toughness change rates from different studies.

Type	Fracture Toughness Change Rate
LIG [23]	69%
Short Kevlar fibers [37]	61.8%
LIG–short Kevlar fibers	381.79%

To further understand the toughening mechanisms of LIG and short Kevlar fibers on the delamination of laminates under Mode II testing, this study employs SEM imaging to observe the fracture surfaces of the specimens.

For the untreated specimens, the fracture surfaces exhibit relatively clean characteristics, with minimal residue of the resin matrix. This phenomenon can be attributed to the extremely smooth surface of the untreated aramid fibers, which hinders the formation of a strong interfacial bond with the resin matrix. Interfacial debonding is identified as the primary cause of failure for the untreated specimens. Additionally, it can be observed that the transverse fibers are arranged in an orderly manner, while the longitudinal fibers show slight signs of damage, accompanied by a small number of fiber breakage, pull-out, and resin residue. This behavior can be explained by the stress distribution characteristics during loading: transverse fibers primarily bear shear stresses, and their orientation is perpendicular to the loading direction, resulting in relatively lower tensile stresses. As the resistance along the transverse fiber direction is minimal, crack propagation tends to occur along this path. In contrast, longitudinal fibers directly face the predominant tensile stresses, making them more susceptible to fracture or damage during loading (Figure 7).

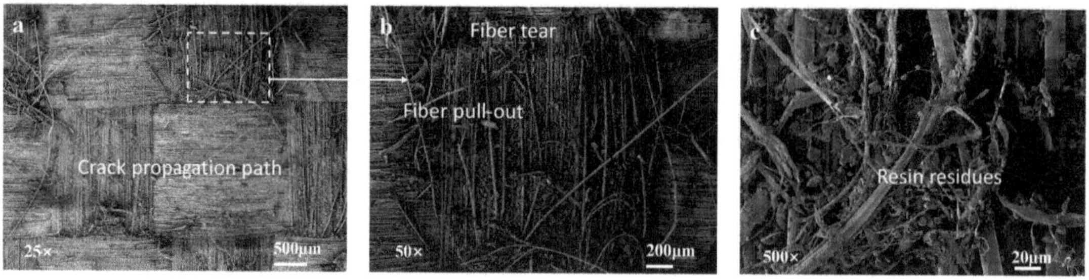

Figure 7. (**a**–**c**) SEM micrographs of the fracture surfaces of untreated specimens.

For the specimens toughened only by short Kevlar fibers, observation of the fracture surface clearly reveals the presence of short fiber breakage bundles in both longitudinal and transverse directions, while the existing aramid fibers showed no significant damage in either direction. These micro-damage mechanisms partially obstruct further crack propagation, thereby effectively enhancing the fracture toughness of the specimens. However, given that no modification treatment was applied to the aramid fiber surfaces, the interfacial bonding between the short Kevlar fibers, the matrix, and the panels remains weak, as evidenced by the minimal resin residue on the fracture surfaces post-failure. This limitation restricts the full realization of the toughening effect (Figure 8).

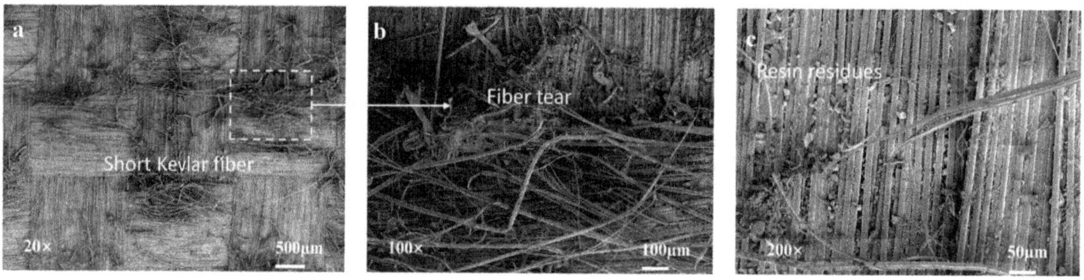

Figure 8. (**a**–**c**) SEM micrographs of the fracture surfaces of only short Kevlar fiber specimens.

In comparison, the fracture surfaces of the specimens only LIG-coated exhibit a significant amount of resin matrix residue, which appears in a plate-like morphology adhered to the surface of the aramid fibers. This surface shows a higher roughness compared to the untreated specimens and those toughened with short Kevlar fibers. This phenomenon can be attributed to the uniform coverage of the LIG coating on the surface of the aramid fibers after treatment, resulting in increased micro-roughness. The enhancement in surface roughness promotes effective wetting between the fibers and the matrix, leading to the formation of a robust interlocking structure. It is noteworthy that the resin matrix on the fracture surfaces displays crack propagation in both the longitudinal and transverse fiber directions, indicating that crack growth is not confined to the longitudinal fiber direction but also involves the transverse direction. This complex crack path increases the length and tortuosity of the fracture trajectory, further dissipating fracture energy and effectively enhancing interlaminar toughness. Additionally, the fiber bridging phenomena observed on the matrix surface reveal the effective connecting capability of the LIG coating between adjacent layers, contributing significantly to the toughening effect (Figure 9).

Figure 9. (**a–c**) SEM micrographs of the fracture surfaces of only LIG-coated specimens.

The fracture surfaces of the specimens containing LIG-coated and toughened short Kevlar fibers clearly reveal the presence of epoxy resin and bundles of short fibers. This phenomenon can be attributed to the uniform distribution of the LIG coating on the surface of the aramid fibers, which promotes effective wetting between the aramid fibers and the resin matrix, consequently significantly enhancing the interfacial bonding strength. Furthermore, the incorporation of short Kevlar fibers leads to frequent fiber pull-out and bridging phenomena, resulting in a more complex crack propagation path. Notably, the short Kevlar fibers exhibit characteristics of longitudinal fracture failure, indicating that the Kevlar fibers, due to their exceptional toughness, effectively absorb a substantial amount of energy through elastic deformation, thereby further enhancing the toughening performance of the material. Additionally, the fracture sites of the short Kevlar fibers clearly show residual resin matrix, further suggesting that the Kevlar fibers effectively form fiber-bridging structures at the interface, which enhances the interfacial toughness and strengthens the adhesion between the matrix and fibers (Figure 10).

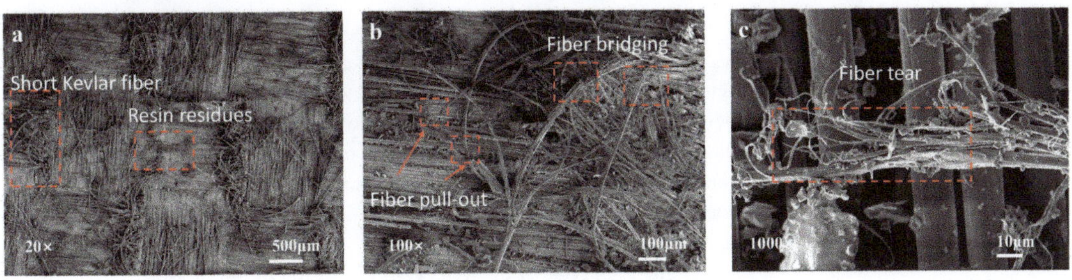

Figure 10. (**a–c**) SEM micrographs of the fracture surfaces of combined specimens.

The LIG-coated aramid fiber laminates serve two key purposes: Firstly, the laser treatment enhances the surface roughness of the aramid fibers, improving the adhesion between the panel and the epoxy resin, thereby increasing interlaminar fracture toughness. Secondly, the inherent piezoresistive properties of graphene enable the monitoring of delamination within the laminate interface. The variations in stress and resistance with strain for aramid fiber laminates containing LIG-coated are illustrated in Figure 11. From the figure, it is evident that there is a correlation between changes in stress and resistance as strain increases. Initially, the relationship between stress and strain is linear, followed by a transition to a nonlinear relationship. Simultaneously, resistance exhibits both linear and nonlinear growth patterns. During the linear phase, the integrity of the laminate structure is maintained, and the conductive pathways within the laminate remain intact, resulting in relatively stable resistance. As the system enters the nonlinear phase, pre-cracks start to propagate, leading to partial disintegration of graphene and affecting the conductive pathways, which accelerates the rate of resistance increase. This indicates a correlation between structural delamination and the rate of resistance change. When the stress reaches a critical value, cracks begin to propagate rapidly, resulting in the failure of the connections within the LIG layer, causing a significant increase in resistance and demonstrating a correlation between crack propagation and resistance change rate. Subsequently, the load experiences a sharp decline, indicating rapid expansion of interlaminar cracks. Concurrently, resistance rises sharply; when the stress stabilizes, the rate of resistance increase correspondingly decreases. Ultimately, bending failure of the laminate leads to a drop in stress, which corresponds to an increased rate of resistance change. In summary, LIG not only effectively monitors the initiation of delamination in laminates but also serves as a reliable indicator for crack propagation and bending failure.

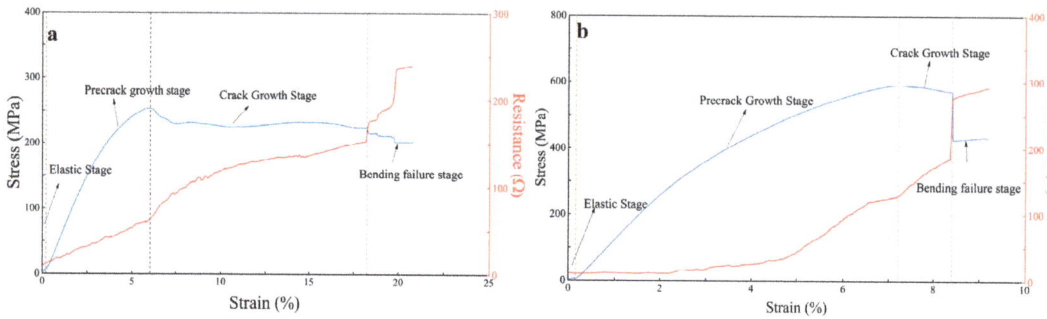

Figure 11. (**a**) Stress/resistance–strain curves of aramid fiber laminate with only LIG-coated; (**b**) Stress/resistance–strain curves of aramid fiber laminates containing LIG-coated and short Kevlar fibers.

3.3. In-Plane Mechanical Properties

The previous studies have confirmed that LIG and short Kevlar fibers significantly enhance the fracture toughness of aramid fiber-reinforced composite laminates. However, many methods for improving interlaminar toughness often come at the expense of in-plane performance [38–40]. Therefore, it is crucial to investigate the impact of LIG and short Kevlar fibers on the in-plane mechanical properties of the composites. This paper examines the effects of LIG and short Kevlar fibers on the in-plane mechanical properties of aramid fiber laminates through tensile testing.

The average load displacement curves for the four types of specimens under tensile testing are shown in Figure 12. It can be observed that all four types of specimens undergo an elastic deformation phase followed by a plastic deformation phase. During the elastic deformation phase, the load displacement relationship increases linearly. Subsequently, in the plastic deformation phase, as the applied displacement load increases, the load exhibits a nonlinear growth trend, and fiber damage begins to occur. After reaching the maximum load that the specimen can withstand, the load starts to decrease, indicating that the specimen has reached its damage limit and ultimately fractures.

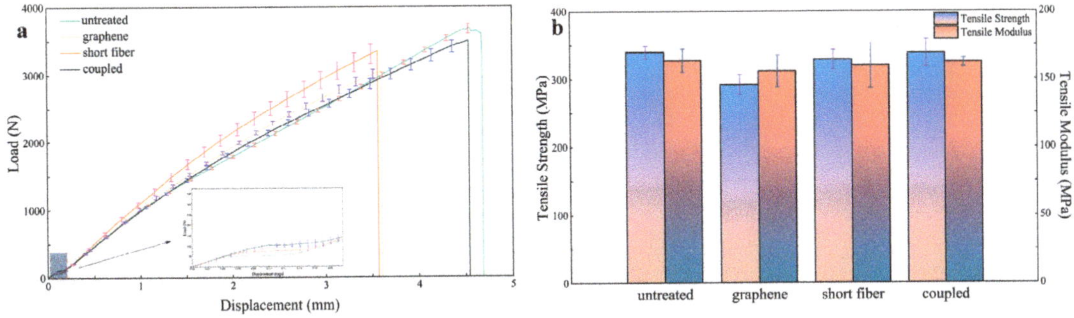

Figure 12. (**a**) Average load displacement curves of the four types of specimens; (**b**) Tensile strength and tensile modulus of the four types of specimens.

The magnified view of the average load displacement curve indicates that the laminated composites containing only LIG-coated aramid fiber and those with a combination of LIG-coated and short Kevlar fibers exhibit a prolonged linear growth phase. This suggests that, compared to the other two types of specimens, they are capable of maintaining a longer elastic phase. This is attributed to the fact that graphene, as a surface coating on aramid fibers, can effectively distribute stress over the fiber surfaces during tensile loading, reducing localized stress concentrations and ensuring that the entire laminate can uniformly withstand external forces during stretching. This capability allows it to endure greater loads without undergoing plastic deformation, thereby extending the elastic deformation phase. Furthermore, the LIG coating can partially constrain the sliding of aramid fibers, enhancing the overall structural integrity of the material at the microscale, resulting in improved linear elastic characteristics during tensile testing. Furthermore, the slope of the linear phase in the load displacement curve represents the tensile modulus of the laminate, with values of 163.65 MPa, 156.01 MPa, 159.94 MPa, and 162.48 MPa, all exhibiting variations within 5% (Figure 12b). This indicates that the incorporation of short Kevlar fibers and LIG did not significantly alter the tensile modulus of the laminate, resulting in comparable slopes during the linear phase of the curve.

An analysis of Table 3 reveals that the ultimate load for the untreated aramid fiber laminate, the laminate containing only LIG-coated, the laminate with only short Kevlar fibers, and the laminate combining both are 3745.63 N, 3220.62 N, 3624.86 N, and 3729.12 N, respectively. It is evident that the ultimate tensile load of the laminate containing only LIG-coated aramid fibers is significantly lower than that of the other three types. This indirectly indicates that the surface modification of aramid fibers using a 12 W laser has caused damage to the in-plane fibers, resulting in the formation of surface defects that compromise the tensile properties of the fibers. Conversely, incorporating short Kevlar fibers into the laminate containing LIG-coated aramid fibers can partially compensate for the reduction in fiber quantity due to laser treatment, ensuring that the overall weight of the fiber fabric does not experience significant loss. Therefore, the laminate that combines LIG-coated with short Kevlar fibers can maintain the in-plane mechanical properties of the structure effectively.

Figure 12b illustrates the tensile strengths of the four types of specimens. The tensile strength of the untreated aramid fiber laminate (339.74 MPa) is comparable to that of the short Kevlar fiber laminate (328.79 MPa) and the laminate combining LIG-coated with short Kevlar fibers (338.24 MPa), with variations within 5%. This indicates that both the laminate containing only short Kevlar fibers and the laminate with LIG-coated and short Kevlar fibers have a limited impact on the tensile properties of the aramid fiber laminate. This limitation is attributed to the relatively low number of short fiber interlayers, which do not significantly affect the thickness of the laminate, thereby resulting in no change in the in-plane mechanical properties of the composite material. Conversely, the tensile strength of the laminate containing only LIG is 292.12 MPa, representing a reduction of 14.02% compared to the untreated laminate. This decrease is attributed to the transformation of aramid fiber surfaces into LIG and gaseous substances upon laser irradiation, leading to a reduction in the number of aramid fibers in both the longitudinal and transverse directions, which in turn decreases the overall density of the aramid fibers and their load-bearing capacity. Therefore, although LIG can significantly improve the fracture toughness of the laminate, this comes at the expense of in-plane mechanical strength. However, the incorporation of short Kevlar fibers effectively mitigates this loss of strength, preserving the in-plane mechanical performance of the composite material.

Table 3. Ultimate load, tensile strength, and tensile modulus of four types of specimens.

Type	Ultimate Load (N)	Tensile Strength (MPa)	Tensile Modulus (MPa)
Untreated	3745.63 ± 98.45	339.74 ± 8.93	163.65 ± 8.46
LIG	3220.62 ± 160.33	292.12 ± 14.54 (−14.02%)	156.01 ± 11.64 (−4.67%)
Short Kevlar fibers	3624.86 ± 156.13	328.79 ± 14.16 (−3.22%)	159.93 ± 16.29 (−2.27%)
LIG–short Kevlar fibers	3729.12 ± 224.55	338.24 ± 20.37 (−0.44%)	162.48 ± 3.28 (−0.71%)

To monitor the damage evolution of the specimens in real time during the tensile testing, load displacement and load were recorded using a universal testing machine for the laminate containing LIG. Simultaneously, the four-probe method was employed to measure the resistance in situ (Figure 13). In the initial stage of displacement loading, the specimens were in the elastic deformation stage, and the stress–strain curve demonstrated a linear relationship; correspondingly, the resistance increased approximately linearly. This is attributed to the laser treatment that transforms the outer fibers of the aramid fibers' surface into piezoresistive graphene. During the application of tensile force, the number of carbon–carbon contacts on the surface of the aramid fibers decreases, resulting in an increase in resistance. At this stage, the laminate experiences minimal damage, maintaining an intact conductive pathway and relatively stable resistance values. As displacement increases, the specimens transition into the plastic deformation phase, where changes in resistance align with stress variations, both exhibiting a nonlinear growth trend. This behavior is due to the emergence of micro-damage, such as cracks in the laminate, which leads to localized breakage of the graphene pathways, causing the rate of resistance increase to accelerate. In this phase, the deformation of the specimens becomes irreversible, and the separation between graphene surfaces is also irreversible. In the later stages of loading, the specimens reach critical load, leading to progressive failure of the fiber reinforcement. The damage intensifies, culminating in macroscopic failure of the laminate, where the conductive pathways on each layer's surface separate, resulting in an increased rate of resistance change and a significant spike in resistance values (Figure 13b). For another specimen, a sudden drop in resistance was observed due to wire fracture caused by the failure during displacement loading (Figure 13a).

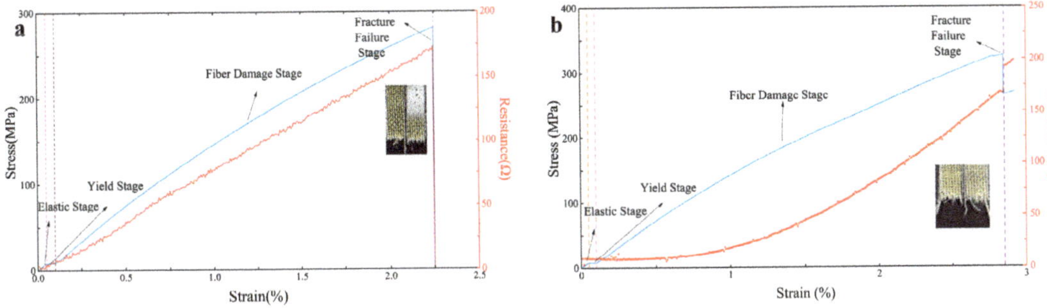

Figure 13. (**a**) Stress/resistance–strain curves of the aramid fiber laminate with only LIG-coated; (**b**) Stress/resistance–strain curves of the aramid fiber laminate containing LIG-coated and short Kevlar fibers.

4. Conclusions

This study investigates the application of LIG coating on the surface of aramid fibers, using short Kevlar fibers as interlayer toughening agents in aramid fiber/epoxy resin composites. The effects of LIG and short Kevlar fibers on the interfacial toughness and damage detection of aramid fiber/epoxy resin composites were systematically explored. The following conclusions were drawn:

(1) Both LIG and short Kevlar fibers significantly improve the interfacial toughness of aramid fiber-reinforced composite laminates. The synergistic effect results in a Mode II fracture toughness of 9460.79 kJ/m^2, representing a substantial 381.60% increase relative to untreated composites. In comparison, LIG alone enhances the toughness by 60.48%, while short Kevlar fibers contribute a 43.00% improvement. When compared to the existing literature on carbon fiber/epoxy composites, the toughening effects of LIG (69% enhancement) and short Kevlar fibers (61.8% enhancement) are notably lower than the 381.60% increase observed in aramid fiber composites. Therefore, the incorporation of LIG and short Kevlar fibers constitutes an effective approach for improving the interfacial toughness of aramid fiber-reinforced composites, significantly enhancing the material's interfacial toughness and structural integrity.

(2) Short Kevlar fibers can compensate for the reduction in in-plane mechanical properties caused by the LIG-coating. The tensile strength of composites containing only the LIG-coated is 292.12 MPa, which represents a decrease of 14.02% compared to the untreated composites. In contrast, the tensile strength of the combined system is 338.24 MPa, with a variation maintained within 5%. These results indicate that the introduction of short Kevlar fibers mitigates the decline in tensile load-bearing capacity resulting from the laser treatment of aramid fiber surfaces, thereby preserving the in-plane mechanical performance of the structure.

(3) The toughening mechanism of aramid fiber-reinforced composites containing LIG and short Kevlar fibers was investigated using SEM. The LIG-coating enhances toughness by increasing the surface roughness of the fibers and promoting fiber-matrix interlocking, while short Kevlar fibers improve toughening effects by introducing complex crack paths and energy absorption mechanisms through fiber pull-out, fiber fracture, and fiber bridging. The combination of both components further enhances the fracture toughness of the laminate by improving interfacial bonding and crack propagation pathways.

(4) LIG can effectively monitor the delamination and tensile damage processes of laminates. Through resistance changes, it allows for real-time tracking of interlayer damage and tensile development. During the Mode II tests, the resistance exhibited

both linear and nonlinear relationships with strain, showing a significant increase in the rate of resistance change at the onset of delamination and during crack propagation. This indicates a correlation between resistance changes and the occurrence of delamination and crack growth. During the tensile tests, resistance initially increased linearly in the elastic phase and then nonlinearly in the plastic phase. A sudden increase in the rate of resistance change was observed when the specimen exhibited tensile fracture damage, suggesting a relationship between resistance changes and the emergence of damage. Thus, LIG provides a novel method for in situ damage detection within structures.

Author Contributions: Conceptualization, B.W.; methodology, B.W.; validation, Q.W. and C.W.; formal analysis, B.W.; investigation, W.T.; resources, W.T. and Q.W.; data curation, W.T.; writing—original draft preparation, B.W. and W.T.; writing—review and editing, C.W.; visualization, W.T. and C.W.; supervision, C.W.; funding acquisition, B.W. All authors have read and agreed to the published version of the manuscript.

Funding: This research was funded by the Shandong Laboratory of Advanced Materials and Green Manufacturing at the Yantai Open Project Foundation (grant number AMGM2022A01), the Shandong Maritime Aerospace Equipment Technology Innovation Center (Ludong University) Open Project Foundation (grant number MAETIC202202), the Shandong Province Key R&D Program—Offshore Satellite Launch and Recovery (grant number 2020CXGC010701), and the Yantai Harbin Engineering University Research Institute School Enterprise Cooperation Foundation (grant number 210F0401004).

Institutional Review Board Statement: Not applicable.

Data Availability Statement: The raw data supporting the conclusions of this article will be made available by the authors on request.

Acknowledgments: We would like to express our gratitude to Wenji Gao for his contributions to the resources and the supervision of this article.

Conflicts of Interest: The authors declare no conflicts of interest.

References

1. Zhang, W.; Liu, S.; Zhang, M.; Tan, Y. Development Status and Application of Aramid Fiber. *J. Text. Sci. Eng.* **2024**, *41*, 86–94. (In Chinese) [CrossRef]
2. Singh, T.J.; Samanta, S. Characterization of Kevlar fiber and its composites: A review. *Mater. Today Proc.* **2015**, *2*, 1381–1387. [CrossRef]
3. Zhu, L.; Guo, K.; Li, Y.; Yu, T.; Zhou, Q. Experimental study on the dynamic behaviour of aluminium foam sandwich plates under single and repeated impacts at low temperature. *Int. J. Impact Eng.* **2018**, *114*, 123–132. [CrossRef]
4. Kim, J.-K.; Mai, Y.-W. *Engineered Interfaces in Fiber Reinforced Composites*; Elsevier: Amsterdam, The Netherlands, 1998; pp. 1–3. [CrossRef]
5. Zeng, Y.; Wan, X.; Zhang, D. Experimental investigation on low-velocity impact response of aramid fiber composites. *China Plast. Ind.* **2019**, *47*, 83–87. (In Chinese) [CrossRef]
6. Ma, X.-M. The Dynamic Behavior of Kevlar Fiber Laminates. Master's Thesis, Taiyuan University of Technology, Taiyuan, 2015. (In Chinese). [CrossRef]
7. Sela, N.; Ishai, O. Interlaminar fracture toughness and toughening of laminated composite materials: A review. *Composites* **1989**, *20*, 423–435. [CrossRef]
8. Daelemans, L.; van der Heijden, S.; De Baere, I.; Rahier, H.; Van Paepegem, W.; De Clerck, K. Nanofibre bridging as a toughening mechanism in carbon/epoxy composite laminates interleaved with electrospun polyamide nanofibrous veils. *Compos. Sci. Technol.* **2015**, *117*, 244–256. [CrossRef]
9. Hassanzadeh-Aghdam, M.; Mahmoodi, M.; Ansari, R. Micromechanics-based characterization of mechanical properties of fuzzy fiber-reinforced composites containing carbon nanotubes. *Mech. Mater.* **2018**, *118*, 31–43. [CrossRef]
10. Zhou, Y.; Pervin, F.; Lewis, L.; Jeelani, S. Fabrication and characterization of carbon/epoxy composites mixed with multi-walled carbon nanotubes. *Mater. Sci. Eng. A* **2008**, *475*, 157–165. [CrossRef]

11. Francesconi, L.; Aymerich, F. Effect of Z-pinning on the impact resistance of composite laminates with different layups. *Compos. Part A Appl. Sci. Manuf.* **2018**, *114*, 136–148. [CrossRef]
12. Jiang, H.; Qian, K.; Zhu, Y.; Zhang, D. Research progress on processing connection and mechanical properties of three-dimensional textile composites. *Acta Mater. Compos. Sin.* **2023**, *40*, 3197–3217. (In Chinese) [CrossRef]
13. Ravandi, M.; Teo, W.; Tran, L.; Yong, M.; Tay, T. Low velocity impact performance of stitched flax/epoxy composite laminates. *Compos. Part B Eng.* **2017**, *117*, 89–100. [CrossRef]
14. Nguyen, M.H.; Davidson, P.; Waas, A.M. Particle-toughened interlayers enhance mechanical response of composite laminates. *Compos. Sci. Technol.* **2019**, *182*, 107761. [CrossRef]
15. Marino, S.; Czél, G. Improving the performance of pseudo-ductile hybrid composites by film-interleaving. *Compos. Part A Appl. Sci. Manuf.* **2021**, *142*, 106233. [CrossRef]
16. Quan, D.; Deegan, B.; Binsfeld, L.; Li, X.; Atkinson, J.; Ivanković, A.; Murphy, N. Effect of interlaying UV-irradiated PEEK fibres on the mechanical, impact and fracture response of aerospace-grade carbon fibre/epoxy composites. *Compos. Part B Eng.* **2020**, *191*, 107923. [CrossRef]
17. Kılıçoğlu, M.; Bat, E.; Gündüz, G.; Yıldırım, M.U.; Urgun, K.; Maviş, B. Fibers of thermoplastic polymer blends activate multiple interlayer toughening mechanisms. *Compos. Part A Appl. Sci. Manuf.* **2022**, *158*, 106982. [CrossRef]
18. Park, B.; Kim, S.C. A study of the interlaminar fracture toughness of a carbon-fiber/epoxy composite containing surface-modified short kevlar fibers. *Compos. Sci. Technol.* **1998**, *58*, 1599–1606. [CrossRef]
19. Zheng, H.; Li, Y.; Tu, H. Research on interlayer properties of short fiber intercalated carbon fiber/epoxy composites. *Acta Mater. Compos. Sin.* **2022**, *39*, 3674–3683. (In Chinese) [CrossRef]
20. Ronglu, Y.; Ousheng, Z. Surface modification of para-aramid fiber and properties of composites thereof. *China Synth. Fiber Ind.* **2022**, *45*, 45–49. (In Chinese) [CrossRef]
21. Ramazan, B.; Tesoro, C. Effect of surface-limited reaction on the properties of Kevlar fibers. *Text. Res. J.* **1990**, *10*, 334–344. [CrossRef]
22. Lin, J.; Peng, Z.; Liu, Y.; Ruiz-Zepeda, F.; Ye, R.; Samuel, E.L.; Yacaman, M.J.; Yakobson, B.I.; Tour, J.M. Laser-induced porous graphene films from commercial polymers. *Nat. Commun.* **2014**, *5*, 5714. [CrossRef]
23. Nasser, J.; Zhang, L.; Sodano, H. Laser induced graphene interlaminar reinforcement for tough carbon fiber/epoxy composites. *Compos. Sci. Technol.* **2021**, *201*, 108493. [CrossRef]
24. Groo, L.; Nasser, J.; Inman, D.; Sodano, H. Fatigue damage tracking and life prediction of fiberglass composites using a laser induced graphene interlayer. *Compos. Part B Eng.* **2021**, *218*, 108935. [CrossRef]
25. Bak, K.M.; Kalaichelvan, K.; Arumugam, V. A novel approach for classification of failure modes in single lap joints using acoustic emission data. *J. Compos. Mater.* **2014**, *48*, 3003–3017. [CrossRef]
26. Bremer, K.; Weigand, F.; Zheng, Y.; Alwis, L.S.; Helbig, R.; Roth, B. Structural health monitoring using textile reinforcement structures with integrated optical fiber sensors. *Sensors* **2017**, *17*, 345. [CrossRef]
27. Chyan, Y.; Ye, R.; Li, Y.; Singh, S.; Arnusch, C.; Tour, J. Laser-induced graphene by multiple lasing: Toward electronics on cloth, paper, and food. *ACS Nano* **2018**, *12*, 2176–2183. [CrossRef]
28. Groo, L.; Nasser, J.; Inman, D.; Sodano, H. Laser induced graphene for in situ damage sensing in aramid fiber reinforced composites. *Compos. Sci. Technol.* **2021**, *201*, 108541. [CrossRef]
29. Steinke, K.; Groo, L.; Sodano, H.A. Laser induced graphene for in-situ ballistic impact damage and delamination detection in aramid fiber reinforced composites. *Compos. Sci. Technol.* **2021**, *202*, 108551. [CrossRef]
30. Naseri, I.; Ziaee, M.; Nilsson, Z.N.; Lustig, D.R.; Yourdkhani, M. Electrothermal performance of heaters based on laser-induced graphene on aramid fabric. *ACS Omega* **2022**, *7*, 3746–3757. [CrossRef] [PubMed]
31. Wang, Y.; Zhao, W. Comparative analysis of data reduction method of ENF test of mode II delamination. *Compos. Sci. Eng.* **2022**, *7*, 81–92. (In Chinese) [CrossRef]
32. Wu, J.; Xu, H.; Zhang, J. Raman Spectroscopy of Graphene. *Acta Chimca Sin.* **2014**, *72*, 301–318. (In Chinese) [CrossRef]
33. Cao, G.; An, F. Effectiveness of the elastic moduli characterization of graphene or other 2D materials via Raman spectroscopy. *Diam. Relat. Mater.* **2024**, *146*, 111201. [CrossRef]
34. Sharma, A.; Dantham, V.R. Observation of reversible and irreversible charge transfer processes in dye-monolayer graphene systems using Raman spectroscopy as a tool. *Spectrochim. Acta Part A Mol. Biomol. Spectrosc.* **2024**, *317*, 124431. [CrossRef] [PubMed]
35. Wang, Y.; Chen, X.; Gong, Y. Experimental Study on Delamination of Composite Laminates with 0/45 Interface under Mode I, Mode II and Mixed-mode I/II Loading. *J. Aeronaut. Mater.* **2018**, *38*, 83–88. (In Chinese) [CrossRef]
36. Kim, B.W.; Mayer, A.H. Influence of fiber direction and mixed-mode ratio on delamination fracture toughness of carbon/epoxy laminates. *Compos. Sci. Technol.* **2003**, *63*, 695–713. [CrossRef]
37. Wang, Y.; Liu, X.; Chen, L.; Shen, W.; Zhu, L. Simultaneously improve the mode II interlaminar fracture toughness, flexural properties, and impact strength of CFRP composites with short aramid fiber interlaminar toughening. *Polym. Compos.* **2022**, *43*, 8437–8442. [CrossRef]

38. Sun, Z.; Shi, S.; Hu, X.; Guo, X.; Chen, J.; Chen, H. Short-aramid-fiber toughening of epoxy adhesive joint between carbon fiber composites and metal substrates with different surface morphology. *Compos. Part B Eng.* **2015**, 77, 38–45. [CrossRef]
39. Wang, J.; Li, C.; Zheng, Q.; Yu, J.; Luo, Z.; Lu, S. Effects of LiCl treatment on surface structure and properties of Kevlar fibers. *Acta Mater. Compos. Sin.* **2016**, 33, 704–713. (In Chinese) [CrossRef]
40. Kim, K.-W.; Jeong, J.-S.; An, K.-H.; Kim, B.-J. A study on the microstructural changes and mechanical behaviors of carbon fibers induced by optimized electrochemical etching. *Compos. Part B Eng.* **2019**, 165, 764–771. [CrossRef]

MDPI AG
Grosspeteranlage 5
4052 Basel
Switzerland
Tel.: +41 61 683 77 34

Polymers Editorial Office
E-mail: polymers@mdpi.com
www.mdpi.com/journal/polymers

www.ingramcontent.com/pod-product-compliance
Lightning Source LLC
Chambersburg PA
CBHW061840210926
51850CB00006B/936